FOOD ENGINEERING

FOOD ENGINEERING

CONCEPTS AND APPLICATIONS

Huaming Chang
Tong Yin

Kruger Brentt
Publishers
2023

Kruger Brentt Publishers UK. LTD.
Company Number 9728962

Regd. Office: 68 St Margarets Road, Edgware, Middlesex HA8 9UU

ISBN: 978-1-78715-094-2

For information on all our publications visit our website at http://krugerbrentt.com/

PREFACE

Food engineering is the application of engineering principles to the storage, processing and distribution of food materials and their bio-products. It requires a sound engineering education, as well as fundamental training in chemistry and food science. As the global population is expected to reach over 9 billion by 2050, there is an increasing need to develop innovative ways to produce food and ensure the food produced can last long, contain as many nutrients as possible per gram, be safe and reach further. Food engineering is the tool to meet these needs of the present and future population. Food Engineering is a relatively new profession and a scientific field involved with food manufacturing and the processing of refined foods. It encompasses the practical application of food science to develop efficient industrial production, packaging, storage, and physical distribution of nutritious and convenient foods that are uniform in quality and safe. The most remarkable difference between food engineering and food science is that the former includes the knowledge of unit operations and processes. However, these operations involve applied food science, just as chemical engineering involves chemistry and thermodynamics. Food science deals primarily with acquiring knowledge to elucidate the course of reactions or changes occurring in foods whether natural or induced by handling procedures. On the other hand, food engineering is the application of principles and the facts of science, engineering, and mathematics to the processing, preservation, storage, and utilization of food.

Food engineering should be differentiated from food technology. In teaching technology, some schools include considerable engineering in the curriculum to train students in the practical application of food science. Graduates are in effect food engineers. Other schools emphasize food science and qualify graduates primarily for research and quality control work. The food technologists from these schools are actually food scientists. Where food processing is involved, understanding only one branch of science is not enough. Scientists and engineers with knowledge of the different fields relevant to various aspects of food manufacturing are needed. Management of different disciplines will enable them to couple principles from traditional engineering fields with principles of food preservation and safety to obtain the most suitable solutions to food processing problems. There have been instances in which a civil, mechanical or chemical engineer has been assigned the task of devising a new procedure for handling food. The process designed often was very efficient from an engineering point of view, but inadequate from a microbiological point of view. For example, allowance was not made for proper cleaning of equipment.

The present book contains eleven -chapters covering all related disciplines. These chapters include 15 chapters. These chapters include Introduction, units and dimensions, fluid flow in food processing, sensible and latent heat changes, energy and controls in food processes , heat transfer in food processin, preservation processes, refrigeration, evaporation, psychrometry, mass transfer, food freezing, membrane separation, dehydration & packaging concepts. Related terminology is given at the end for ready reference. It is written in a lucid style with the most recent advances, and with theoretical discussions being supplemented with illustrations, and tables for easy understanding of the subject. This book describes the principles as well as the equipment used, food preservation, post-harvest handling, and packaging techniques. The book is also supported with a comprehensive glossary and index in the end. This book is written for food, mechanical, chemical, and electrical engineers, students as well as food scientists and technologists working in the food industry.

We are grateful to all those persons as well as various books, manuals, periodicals, magazines, journals etc. that helped in the preparation of this book. In spite of the best efforts, it is possible that some errors may have occurred into the compilation and editing of the book. Further queries, constructive suggestions and criticisms for the improvement of the book are always welcome and shall be thankfully acknowledged.

Huaming Chang

Tong Yin

CONTENTS

CHAPTER 1
INTRODUCTION

Food engineering is one of the most multidisciplinary fields in engineering. It combines engineering principles with other fields like microbiology, physical sciences, chemistry, and more. Food engineers are responsible for ensuring the safe and efficient processing, packaging and delivery of food to every store shelf in the world. Needless to say, it is a vital job for the entirety of humanity. Food Engineers take engineering principles and apply them to the storage, processing, and distribution of food materials and other bio-products. They are the people responsible for ensuring that store shelves remain stocked with arrays of canned, bagged, and boxed foodstuffs that are nutritious and safe to eat and drink.

Food Engineering is the study of processing; packaging and delivery of food in a safe and efficient way. Food engineering principle applies to all canned, bagged and boxed food to maintain their nutrition throughout the supply chain till it reaches the consumer.

Food engineering encompasses machines used in the manufacturing plant to storage systems on transport trucks. By designing and operating food processing plants food engineers control the health and safety of food production. They also work for waste management, genetic modification of foods and pharmaceutical health care industries. Design of processing, packaging, handling of storage equipment, machinery, scale-up of prototype food processes, regulation, protection of the public health, product research development and technical sales all come under the area of food engineering.

Food Engineering is a specialized sub field within Agricultural Engineering and is focused on the application of Engineering to the production and distribution of food. In today's world, food production is a significant issue and Food Engineers play an important role in addressing that.

Food that is grown or processed must follow strict safety health and standards to ensure they are safe and that people do not get sick (or worse) from consuming produced food. Food Engineers control the health and safety of food production by designing and operating food processing plants. They also involve themselves in waste management and the genetic modification of foods. Food Engineers generally work for food manufacturing companies, but they can also work in the pharmaceutical and health care industries.

Fig. 1

Foods represent a complex mixture of hundreds of components, which contribute to their characteristic flavor. Nowadays, assuring the food's flavor stability represents an issue of food industry involved actors, becoming one of the main technological challenges. An educated consumer has increasing demands related to food quality perception over the last years. As a consequence, the food flavor instability is a critical quality issue of food industry.

Achieving food's quality in respect to its flavor and stability is a complex process, especially by considering the large variety of flavor compounds and diversity of food accepted by consumers. The knowledge of all mechanisms involved in food flavor changes during processing and storage period is compulsory. Ideally, food should have the same flavor from the moment of packaging to the moment it gets to the consumer.

Unfortunately, food is exposed to chemical reactions starting with the moment it is taken out from the plant. Given the actual trend of big food industry producers to accede to the global food market, this involves longer durations on which the product passes until it arrives to the consumers, contributing in increasing risk of its quality damage. New innovative cost-effective technologies are in discussion, capable of satisfying the actual consumer's quality demands.

Foods were thermally treated for cooking purposes since many centuries ago, firstly to modify and preserve their organoleptic and nutritional characteristics.

Anthropologists agree that thermal cooking of food had a great impact on phenotypical properties, intellectual development of mankind, and later to the society, depending on economy, culture, and geography. The transfer from home cooking to industrialized processes began in the late nineteenth century, firstly aiming food preservation and later, after the Second World War, focusing food safety and quality issues. Food engineering as a study discipline was first introduced in West European countries willing to offer a curriculum for actors in the food industry, including chemical engineering aspects, food microbiology, and biology. Nowadays, there is a difference between food technology and food engineering. The first treats the application of methods of food preservation and food processing, while food engineering is a more complex area focusing on a combination between food and applied sciences such as microbiology, physical sciences, chemistry, and engineering.

Thermal food processing is one of the most efficient methods of preservation. Beneficial effects include inactivation of food pathogens, natural toxins, and other unwanted changes, prolonging the shelf-life, adjusting the food digestibility, increasing the bioavailability of nutrients, and improving the functional properties. Still, thermal processing usually comes with unwanted changes in food composition such as decrease in nutrients, formation of toxic compounds, or other influencing the organoleptic properties. Continuously, heat treatment has to be improved according to the latest trends in food research, focusing on the increasing beneficial effects and counteracting the undesired effects.

Active packaging concept gains more interest due to its effectiveness in antimicrobial protection insurance, especially to pathogenic microorganisms. Still open for discussion, the subject refers to the influence of food-packaging polymers on food products after the incorporation of protective microorganisms. Actually, this is one of the biggest challenges when considering transposing these biomaterials to industrial scale use.

The past decade offered a great opportunity in valorizing the results in active biofilms characterized by antimicrobial and antifungal activities in order to extend the food shelf-life and to reduce the addition of food additives. As so, biologically active polysaccharides such as kefiran or chitosan proved an incredible potential to food industry by their high antimicrobial activity against pathogenic and spoilage microorganisms. The positive impact relates also to better environmental practices by avoiding the conventional packaging materials such as plastics and reducing the food waste.

Advanced processing methods tend to preserve the characteristic properties of food including their organoleptic and nutritional qualities better when compared to the conventional food processing methods and, in some cases, are even more cost-effective. There is a clear rise in the populations suffering from food allergies, especially infants and children. Though this fact is widely attributed to the changing livelihood of population in both developed and developing nations and to the introduction of new food habits with advent of novel foods and new processing techniques, their complete role is still uncertain. As so, the return to ancient natural food recipes could be an alternative for reinterpretation of food in a more safety and healthier way.

For example, cereals are traditionally processed into foods and beverages through fermentation worldwide. With lactic acid fermentation, as the most known ancient preserving method, studies found that valuable compounds are formed during the process. Additionally, lactic acid fermentation enhances food safety by reducing the toxic compounds and producing antimicrobial factors, which facilitate the inhibition and elimination of food pathogens. Many health-enhancing attributes are related to the consumption of lactic acid fermented products.

Nowadays, demands for nondairy substitutes with high acceptance and functionality, given by pleasant flavor and nutrient content (vitamins, minerals, antioxidants, prebiotics, and probiotics), respectively, are increasing. Cereal-based beverages have a great potential to fulfill this demand. As part of the worldwide traditional cuisine and culture, these products are often transposed to industrial re-interpretation as novel food products with various beneficial microorganisms (especially in the case of lactic acid fermented products) and functional properties designed for a balanced nutrition and life quality improvement. As so, through intensive research, focusing on this field and industrial production, these novel foods are taken beyond their national reputation. By applying novel trend technology, food products are designed for a balanced nutrition, easy and safe to store, prepare, and consume, and able to supply all the needed vitamins and nutrients.

The evolution of food industry offered many opportunities of using low-cost food additives and ingredients. Beverage industry is the most exposed to using caloric sweeteners, sucrose, high-fructose corn syrup, fruit juice concentrates, etc., which in latest studies proved their harmful health effects. Products such as soft drinks, carbonated soft drinks, fruit ades, fruit drinks, sports drinks, energy and vitamin water drinks, sweetened iced tea, cordial, squashes, and lemonade are attributed as the major sugar contributors in the USA for their consumption after each meal, as a supplementary source of calories. This intake severely contributes to obesity and obesity-related diseases in children and adults. As a result of novel food regulation, past years' research focused on the reduction of sugar in food and beverages along with avoiding the addition of other artificial sweeteners. Some prospects were made and proved the positive effects of a predicted reduction of sweeteners in food.

Wild edible plants as part of the cultural and genetic heritage are harvested and processed into value-added products especially in the Eastern Europe. They represent important sources of nutrients and health-promoting compounds, being consumed as remedies especially in rural and suburban areas. However, these products lack recognition as important contributors to human diet in developed areas. Current research promotes the using of wild edible plants in human diet, while several strategies present their importance for social, economic, and agro-ecological development. Given the growing interest for a natural alternative of products obtained by chemical synthesis, bioactive compounds found increasing application in food industry, cosmetics, pharmaceutical industry, and even in agriculture.

Sea-buckthorn (*Hippophaë rhamnoides*), hawthorn (*Crataegus monogyna*), wild grown European blackberry (*Rubus fruticosus* L.), cornelian cherry (*Cornus mas*), blackthorn (*Prunus spinosa*), dog rose (*Rosa canina*), and bird cherry or hackberry (*Prunus padus*) are only few of the wild fruits with great potential for food industry as a valuable source of bioactive compounds, mainly phenolic compounds.

Novel bioactive compounds obtained by lab-scale extraction techniques such as high-pressure processing, pulsed electric field on berry fruits, grape pomace and food industry byproduct extraction, microwave-assisted extraction, ultrasonically assisted diffusion processes, supercritical fluid extraction, pressurized hot water extraction, gas-assisted extraction, and enzyme-assisted extraction were transposed to industrial application such as juice industry and oilseed processing industry. Extraction-dependent factors—temperature, infusion duration, tea, and water ratio—were also tested in tea brewing in order to deliver important product information to consumers.

Modeling of processes through mathematics is widely used in food industry. Recently, an experimentally validated multiscale modeling for coffee extraction was developed.

Food loss is a major concern to all economic, environmental, and social aspects. In recent years, several modeling of waste treatments depending on food characteristics were proposed. In order to assess the stability of a food product during storage period, degradation kinetics modeling was elaborated using different indicators such as the ascorbic acid content and color intensity loss.

Given the fast-growing engineering fields, namely the food industry with novel food processes technologies, novel ingredients, advanced enzyme production and applications, and other complementary technologies, this book will disclose the state of the latest trends in food engineering.

Physics, chemistry, and mathematics are essential in gaining an understanding of the principles that govern most of the unit operations commonly found in the food industry. For example, if a food engineer is asked to design a food process that involves heating and cooling, then he or she must be well aware of the physical principles that govern heat transfer.

The engineer's work is often expected to be quantitative, and therefore the ability to use mathematics is essential. Foods undergo changes as a result of processing; such changes may be physical, chemical, enzymatic, or microbiological. It is often necessary to know the kinetics of chemical changes that occur during processing.

Such quantitative knowledge is a prerequisite to the design and analysis of food processes. It is expected that prior to studying food engineering principles, the student will have taken basic courses in mathematics, chemistry, and physics. In this chapter, we review some selected physical and chemical concepts that are important in food engineering.

CHEMISTRY OF FOOD

Food chemistry deals with the biochemical nature of food, its properties and how they are processed in the body. It involves the study of chemical components from such as proteins, vitamins, minerals, lipids, amino acids, fats, carbohydrates and many more. The study also involves how food properties change upon certain food processing techniques and ways to prevent it or enhance. For example apple can be prevented from browning by application of acidic substances such as lemon juice. An example for enhancement includes addition of microorganisms that convert lactose to lactic acid.

The effect of different storage conditions may lead to change in physical and chemical properties of food products. Such as high temperature causes denaturation of proteins in food and temperature also causes break down of long chain fatty acids in oils. Thus making it unfit for consumption. Food chemistry deals with all these physical and chemical changes in food and ways to prevent them.

DIETARY FIBER

Dietary fiber is a type of carbohydrate found in cereals, grains, nuts, lentils, peas and vegetables. Dietary fiber based on its physical properties is classified in to soluble and insoluble fiber (resistant starch). It cannot be digested by our body enzymes and reaches the large bowel undigested. It gets fermented by bacteria in the large intestine and break down in to methane, hydrogen, carbon dioxide and short-chain fatty acids. It supports the growth of friendly bacteria needed to maintain a healthy gut, slows down the digestion time and makes one feel fuller for longer. It also helps to control blood sugar levels by slow absorption of glucose in to the blood stream. It helps in maintaining healthy heart by reducing absorption of cholesterol.

DIETARY SUPPLEMENTS

Dietary supplements contain one or more of the dietary ingredients such as mineral, vitamin, herbs, amino acids and other botanical ingredients. They are available in the form of tablets, capsules, gummies, energy bars and drinks. A dietary supplement helps to balance the daily requirement of nutrients if not completely available by food. Some of the dietary supplements are meant to manage health conditions and overall health. Such as calcium and vitamin D to maintain bone health in women and omega-3 fatty acids for heart health. The dietary supplements come with label which lists all the active ingredients and other ingredients amount per serving.

Fig. 2. Dietary Supplements

Excessive intakes of these dietary supplements are not recommended as they also have several side effects. For example Vitamin K supplement reduces the effect of Coumadin drug. Vitamin C and E (anti-oxidant supplements) might reduce the effect of some type of chemotherapy treatment.

FOOD NANOTECHNOLOGY

Nanotechnology is used in food industry to enhance delivery of food ingredients. It is defined as the creation of materials in nanometer scale. Polymeric nanoparticles, lipid-based nanoparticles such as liposomes, Nano emulsions, dendrimers, carbon-based Nano carrier's and quantum dots are examples of Nano carriers used in food nanotechnology. It is mainly applied in food packaging increasing the shelf life of the product, detecting the bacteria in packaging and food safety. Manufacturing food at nanometer level helps to provide lower fat foods and also incorporates much of the nutrients in a small package. Nanotechnology in food also includes several types such as Nano emulsification and Nano encapsulation. An example of Nano emulsification is mayonnaise which is a blend of oil and water leading to thick paste. Furthermore fatty content in mayonnaise or other food can be lowered by manipulating the droplets structure at the nanoscale level.

FOOD PRESERVATION

Food Preservation is a process that prevents the growth of microorganisms, bacteria, fungi and thus increases the shelf life of the product. It also slows down enzymatic browning and oxidation of fats that leads to rancidity. Thus science of food processing ensures that the food is not contaminated by chemicals or pathogenic microorganisms and does not lose its original color, flavor, texture and nutritive value. Traditional techniques used for food processing are curing, cooling, vacuum packaging, freezing, boiling, potting, heating, sugaring, pickling, lye, canning, jellying, bottling, jugging, burial, confit and fermentation. Modern techniques include Pasteurization, Artificial food additives, nonthermal plasma, Freeze drying, Hurdle technology, Irradiation, Modified atmosphere, High-pressure food preservation, bio-preservation and Pulsed electric field electroporation.

FOOD SAFETY

Food safety includes the guidelines for handling, cooking, and preservation of food to prevent foodborne illnesses caused by microbes such as fungi, bacteria, parasites, and viruses. Food safety is important to prevent food poisoning, diarrhea, vomiting, stomach aches, muscle aches, and many more. This is what makes food safety important as it assures that the food is safe throughout the food chain from preparation, distribution, storage and harvesting. Food safety refers to the absence, or safe acceptable levels, of hazards in food that may harm the health of consumers.

As food is a medium for growth of microorganisms it has to stored and packed properly to avoid entry of pathogens. The final products of food samples are tested for the presence of microbes, flavor, color, and any physical-chemical changes. If the final products pass the entire test and has foreign substances within acceptable limits, required shelf life it gets approved for distribution among the consumers.

NUTRIGENOMICS

Nutrigenomics is the study of interaction between nutrition, health and genes. It helps to determine body nutritional requirement and allergy conditions if any to any of the food ingredients. Amount of fats, carbohydrates, lipids, nutrients present in the body is known by this study and one can take care of what is lacking or more than required in our body to maintain overall health. Nutrigenomics corresponds to the use of epigenomics, biochemistry, nutrition, metabolomics, proteomics, genomics, transcriptomics, and physiology to study relation between nutrition and genes. The food we eat has an impact on our body this effect on genome, proteome and metabolome is studied in nutrigenomics, based on which the diet of a person is planned to protect genes from damage.

NUTRITION

Nutrients are substances that allow your body to make energy, build and maintain tissues, and regulate bodily processes. Nutrition, nourishment, or aliment, is the supply of materials-food -required by organisms and cells to stay alive. In science and human medicine, nutrition is the science or practice of consuming and utilizing foods. As the body can't make Essential nutrients; these nutrients must come from food.

Essential nutrients are of two types: micro and macro

- Micronutrients: Vitamins, minerals.
- Macronutrients: Carbohydrate, proteins, fat.

As molecular biology, genetics advance and biochemistry, nutrition has become more focused on metabolism and metabolic pathways; to lead a healthy life style good nutrition is important. Nutrition is the intake of food and liquids that aid in your body survival and functioning. Proteins, Vitamins and Minerals, Carbohydrates, Roughage, Water are the important nutrients required for the body.

NUTRITIONAL DEFICIENCY

Nutritional deficiency occurs when the body doesn't absorb or get enough nutrients such as vitamins and minerals. Sometimes it can result from an imbalanced diet or certain health conditions. The body requires many vitamins and minerals to stay healthy and function properly. The vitamins and minerals are crucial for both body development and preventing disease. These vitamins and minerals are often referred to as micronutrients. They are not produced naturally in the body, so you have to get them from your diet.

Nutritional deficiencies are more like to occur due to inadequate or poor food choices. There are a number of Deficiencies like Iron deficiency, Vitamin A deficiency, Thiamine (vitamin B-1) deficiency, Niacin (vitamin B-3) deficiency; Folate (vitamin B-9) deficiency, Cobalamin (vitamin B-12) deficiency, Vitamin D deficiency, Calcium deficiency etc. These deficiencies can lead to a variety of health problems such as anemia, digestion problems, skin disorders, stunted or defective bone growth, and even dementia. The treatment for a nutritional deficiency depends on the type and severity of the deficiency.

NUTRITIONAL TOXICITY

Nutrient toxicity condition is also described as nutrient deficiency. Intake of nutrients obtained in food found in plants and animals for proper growth, metabolism and repair is known as nutrition. Providing more or excess fertilizers to the nutrient uptake may leads to nutritional toxicity. The deficiency of nutrient in plants cause to Stunted growth (decreased growth), Chlorosis (lack of chlorophyll), Interveinal chlorosis (yet veins remain green), Necrosis (death of plant tissue), Burning (severe localized yellowing, scorched appearance), Immobile nutrient (stops the flow of nutrients from one body parts to another) and Mottling (spotted, irregular patterns on leaves). By using different elements like calcium, iron, manganese, potassium, magnesium, molybdenum, copper, sulfur and zinc as fertilizers can also changes the pH fluctuations in the plants and leads to more toxicity. In this cases, the nutritional values changes and mostly leads to nutritional toxicity and even effects the regular diets in living organisms. S

CHAPTER

UNITS AND DIMENSIONS 2

Dimensions in mathematics are the measure of the size or distance of an object or region or space in one direction. In simpler terms, it is the measurement of the length, width, and height of anything. In Physics it can also mean any physical measurement such as length, time, mass, etc.

Dimensions are generally expressed as:

- Length
- Breadth
- Width
- Height or Depth

Primary dimensions, such as length, time, temperature, and mass, express a physical entity. Secondary dimensions involve a combination of primary dimensions (e.g., volume is length cubed; velocity is distance divided by time).

Dimension is one of the most fundamental concepts of geometry, so fundamental that an adequate axiomatic definition has only been produced in the last century. Intuitively we all understand that the dimension of a point, line (or curve), surface, and solid all differ and are ordered. The notion of dimension has been extended beyond the three dimensions that we are familiar with from experience to higher dimensions and even infinite dimensions.

These abstract higher dimensions are used in both formal mathematics and physics (e.g. the phase space, Hilbert space), as well as in other sciences like computer science and data science. Going into a truly rigorous point-set topology (also called set-theoretic geometry) demonstration of dimension is beyond the scope of this lesson, though we will work our way to a intuitive and clear understanding of the same concepts.

ENGINEERING UNITS

Throughout the engineering industries, many different definitions and units have been proposed and used for mechanical and thermal properties.

The problems this caused led to the development of an agreed international system of units. In the SI system there are seven well-defined base units from which the units of other properties can be derived, and these will be used throughout this chapter.

The SI base units include length (in metres), mass (in kilograms), time (in seconds) and temperature (in kelvin). The first three will hopefully need no further explanation, while the latter will be discussed in more detail later.

The other SI base units are electric current (in amperes), amount of substance (in moles) and luminous intensity (in candela). These may be familiar to readers with a background in electronics, chemistry and physics respectively, but have little relevance to steam engineering nor the contents of The Steam and Condensate Loop.

SI BASE UNITS

The core of the SI system is a short list of base units defined in an absolute way without referring to any other units. The base units are consistent with the part of the metric system called the MKS system. The International System of Units (SI) is founded on seven base units.

There are just seven base units

Quantity	Unit	Symbol
length	metre	m
mass	kilogram	kg
time	second	s
temperature	kelvin	K
amount of substance	mole	mol
electric current	ampere	A
luminous intensity	candela	cd

Prefixes are used to indicate powers of ten. They mean that with only one unit of measurement you can express any measurement from the smallest to the largest. They each have a short name and symbol. Large factors have upper case symbols, and small factors lower case symbols.

For example:

giga	1 000 000 000	G
mega	1 000 000	M

kilo	1 000	k
milli	0.001	m
micro	0.000 001	μ

- **Metre** - is the length of a path traveled by light in vacuum during a time interval of 1/299792458 of a second
- **Kilogram** - is equal to the mass of the international prototype of the kilogram
- **Second** - is the duration of 9192631770 periods of radiation corresponding to the transition between the two hyperfine levels of the ground state of caesium-133 atom
- **Ampere** - is the constant current in two parallel conductors of infinite length with negligible circular cross section and placed 1 metre apart in vacuum - that produces a force equal to 2 10^{-7} newton per metre of length between the conductors
- **Kelvin** - is the 1/273.16 fraction of the thermodynamic temperature of the triple point of water
- **Mole** - is the amount of substance which contains as many elementary entities as there are atoms in 0.012 kilogram of carbon-12
- **Candela** - is the luminous intensity in a given direction of a source that emits monochromatic radiations of 540 10^{12} hertz frequency with a radiant intensity of 1/683 watt per steradian in the direction

STYLE CONVENTIONS AND RULES FOR USING SI UNITS

General principles for the writing of unit symbols and numbers were first proposed in 1948. These were subsequently adopted and elaborated by ISO, the international standards organization.

- Unit symbols are printed in roman (upright) type, irrespective of how the rest of the text is printed.
- Unit symbols are unaltered in the plural.
- Unit symbols are written without a final full stop (period) except for normal punctuation such as at the end of a sentence.
- Unit symbols are placed after the numerical value, leaving a space between the value and the symbol. For example 5 V not 5V.
- Unit symbols are generally written in lower case letters, except when the name of the unit is derived from a proper name. (Note that when the name of a unit which is derived from a proper name is written out in full, such as ampere or herz, the name is not capitalised. The only exception to this is Celsius.)

- The the given SI unit symbol should be used. The symbol for second, for example, is s. To use sec or secs is incorrect.
- Unit symbols and unit names shouldn't be mixed. Metre per second, not metre/second or metre/s.

DERIVED UNITS

Derived units are algebraic combinations of the seven base units and the two supplementary units with some of the combinations being assigned special names and symbols.

Table 1. Derived units

Quantity	Name of Unit	Symbol	Expression in terms of SI base units	Expression in terms of other units
Plane angle	radian	rad		
Solid angle	steradian	sr		
Adsorbed radiation	gray	Gy	$m^2 s^{-2}$	J/kg
Electrical capacitance	farad	F	$m^{-2} kg^{-1} s^4 A^2$	C/V
Electrical charge	coulomb	C	A s	
Electrical conductance	siemens	S	$m^{-2} kg^{-1} s^3 A^2$	A/V
Electrical inductance	henry	H	$m^2 kg s^{-2} A^{-2}$	
Electrical potential	volt	V	$m^2 kg s^{-3} A^{-1}$	W/A
Electrical resistance	ohm	Ù	$m^2 kg s^{-3} A^{-2}$	V/A
Force	newton	N	$kg ms^{-2}$	
Frequency	hertz	Hz	s^{-1}	
Illuminance	lux	lx	m^{-2}cd sr	lm/m^2
Luminous flux	lumen	lm	cd sr	
Magnetic flux	weber	Wb	$m^2 kg s^{-2} A^{-1}$	V s
Magnetic flux density	tesla	T	$kg s^{-2} A^{-1}$	Wb/m^2
Power or radiant flux	watt	W	$kg m^2 s^{-3}$	J/s
Pressure	pascal	Pa	$kg/(m s^2) = (N/m^2)$	
Radioactivity	becquerel	Bq	s^{-1}	
Temperature relative to	degree Celsius	°C	K	
273.15 K				
Work, energy, heat	joule	J	$m^2 kg s^{-2}$	N m

SI DERIVED UNITS DESCRIBED IN TERMS OF ACCEPTABLE SI UNITS

Derived units are algebraic combinations of the seven base units and the two supplementary units with some of the combinations being assigned special names and symbols.

Table 2. SI Derived Units

Quantity	Description	Symbol	Expression in terms of SI base units
acceleration	meter per second squared	m/s^2	m s^{-2}
area	square meter	m^2	m^2
coefficient of heat transfer (often used symbol *h* or *U*)	watt per square meter Kelvin	W/(m^2 K)	kg s^{-3}K^{-1}
concentration (of amount of substance)	mole per cubic meter	mol/m^3	mol m^{-3}
current density (often used symbol *r*)	ampere per square meter	A/m^2	A m^{-2}
density (mass density)	kilogram per cubic meter	kg/m^3	kg m^{-3}
electrical charge density	coulomb per cubic meter	C/m^3	m^{-3} s A
electric field strength	volt per meter	V/m	m kg s^{-3} A^{-1}
electric flux density	coulomb per square meter	C/m^2	m^{-2} s A
energy density	joule per cubic meter	J/m^3	m^{-1} kg s^{-2}
force	Newton	N or J/M	m kg s^{-2}
heat capacity	joule per Kelvin	J/K	m^2 kg s^{-2} K^{-1}
heat flow rate (often used symbol Q or *q*)	watt	W or J/s	m^2 kg s^{-3}
heat flux density or irradiance	watt per square meter	W/m^2	kg s^{-3}
luminance	candela per square meter	cd/m^2	cd m^{-2}
magnetic field strength	ampere per meter	A/m	A m^{-1}
modulus of elasticity	giga Pascal	GPa	10^{-9} m^{-1} kg s^{-2}
(or Young's modulus)			
molar energy	joule per mole	J/mole	m^{-2} kg s^{-2} mol^{-1}
molar entropy (or molar heat capacity)	joule per mole Kelvin	J/(mole K)	m^{-2} kg s^{-2} K^{-1} mol^{-1}
moment of force (or torque)	Newton meter	N m	m^2 kg s^{-2}
moment of inertia	kilogram meter squared	kg m^2	kg m^2
momentum	kilogram meter per second	kg m/s	kg m s^{-1}

Quantity	Description	Symbol	Expression in terms of SI base units
permeability	Henry per meter	H/m	m kg s^{-2} A^{-2}
permitivity	farad per meter	F/m	m^{-3} kg^{-1} s^4 A^2
power	kilowatt	kW	10^{-3} m^2 kg s^{-3}
pressure (often used symbol *P* or *p*)	kilo Pascal	kPa	10^{-3} m^{-1} kg s^{-2}
specific energy	joule per kilogram	J/kg	m^2 s^{-2}
specific heat capacity (or specific entropy, often used symbol c_p,c_v or *s*)	joule per kilogram Kelvin	J/(kg K)	m^2 s^{-2} K^{-1}
specific volume	cubic meter per kilogram	m^3/kg	m^3 kg^{-1}
stress	mega Pascal	MPa	10^{-6} m^{-1} kg s^{-2}
surface tension	Newton per meter	N/m	kg s^{-2}
thermal conductivity (often used symbol *k*)	watt per meter Kelvin	W/(m K)	m kg s^{-3} K^{-1}
torque	Newton meter	N m	m^2 kg s^{-2}
velocity (or speed)	meters per second	m/s	m s^{-1}
viscosity, absolute or dynamic (often used symbol *m*)	Pascal second	Pa s	m^{-1} kg s^{-1}
viscosity, kinematic (often used symbol *n*)	square meter per second	m^2/s	m^2 s^{-1}
volume	cubic meter	m^3	m^3
wave number	1 per meter	1/m	m^{-1}
work (or energy heat, often used symbol *W*)	joule	J or N m	m^2 kg s^{-2}

SUPPLEMENTARY UNITS

The supplementary unitis referred as a dimensionless unit which is employed with the fundamental units to create the derived units. The supplementary units are utilized in two major geometric variables such as phase angle and solid angle.

Example

The geometric variable "length" has a unit "radian".

Supplementary SI units

Table 3. Supplementary Units

Parameter	SI Unit	Abbreviation	Definition
Plane Angles	radian	rad	A plane angle with vertex at the center of a circle that is subtended by an arc equal in length to the radius.
Solid Angles	steradian	sr	The solid angle with vertex at the center of a sphere that is subtended by an area of the spherical circle equal to that of a square with sides equal in length to the radius.

The international system of units represented as SI units is the new method of metric system which is the mostly used unit system. It is employed in all research and development applications and it is defined to be a coherent system of units. This system is developed based on the seven fundamental units and two supplementary units of physical variables.

Geometric variables	Units
Phase angle	radian, rad
Solid angle	steradian, sr

SYSTEM AND PROPERTIES

System

A careful description of the system is vital in engineering analysis. A region prescribed in space or a finite quantity of matter is called a system and that is enclosed by an envelope, which is stated to be the boundary of the system. The boundary of a system can be real, such as walls of a tank, or it can be an imaginary surface that encloses the system. For example, in Fig. 1, the boundary in system A is along the walls of a storage tank; thus it does not include the pipe and the valve. However, in system B, the boundary envelops the tank, valve and the pipe. The composition of the system is described by the components present inside the system undary.

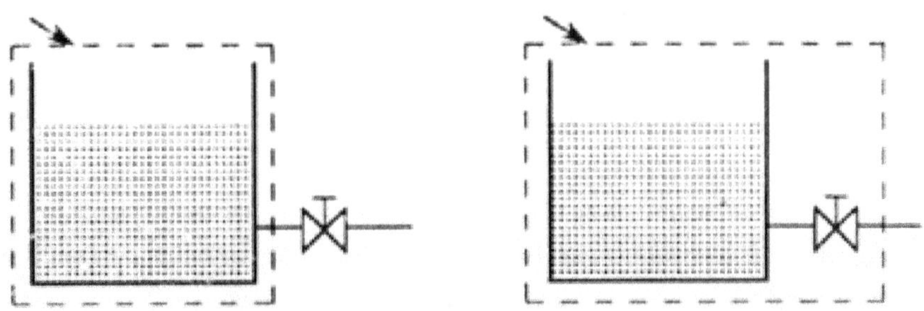

Fig. 1: Examples of systems and their boundaries

Both closed and open systems are often encountered. In closed systems, the boundary of the system is impervious to any flow of matter. In an open system, heat and/or matter can flow into or out of the system. For example, a system boundary that contains only a small section of the wall is impervious to the flow of matter, and may be considered a close system. On the other hand, system B in Fig. 1 is an open system since both heat and liquid can flow through the system.

Properties

Properties are those observable characteristics, such as pressure and temperature, which define the equilibrium state of a thermodynamic system. Properties do not depend on how the state of a system is attained: they are only functions of the state of a system. Thus, properties are independent of the process by which a system attained a certain state.

Intensive Properties

Intensive properties do not depend on the size of a system, such as temperature, pressure and density.

Extensive Properties

An extensive property depends on the size of the system: for example, mass, length, volume, energy. This definition implies that an extensive property of a system is a sum of respective partial property values of the components of a system. These properties, one of which may be mass are required to uniquely give an extensive property of a single component system.

The ratio of two extensive properties of a homogenous system is an intensive property. For example the ratio of two extensive properties mass and volume is density, which is an intensive property.

The state of a system is defined by independent properties. Once the properties become fixed, when the state of the system is defined, they are called dependent properties.

DENSITY

Density is defined as the mass of substance divided by the volume occupied. Its unit in SI is the kilogram per cubic meter (kg m^{-3}). At 5°C water has a density of 1.00 g/ml or 10^3 kg m^{-3}.

The addition of solids, e.g. minerals, sugar, protein will increase the density, whereas oil and fat will decrease the density. The density of fluid is usually measured with a hydrometer. The density is temperature dependent, so temperature should always be recorded. The density of milk usually falls within the range of 1028 - 1035 kg m^{-3} depending on the composition. It is generally measured with a special hydrometer known as a lactometer and the result can be used to estimate total solids.

The densities of the solid constituents are regarded as fat (930), water (1000), and milk solids-not-fat (1035 kg m^{-3}). Fat contents range between 1 and 10 per cent and the total solids determination is based on the following equation.

$$T = 0.25\ \rho + 1.2.2\ F + 0.72 \qquad (1)$$

where

T=total solids [w/w],

ρ = 1000 (density in g/ml).

F=fat percentage.

Fat is determined separately usually by the Gerber method.

Researcher presents relationship for the density of whole milk and cream over the temperature range 0 - 150°C.

Whole milk; $\rho = 1033.7 - 0.2308\ \theta - 2.46 \times 10^{-3}q^2$ (2)

Cream (20 % fat) ; $\rho = 1013.8 - 0.3179\ \theta - 1.95 \times 10^{-3}q^2$ (3)

where θ = temperature, C.

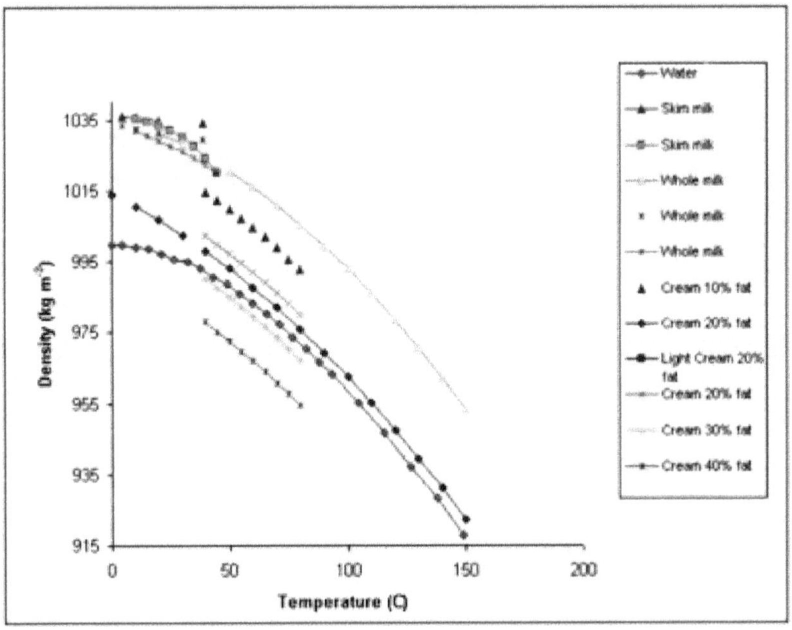

Fig. 2. Density of Fluid Milk Products

Density is useful for monitoring changes occurring during processing. e.g. evaporation, or for checking whether extraneous water has entered the product. When dealing with solids, it is necessary to differentiate between solid density and bulk density, particularly with particulate matter and powder. The solid or particle density is mass of solid/volume of solid, and it will take into account the presence of air within the solid. For particulate matter, it can be determined either by flotation using liquids of known specific gravity or by using a density bottle.

The density of milk varies between 1.027 and 1.035 g/cm^3 as a function of the type and quantity of the dispersed particles (emulsified, colloidal or soluble). With increasing fat content, the density decreases; density increases with an increasing protein, lactose and mineral content. Density is influenced by temperature, average values for densities at 20°C and 1.013 bar (standard atmospheric pressure) in g/cm^3 are:

Whole milk	1.027-1.032	Cream	1.0035-1
Whey	1.025-1.027	Skimmed milk	1.033-1.036
Buttermilk (heat serum)	1.0214		

Density is technologically important when calculating the raw milk quantity, when determining adulterations, for automatic fat standardization, for the determination of water in buttermilk and when manufacturing concentrates for longlife products.

A rapid method for the determination of density in milk uses the lactodensitometer.

The solid density is important in separation processes. e.g. centrifugation of cheese fines, cyclone operation and the pneumatic or hydraulic transport of powders and particulate matter.

Bulk density is an important property, particularly for the transportation and storage of bulk particulate material e.g. fruit, grain, powders. It is the mass of material divided by the total volume occupied.

In most cases, it is important to have a high bulk density. The bulk density of milk powder is affected by processing condition. The bulk density increases as the total solids increases. The particle density increases as the total solids increase indicating that less air is incorporated into the particles at higher total solids. Injection of air or nitrogen into the product immediately before atomisation may reduce the bulk density and agglomeration achieved by a re-wetting process substantially decreases the bulk density.

The method of atomisation will also affect the bulk density. Early designs of atomiser wheel produced powder 0.45 - 0.55 g/ml, whereas later designs typified by the vaned wheel produced bulk densities of ·0.55- 0.65 g/ml; later designs have used steam to occlude air from the fluid. Jet nozzles can produce powders with bulk densities as high as 0.83 g/ml.

The porosity is the fraction of the total volume which is occupied by air between particles. It does not take into account air within the particles.

Porosity = $(r_s - r_b) / r_s$

where

r_s = solid density and

r_b = bulk-density

The density of whole milk powder ranges from 43 to 51 per cent.

BOILING AND FREEZING POINT

The boiling point for milk is 100.2°C, slightly higher than for pure water. The boiling point of both cow and buffalo milk ranges from 100.2°C to 101°C with an average of 100.5°C. The boiling point ranges slightly with the percentage of solids present in milk.

The freezing point of cow milk and buffalo milk ranges from – 0.535°C to – 0.59°C with an average of – 0.553°C depending on the lactose, proteins and mineral content.

The average freezing point of raw milk is at -0.526 C; of pasteurized milk in the range -0.517 to -0.521 C. This is influenced by the dissolved lactose and the similarly dissolved ions of the milk salt and a few other compounds with a relatively low molecular mass.

This relationship permits to detect adulteration with water or with additives e.g. salts, disinfectants or neutralizing agents.

The freezing point of raw milk has a lower value of -0.515°C, values of < -0.5°C indicate adulteration with water. Values of > -0.62°C indicate adulteration with salts.

The freezing point can be modified by gassing/ degassing of milk, lactose splitting or pH modification.

The presence of dissolved substances elevates the boiling point of a solution.

Dissolved substances lower the freezing point of a solution, since milk is a solution containing salts and sugars its freezing point is lower than that of water.

The fortification of milk with milk powder or lactose lowers the freezing point.

Table 4 Bulk Density of Selected Food Materials

Material	Bulk density (kg/m^3)
Beans, cocoa	1073
Beans, soy, whole	800
Coconut, shredded	320–352
Coff ee beans, green	673
Coff ee, ground	400
Coff ee, roasted beans	368
Corn, ear	448

Material	Bulk density (kg/m³)
Corn, shelled	720
Milk, whole dried	320
Mustard seed	720
Peanuts, hulled	480–720
Peas, dried	800
Rapeseed	770
Rice, clean	770
Rice, hulled	320
Sugar, granulated	800
Wheat	770

CONCENTRATION

Concentration is a measure of the amount of substance contained in a unit volume. It may be expressed as weight per unit weight, or weight per unit volume. Normally, concentration is given in percentage when weight per unit weight measurement is used. Thus, a food containing 20% fat will contain 20 g of fat in every 100 g of food.

Concentration values are also expressed as mass per unit volume—for example, mass of a solute dissolved in a unit volume of the solution. Another term used to express concentration is **molarity**, or molar concentration. Molarity is the concentration of solution in grams per liter divided by the molecular weight of the solute. To express these units in a dimensionless form, **mole fraction** may be used; this is the ratio of the number of moles of a substance divided by the total number of moles in the system.

Thus, for a solution containing two components, A and B, with number of moles n_A and n_B, respectively, the mole fraction of A, X_A, is

$$XA = \frac{n_A}{n_A + n_B} \quad (4)$$

Concentration is sometimes expressed by **molality**. The molality of a component A in a solution is defined as the amount of a component per unit mass of some other component chosen as the solvent. The SI unit for molality is mole per kilogram.

A relationship between molality, M A, and mole fraction, X A, for a solution of two components, in which the molecular weight of solvent B is M B, is (5)

Both molality and mole fraction are independent of temperature.

Develop a spreadsheet on a computer to calculate concentration units for a sugar solution. The sugar solution is prepared by dissolving 10 kg of sucrose in 90 kg of water. The density of the solution is 1040 kg/m³. Determine

a.	concentration, weight per unit weight
b.	concentration, weight per unit volume
c.	°Brix
d.	molarity
e.	mole fraction
f.	molality
g.	Using the spreadsheet, recalculate (a) to (f) if (1) the sucrose solution contains 20 kg of sucrose in 80 kg of water, and density of the solution is 1083 kg/m3; (2) the sucrose solution contains 30 kg of sucrose in 70 kg of water, and density of the solution is 1129 kg/m3.

THERMAL PROCESSING

Since many of the processes utilized to preserve food products depend on the addition of thermal energy, it is important to understand the principles associated with food preservation through the addition of thermal energy. The design of a thermal process to achieve food preservation involves two principles: (a) the use of elevated temperatures to increase the rate of reduction in the microbial population present in the raw food material and (b) the transfer of thermal energy into the food product required to achieve the desired elevated temperatures.

The information in this section will address the typical parameters used to quantify the influence of elevated temperatures on reduction of microbial populations.

Influence of Elevated Temperatures on Microbial Populations

When microbial population in a food is exposed to an elevated temperature, changes in individual microbial cells within the population and a reduction in the viability of the microorganisms result in a reduction of the population with time of exposure when quantified by standard microbiological procedures. The reduction in population occurs in a logarithmic manner with increasing time at a given constant elevated temperature.

Decimal Reduction Time

The decimal reduction time (D Value) is defined as the time necessary for 90% reduction in the microbial population. When the microbial population is plotted against corresponding heating time on semi log coordinates, the D value is the time required for a one log cycle reduction in the number of microorganisms. The initial microbial population has no influence on the D value since the magnitude is directly related to the slope of the straight line. Exposure of the microbial population to higher temperatures results in a decrease in the D value. In fact, a plot of decimal reduction time as a function of temperature on semi log coordinates results in a linear relationship (Fig. 3). This curve is known as thermal resistance curve.

Based on the definition, the following equation can be usedlog

N_0 - log N = t/D (6)

or $N/N_0 = 10^{-t/D}$ (7)

where N_0 is the initial number of microorganisms (t=0), N is the number of microorganisms surviving after heating for time 't' at a temperature and D is decimal reduction time for the microorganism under conditions of heating.

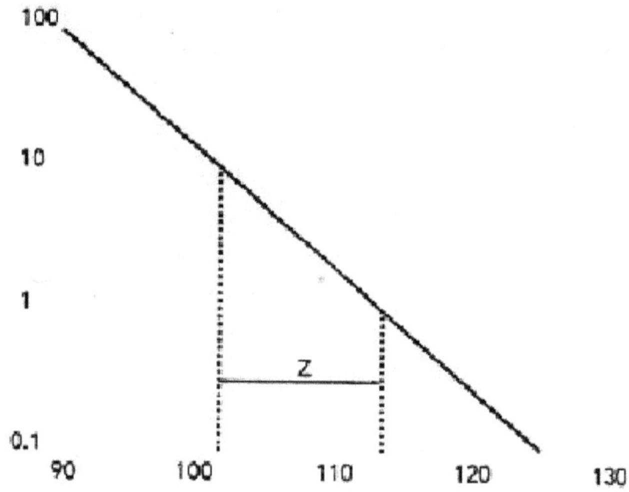

Fig. 3: Thermal resistance curve of microbial population

Thermal Resistance Constant

The temperature increase required to cause a one log cycle reduction in the decimal reduction time is defined as the thermal resistance constant (z). The thermal resistance constant, or z-value, is a second quantitative parameter normally used to quantify the thermal process required for a given microbial population. A large z-value suggests that a given increase in temperature of exposure for the microbial population results in a small change in decimal reduction time. In most situations, this observation would indicate that the microbial population contains vegetative cells or microbial spores; the spores exhibit greater heat resistance or higher z-value than vegetative populations which are characterized by a lower z-value. It is important to note that the complete characterization of the impact of elevated temperature on microbial population requires reference to both the decimal reduction time (D) and the thermal resistance constant (z). The z-value as shown in Fig. 3 can be expressed by the following equation:

$$z = \frac{T_2 - T_1}{\log D_{T1} - \log D_{T2}} \qquad (8)$$

where D_{T1} and D_{T2} are decimal reduction times for micro-organisms at temperatures T_1 and T_2, respectively and 'z' is thermal resistant constant.

Thermal Death Time

The third quantitative parameter found frequently in thermal processing is the thermal death time, or F-value, which finds use in the actual thermal process. Thermal death time is defined as the time required for achieving a stated reduction in the microbial population at a given temperature. The key part of the definition is the stated reduction, which may be the reduction in the population of microbial pathogens required to establish product safety. Alternatively, the stated reduction may be the reduction in the population of a vegetative microorganism causing product spoilage, and the stated reduction is that required to achieve the desired product shelf-life.

A thermal death time curve would appear to be very similar to a thermal resistance curve (Fig. 3), but would cover the entire range of times required to achieve the desired or stated reduction in microbial population. A typical thermal death time curve is presented in Fig. 4. Note that the curve describes the entire reduction in microbial population from time zero until the population has been reduced to a defined level of microbial survivors. This point has been defined as the thermal death time, or F-value.

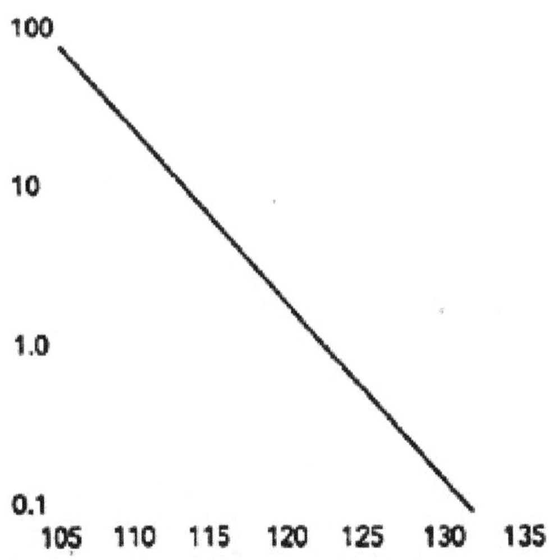

Fig. 4. Thermal death time curve

The similarity in the thermal resistance and the thermal death time curve is evident from Figures 3 and 4. Each log cycle reduction in the microbial population on the thermal resistance curve represents a decimal reduction time, or D-value. It follows that F-values may be expressed as multiples of D-values. The most common of these relationships is F = 12D for Clostridium botulinum in commercial sterilization of low acid foods.

It must be emphasized that any given F-value will apply to a single elevated temperature. In other words, different elevated temperatures result in different F-values or times required achieving the same stated level of reduction in microbial population.

MOISTURE CONTENT

Food is any substance consumed to provide nutritional support for the body. It is usually of plant or animal origin, and contains essential nutrients, such as carbohydrates, fats, proteins, vitamins, or minerals. All foods content solids, water and other chemicals. The moisture contained in a material comprises all those substances which vaporize on heating and lead to weight loss of the sample. The weight is determined by a balance and interpreted as the moisture content. According to this definition, moisture content includes not only water but also other mass losses such as evaporating organic solvents, alcohols, greases, oils, aromatic components, as well as decomposition and combustion products. The moisture content also called as moisture assays is one of the most important analyses performed on most of the food products. Table 1 gives the general idea about the moisture content of different foods. Water activity measurements parallel to the moisture content is also an important parameter for quality and stability of food.

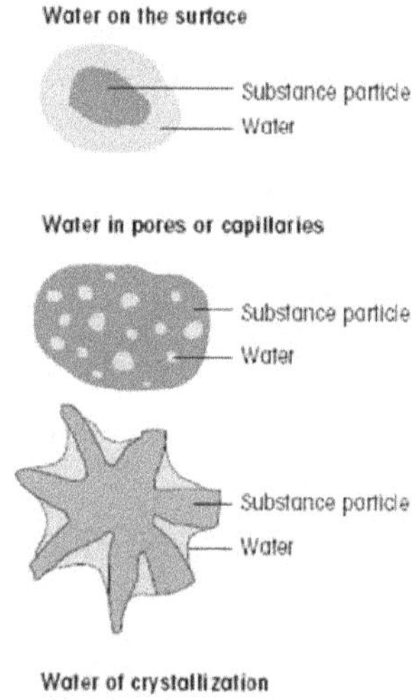

Fig 5 Types of bonding of moisture in food products

Types of bonding of moisture in the product

The moisture in food can be present in different forms which are decided by type of bonding with solids (Fig 5). It is available in following forms:

Free water: water on the surface of the test substance and it retains its physical form **Absorbed water:** water in large pores, cavities or capillaries of the test substance **Water of hydration:** Occluded in lattice ions or water of crystallization coordinately bonded to ions.

Esimation of Moisture Content

Moisture content is the quantity of water contained in a food material. Moisture content is used in a wide range of scientific and technical areas, and is expressed as a ratio, which can range from 0 (completely dry) to the value of the materials' porosity at saturation. It can be given on a volumetric or mass (gravimetric) basis. Moisture content is expressed as a percentage of moisture based on total weight (wet basis) or dry matter (dry basis). Wet basis moisture content is generally used. Dry basis is used primarily in research. The moisture content is expressed by following formulae.

Based on the different forms of moisture present in the food the method used for measurement of moisture may estimate more or less moisture content. Therefore, for different food products *Official Methods* of moisture measurement have been given by agencies like AOAC (Association of Official Analytical Chemists), AACC (American Association of Cereal Chemists) and BIS (Bureau of Indian Standards).

Importance of Moisture Content in Foods

Proper moisture content is essential for maintaining fresh, healthy foods. If a food is too moist or too dry, it may not be suitable to eat and will not taste as good as it would if it had the correct moisture content. Most of the food products contain moisture. The moisture content per cent is seldom of interest. Rather, it shows whether a product intended for trade and production has standard characteristics such as:

- Storability
- Agglomeration in the case of powders
- Microbiological stability
- Flow properties, viscosity
- Dry substance content
- Concentration or purity
- Commercial grade (compliance with quality agreements)
- Nutritional value of the product
- Legal conformity (statutory regulations governing food)

In addition to above characteristics, the determination of moisture content plays important role commercially with respect to following aspects:

- **Freshness:** Fresh, ripe fruits and vegetables are moist to the touch. As they age and begin to rot, some dry out and some pick up excess moisture and begin to mold.
- **Labeling:** Food industries require a minimum or maximum percentage of moisture on certain foods in order for them to be packaged and labeled. If they don't fit to these standards, the foods cannot be sold.
- **Cost:** In processed foods, the percentage of water in a product can determine its final price. Generally, a product with more water will cost less.
- **Processing:** Biologists and manufacturers need to know the moisture content of food to ensure that it's processed and packaged in a safe, stable way.
- **Quality:** Moisture content determines the way most foods taste, feel and look. It is one of the important ways to measure food quality.
- **Shelf life:** Shelf life of product depends on its moisture content at the time of packaging and rate of moisture gain during storage which is also called as sorption isotherm study.

HIGH PRESSURE PROCESSING

Introduction

High hydrostatic pressure processing (HPP), a relatively new technology to the food industry inactivates microorganisms without causing significant flavour and nutritional changes to foods. On the other hand, the effectiveness of thermal processing technologies explains why it remains as the prevailing method to achieve microbial safety and the inactivation of enzymes and microorganisms responsible for food spoilage. However, the high temperatures used in these processes cause significant chemical changes in foods. Particularly important are thermal degradation reactions leading to off-flavours, destruction of nutrients and other product quality losses.

For example, high-temperature short-time (HTST) pasteurization treatments (72°C for 15 s) impart a slight cooked, sulphurous note that has become acceptable to milk consumers but its refrigerated shelf life is only approximately 20 days. Ultra pasteurization (UP), a process similar to HTST pasteurization using more severe treatments (e.g. 1 s at 89 p C, 0.1 s at 96 p C or 0.01 s at 100 p C) lowers flavour quality and causes more nutrient damage but yields milk with a refrigerated shelf life of approximately 30 days. Pressure treatments of 400 MPa for 15 min or 500 MPa for 3 min at room temperature achieves microbiological reductions similar to thermal pasteurization but it is not used commercially because long pressure processing times are not financially viable. HPP treatments (586 MPa for 3 and 5 min) at moderate temperature (55°C) extend the refrigerated shelf life of milk to over 45 days while retaining milk volatile profiles similar to those observed after conventional HTST treatments. Finally, ultra high temperature (UHT) processing (135–150°C for 3–5 s) yields milk that is stable at room temperature for 6 months; however, this process induces strong 'cooked' off-flavour notes thus limiting its consumer acceptance in important markets.

Future advances are expected from the synergistic effects of using high pressure and high temperature combinations in the rapidly evolving pressure-assisted thermal processing technology (PATP). PATP is not yet a commercial application and will require more complex safety validation procedures than HPP, particularly for the case of low-acid foods (pH under 4.5). PATP conditions are sufficiently severe to achieve the inactivation of bacterial spores and recent studies suggest that pressure can lower the degradation rate of product quality caused by high temperature treatments.

High pressure processing at refrigeration, ambient or moderate heating temperature allows inactivation of pathogenic and spoilage microorganisms in foods with fewer changes in food quality as compared to conventional technologies. Pressure acts by disrupting mainly hydrogen bonds without affecting covalent bonds. Therefore, high pressure processing (HPP) treatments at low (approximately 0–30°C) and moderate (approximately 30–50°C) temperature cause minimum losses in quality factors associated with small molecules such as vitamins, pigments and volatile flavours. Research has confirmed that the sensory characteristics of HPP products make them often indistinguishable from untreated controls. Five decimal reductions in pathogens including *Salmonella typhimurium, S. enteritidis, Listeria monocytogenes, Staphylococcus aureus* and *Vibrio parahemolyticus* can be achieved by HPP.

Principle

The high pressure technique is essencially additive-free, mostly non-thermal or involves reduced heat treatments. Based on the pascal or isostatic principle, the hydrostatic pressure at a given point is the same in all directions and pressure is transmitted uniformly and immediately through the pressure transferring medium. Thus, the effects of pressure are independent on product size and geometry. It is often stated that HP processing is a uniform way of processing foods.

The effectiveness of a high pressure treatment is influenced by various intrinsic and extrinsic factors. Treatment time, pressurization or decompression rate, temperature and the number of pulses are critical to the effectiveness of the process. Moreover, the factors which include the effect of pressure on water, adiabatic heating and heat dissipation, food composition and the physiological states of microorganisms to be inactivated must be taken into account when optimising pressure treatments for the production of safe, quality foods.

Effects of HPP

- Microorganism inactivation
- Modification of biopolymers including enzyme activation and inactivation, protein denaturation and gel formation.
- Quality retention (colour, flavour, nutritional value)
- Modification of physiochemical properties of water

General Description of HP Equipment For Food Industry

The main components of an HP system are a pressure vessel, a pressure generation system, a temperature control device and a material handling system. Most pressure vessels are made from a high tensile steel alloy 'monoblocs' (forged from a single piece of material), which can withstand pressure of 400-600 MPa. For high pressures, pre-stressed multilayer or wire-wound vessels are used. In operation, after all air has been removed, a pressure-transmitting medium (either water or oil) is pumped from a reservoir into the pressure vessel using a pressure intensifier until the desired pressure is reached. Temperature control in commercial operations can be achieved by pumping a heating/cooling medium through a jacket that surrounds the pressure vessel. This is satisfactory in most applications as a constant temperature is required but if it is necessary to change the temperature regularly, an internal heat exchanger is fitted.

There are two methods of processing foods in high pressure vessels: in-container processing and bulk processing. Since foods reduce in volume at the very high pressure used during processing, there is considerable stress and distortion to the package selection is an important issue in using this method. Materials handling for in-container processing is achieved using equipment similar to that used to load/unload batch retorts. Bulk handling of liquids is simpler, requiring only pumps, pipes and valves.

There are two main types of High pressure equipments:

- **Batch type:** A batch press can be used for any kinds of food in flexible packages, such as pouches, cups, or bulk bags. With the food already packed in the final consumer package at the processing stage, the risk of contamination is eliminated. The food packaged, are placed in the pressure vessel where they are isostatically compressed.

- **Continuous type**: Continuous systems can be used for pumpable food. The system is installed with other equipment, and in the end the liquid food reaches an asceptic or clean filler. Thus any kind of consumer package can be used. Top of-the-line, high quality juice may be perceived as more valuabe if sold in glass bottles, rather than PET or other plastic that would require for batch cycling.

The volume in a pressure vessel for continuous use is better utilized that in a batch press, where there is dead space between the food packages. Thus the output volume is large despite the fairly small dimensions of the vessels used.

The principle of the continuous system is comparable to a four stroke engine. A valve is opened at the top of the press cylinder to let the product in and then closed. A floating piston inside the cylinder acts as a movable divider, and separates the water from the product. The pressure is the same on both the sides of the floating piston. When full pressure has been reached, it is held for a short period of time. After the hold time is over, the water is released through the bottom of the vessel and thus the vessel is decompressed. Another valve is then opened at the top, and the product leaves the cylinder when water is pumped in at the vessel filling, one for holding, and one for emptying the product. As the vessels take turns delivering the high pressure cycled product. There is an almost continuous output of product. With a balance tank in line with the system, the output will be continuous.

Advantages of HPP

- Retention of flavour and texture of the product
- Increase in Microbiological safety and shelf-life
- Low energy consumption
- Minimal heat input
- Minimal effluent and losses
- Uniform isostatic pressure & adiabatic temperature distribution
- Combination with heat gives better effects

Applications of HPP

- Milk treated at pressures of upto 500 MPa for few minutes has been shown to have a shelf-life at least equivalent to HTST pasteurized milk. Most vegetative cells, including non-sporeforming thermodurics, can be eleminated.
- HHP treatment (200 MPa, 10 min) after acidification (rise of acidity after acidification) in yogurt, increases the water binding capacity of whey proteins.

- The cheese yield is not influenced when milk treated at pressure <250 MPa, but at 600-800 MPa, it gets increased by up to 25% with increase in moisture content in curd and decrease in protein content in whey.
- Cheese Ripening can be accelerated by using the High Pressure treatment, which avoids the usage of elevated temperatures, addition of cheese slurries or exogenous enzymes or by the use of adjunct starters.

Cheddar cheese, when exposed to HP from 5 to 300 MPa for 3 days at 25°C, shows free amino acid levels of 26.5 mg/g at 50 MPa compared to 21.3 mg/g in the 6-month old cheese (which had not been HP treated). The taste of both the cheese were described as "excellent". This shows a considerable reduction in the ripening times of the cheese, attained through the application of HPP.

High pressure sterilization is possible by starting at elevated temp. e.g. 60–90°C, & using the adiabatic compression for rapid heating to higher temperatures. High pressure sterilisation is a combined process where both pressure and temperature contribute to sterilisation by the inactivation of spores and enzymes. The result is a shelf stable product, and in many cases a higher general quality than those products obtained using conventional processing.

The standard atmospheric pressure is defined as the pressure produced by a column of mercury 760 mm high.

Zero pressure characterizes absolute vacuum. When pressure is measured relative to absolute vacuum, it is called **absolute pressure** .

However, when we use a pressure measuring device such as a pressure gauge, it is usually calibrated to read zero at one atmospheric pressure. Therefore, these devices are actually reading the difference between the absolute pressure and the local atmospheric pressure. A pressure measured by a gauge is often called **gauge pressure**, and it is related to the atmospheric pressure based on the following expression:

$$P_{absolute} = P_{gauge} + P_{atmosphere}$$

(for pressures greater than Patmosphere)

$$P_{vacuum} = P_{atmosphere} - P_{absolute}$$

(for pressures below Patmosphere)

A visual description of the relationships between the various terms used to define pressure is given in Figure 6.

In fluid flow problems, two additional terms are often encountered, namely, static pressure and impact pressure. **Static pressure** is the pressure measured by a device if it is moving with the same velocity as the fluid velocity. **Impact pressure** is force per unit area perpendicular to the direction of flow when the fluid is brought reversibly to rest.

The pressure of a fluid is measured using a variety of instruments, including a Bourdon tube, a manometer, and a pressure transducer. A Bourdon tube is shown in Fig. 6. It consists of an oval-shaped arm ABCD. An increase in the internal pressure extends the arm, and the movement of the pointer on the dial is calibrated to indicate the pressure.

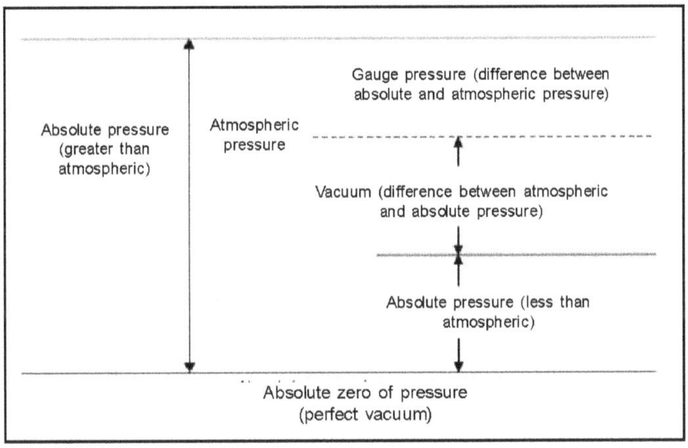

Fig. 6. Illustration of the relationships between the terms used to define pressure.

THE IDEAL GAS LAW

This independent behaviour makes gases do interesting things and those things can be described using the ideal gas law. As the name says, this law describes the behaviour of an ideal gas.

In reality, most gases are not ideal. Nevertheless, those basic concepts hold true well enough in most foods to help you explain that choux pastry, souffle or proofing bread dough.

The ideal gas law describes how gases behave under varying temperatures, pressures and volumes. It can be expressed by quite a simple formula:

$$p * V = n * R * T$$

Where:

- p = pressure (the unit is Pa)
- V = volume (unit = m^3)
- n = number of molecules (expressed in the unit 'moles')
- R = gas constant, this is a fixed number that'll always be the same (8,314 J/molK)
- T = temperature (unit = Kelvin)

This formula tells you that if you multiply the pressure of the space that the gas is with the volume of that space, that outcome is identical to a multiplication of the number of gas molecules present, a specific constant and the temperature.

It's use though for cooks is mostly for when you start comparing scenarios. If you have a gas in a specific space (e.g. an air bubble in your bread dough). You can assume for now that the amount of molecules is constant, as is the gas constant.

In the oven the temperature increases. As such, $n * R * T$ becomes a larger value. As such, $p * V$ should also become larger. In this case that will result in an increase in volume, the gas within that bubble expands because of the heat.

These types of changes in temperature occur in food all the time and even the number of molecules may change as may the volume! The following examples will bring some of those scenarios to life.

Why aren't 'real' gases ideal?

In the ideal gas law scientists make a few assumption about that gas. First of all, you assume that the molecules that make up a gas do not have any attractions between one another. They don't stick to one another, don't push or pull at one another. Second of all, you assume that the gas molecules themselves don't actually take up any space.

In real life both assumptions aren't correct. However, when you stay close to moderate temperatures such as room temperature and atmospheric pressure, a lot of gases behave quite ideal. For widely different pressures and temperatures this assumption doesn't hold up well, but in food we tend to stay pretty close to these 'normal' conditions, making the ideal gas law a useful tool!

Ideal gas law example 1 – Popping popcorn

Imagine a vessel (or a huge popcorn corn kernel) of 1 m^3 at a temperature of 300K (27°C) and a pressure of 100.000 Pa. Using these three variables you can calculate how many moles of a gas sit in this huge corn kernel:

$p * V = n * R * T$

This can be re-written to:

$n = (p * V) / (R * T)$

Which then gives:

$n = (100.000 * 1) / (8,314 * 300) = 40$ moles

Now we're going to take this imaginary corn kernel and we'll heat it to 400K (127°C). Remember that corn kernels are really strong and rigid. It doesn't deform under pressure and there is no way molecules can go in our out. (In reality more gas molecules will be formed due to the evaporation of water inside the kernel, but we'll neglect those for now.)

As such, we know that in the formula three things stay the same: n, V and R.

So, in order to even out the left and right side of the formula, the pressure has to change. And that is exactly what happens when you're heating a corn kernel:

$p * V = n * R * T$

This can be re-written to:

$p = (n * R * T) / V$

Which then gives you:

$p = (40 \times 8{,}314 \times 400) / 1 = 133.333$ Pa.

Remember how the pressure was 100.000 Pa at the start. In other words, it has changed considerably! The pressure build up can only take place if your vessel or kernel is strong enough to handle the pressure. Even a popcorn kernel will break at some point, when the pressure is just to high in there for the gas to be contained.

PHASE DIAGRAM OF WATER

Water is considered to be a pure substance. It has a homogeneous and invariable chemical composition, even though it may undergo a change in phase. Therefore, liquid water or a mixture of ice and liquid water, and steam, a mixture of liquid water and water vapor, are pure substances.

When a substance occurs as a vapor at the saturation temperature and pressure, it is called *saturated vapor*. Saturation temperature is the temperature at which vaporization takes place at a given pressure. This pressure is called the *saturation pressure*. Thus, water at 100°C has a saturation pressure of 101.3 kPa. When the temperature of the vapor is greater than the saturation temperature at the existing pressure, it is called *superheated vapor* .

When a substance occurs in a liquid state at its saturation temperature and pressure, it is called a *saturated liquid*. If at the saturation pressure, the temperature of the liquid is lowered below the saturation temperature, it is called a *subcooled liquid*. In the case where, at the saturation temperature, a substance exists partly as liquid and partly as vapor, the ratio of the mass of water vapor to the total mass of the substance is expressed as the quality of the vapors. For example, if steam has 0.1 kg water and 0.9 kg vapor, the quality of steam is 0.9 divided by 1.0 (which represents the total mass of steam); thus steam quality equals 0.9 or 90%.

Difference between General Phase Diagram and Phase Diagram for Water

There is one significant difference between the general phase diagram and the water phase diagram. The slope of the line connecting the solid and liquid states in the water diagram is negative rather than positive. Water is an unusual substance in that its solid state is less dense than its liquid state. In liquid water, ice floats. As a result, a change in pressure has the opposite effect on those two phases. When ice is near its melting point, it can be converted into liquid water by applying pressure. In the liquid phase, water molecules are actually closer together than in the solid phase.

Curves

OA- Water & Vapor

OA is the vaporisation curve. The curve OA comes to an end at A. It has a critical pressure of 218 atm and a temperature of 374!. It represents the vapour pressure of a liquid at various temperatures.

Along the curve, two phases of water and water vapours coexist in equilibrium.

The vapour pressure is one atmosphere. The corresponding temperature in degrees Celsius is the boiling point of water, which is 100!.

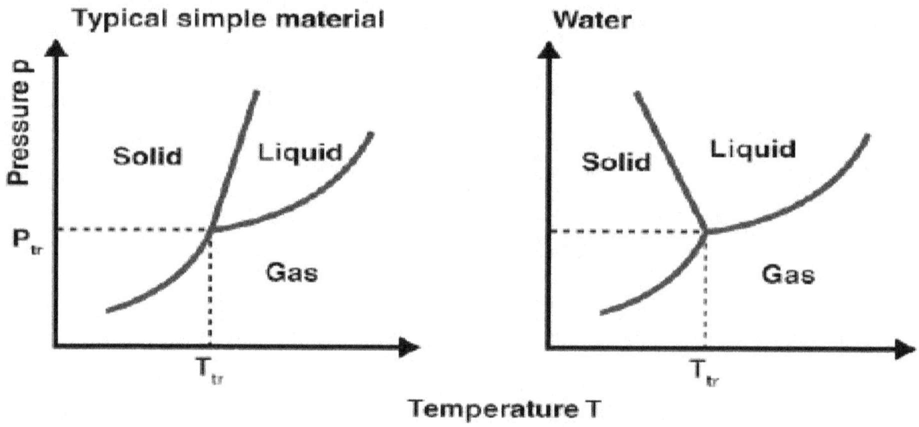

Fig. 7. General Phase Diagram

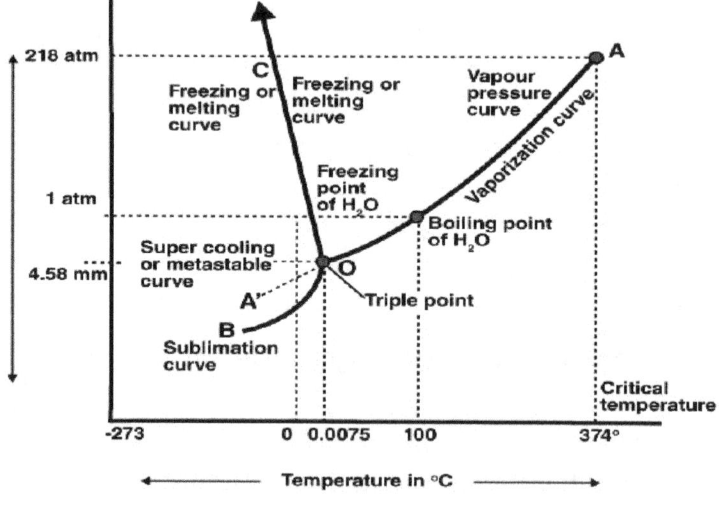

Fig. 8. Phase Diagram for Water

OB- Ice and Vapor

The OB curve is a sublimation curve. The curve OB comes to an end at B, which is the absolute zero temperature of -273!. It depicts the vapour pressure of solid ice at various temperatures.

In equilibrium, the two phases of solid ice and water vapour coexist.

OC- Ice & Water

The OC curve is the fusion curve. The OC curve comes to an end at C, the critical pressure.

In equilibrium, the two phases of solid ice and liquid water coexist. The curve shows that the melting point of ice decreases as pressure increases. At one atm, the line intersects the curve at 0!.

CONSERVATION OF MASS AND ENERGY

The law of conservation of mass states that mass can neither be created nor destroyed. Thus in a processing plant, the total mass of material entering the plant must equal the total mass of material leaving the plant, less any accumulation left in the plant. If there is no accumulation, then the simple rule holds that "what goes in must come out". Similarly all material entering a unit operation must in due course leave.

For example, if milk is being fed into a centrifuge to separate it into skim milk and cream, under the law of conservation of mass the total number of kilograms of material (milk) entering the centrifuge per minute must equal the total number of kilograms of material (skim milk and cream) that leave the centrifuge per minute.

Similarly, the law of conservation of mass applies to each component in the entering materials. For example, considering the butter fat in the milk entering the centrifuge, the weight of butter fat entering the centrifuge per minute must be equal to the weight of butter fat leaving the centrifuge per minute. A similar relationship will hold for the other components, proteins, milk sugars and so on.

The law of conservation of energy states that energy can neither be created nor destroyed. The total energy in the materials entering the processing plant, plus the energy added in the plant, must equal the total energy leaving the plant.

This is a more complex concept than the conservation of mass, as energy can take various forms such as kinetic energy, potential energy, heat energy, chemical energy, electrical energy and so on.

During processing, some of these forms of energy can be converted from one to another. Mechanical energy in a fluid can be converted through friction into heat energy. Chemical energy in food is converted by the human body into mechanical energy.

Note that it is the sum total of all these forms of energy that is conserved.

For example, consider the pasteurizing process for milk, in which milk is pumped through a heat exchanger and is first heated and then cooled. The energy can be considered either over the whole plant or only as it affects the milk. For total plant energy, the balance must include: the conversion in the pump of electrical energy to kinetic and heat energy, the kinetic and potential energies of the milk entering and leaving the plant and the various kinds of energy in the heating and cooling sections,as well as the exiting heat, kinetic and potential energies.

To the food technologist, the energies affecting the product are the most important. In the case of the pasteurizer, the energy affecting the product is the heat energy in the milk. Heat energy is added to the milk by the pump and by the hot water passing through the heat exchanger. Cooling water then removes part of the heat energy and some of the heat energy is also lost to the surroundings.

The heat energy leaving in the milk must equal the heat energy in the milk entering the pasteurizer plus or minus any heat added or taken away in the plant.

Heat energy leaving in milk = initial heat energy
+ heat energy added by pump
+ heat energy added in heating section
- heat energy taken out in cooling section
- heat energy lost to surroundings.

The law of conservation of energy can also apply to part of a process. For example, considering the heating section of the heat exchanger in the pasteurizer, the heat lost by the hot water must be equal to the sum of the heat gained by the milk and the heat lost from the heat exchanger to its surroundings.

From these laws of conservation of mass and energy, a balance sheet for materials and for energy can be drawn up at all times for a unit operation. These are called material balances and energy balances.

SCIENCE OF THERMODYNAMICS

Thermodynamics is directly related to the field of transport of phenomena disciplines, as in heat and material transfer. Thermodynamics plays a very important role in food engineering, for example, in the selection of the type of refrigerant used in a cooling system, in the humidity control of air, or in the final moisture content of a product. Many phase changes that appear during the processing or storage of food can be described through the principles of thermodynamics.

Also, the heat requisites of a heat exchanger employed in pasteurization processes or the equilibrium composition of a multicomponent mixture being separated can be determined through thermodynamic analysis. Furthermore, it is possible to determine the degradation rate of a nutrient with the aid of chemical kinetics. These are some of the examples highlighting the importance of thermodynamics in food processing. The purpose of this discussion is to present, in a concise form, thermodynamic concepts and functions that serve as background for the study and comprehension of food engineering processes.

In examining engineering processes, we are concerned mostly with a macroscopic view. The branch of thermodynamics that deals with this macroscopic approach is called *classical thermodynamics*. Another branch of thermodynamics, called *statistical thermodynamics*, is concerned with what happens at a molecular level, and the average behavior of a group of molecules is considered.

In food engineering, many of the processes of concern to a food engineer are applications of thermodynamics. For example, we may need to calculate heat and work effects associated with a given process. In other instances, the maximum work obtainable from a process may be the key calculation, or we may need to determine how to carry out a process with minimum work. Furthermore, we will encounter the need to determine relationships that exist between various variables of a system when it is at equilibrium.

When we conduct experiments and then want to know the behavior of a given system, the laws of thermodynamics are most useful in the analysis. Based on an experimental foundation, classical thermodynamics is concerned with the macroscopic properties of a system. These properties may be directly measured or they may be calculated from other properties that are directly measurable. For example, we can measure pressure of a gas enclosed in a chamber by attaching a pressure gauge to the chamber.

Thermodynamics also helps us in determining the potential that defines and determines the equilibrium. By knowing potential, we can determine the direction a process will undertake. Although thermodynamics may not tell us how long a process will take to arrive at its final state, it does help in determining what the final state will be. Thus, time is not a thermodynamic variable, and we will rely on other procedures to determine the rate of a process—another important area of calculations for food engineers.

WATER ACTIVITY AND MASS TRANSFER

Introduction

Drying is defined as the removal of moisture from a product, and in most practical situations the main stage during drying is the internal mass transfer. The mechanisms of water transfer in the product during the drying process can be summarized as follows:

- Water movement due to capillary forces
- Diffusion of liquid due to concentration gradients
- Surface diffusion
- Water vapor diffusion in pores filled with air
- Flow due to pressure gradients
- Flow due to water vaporization–condensation

In the pores of solids with rigid structure, capillary forces are responsible for the retention of water, whereas in solids formed by aggregates of fine powders, the osmotic pressure is responsible for water retention within the solids as well as in the surface. The type of material to be dried is an important factor to consider in all drying processes, since its physical and chemical properties play a significant role during drying due to possible changes that may occur and because of the effect that such changes may have in the removal of water from the product.

A hygroscopic material is one that contains bound water that exerts a vapor pressure lower than the vapor pressure of liquid water at the same temperature. It is expected that products made mainly of carbohydrates will behave in a hygroscopic way, since the hydroxyl groups around the sugar molecules allow formation of hydrogen bonds with water molecules. The interaction between the water molecules and the hydroxyl groups causes solvation or solubilization of sugars. In water soluble proteins, as in most of the globular proteins, the polar amino acids are uniformly distributed in the surface, while the hydrophobic groups are located towards the inside of the molecule. This arrangement allows formation of hydrogen bonds with water, which explains the solubility of this type of proteins.

Dehydration involves the simultaneous transfer of heat, mass and momentum in which heat penetrates into the product and moisture is removed by evaporation into an unsaturated gas phase. Owing to the complexity of the process, no generalized theory currently exists to explain the mechanism of internal moisture movement. Although it is now accepted that in most practical situations of air drying of foods the principal rate-determining step is internal mass transfer, there is no agreement on the mechanism of internal moisture movement. In the case of capillary-porous materials such as fruits and vegetables, interstitial spaces, capillaries and gas-filled cavities exist within the food matrix and water transport takes place via several possible mechanisms acting in various combinations. The possible mechanisms include:

- liquid diffusion caused by concentration gradients,
- liquid transport due to capillary forces,
- vapour diffusion due to shrinkage and partial vapour-pressure gradients (Stefan's law),

- liquid or vapour transport due to the difference in total pressure caused by external pressure and temperature (Poiseuille's law),
- evaporation and condensation effects caused by differences in temperature,
- surface diffusion in liquid layers at the solid interface due to surface concentration gradient,
- liquid transport due to gravity.

Most foods are classified as capillary porous rigid or capillary porous colloids. Therefore, it is often proposed that a combination of capillary flow and vapour diffusion mechanisms should be used to describe internal mass transfer. Water activity, rather than moisture content, influences biological reactions. In the regions of water adsorption on polar sites or when a mono-molecular layer exists, there is little enzyme activity. Enzyme activity begins only above the region of mono-molecular adsorption. When the moisture content of a substrate is reduced below 10 %, microorganisms are no longer active. It is necessary however to reduce the moisture content to below 5 % in order to preserve nutrition and flavour.

Water Activity

Water activity and moisture content of the material are very important in food unit operations. During processing and storage, many chemical and physical factors are influenced by the water activity and moisture content level. Chemical changes that are enhanced by water activity include enzymatic reactions, nonenzymatic browning, and microbial activity. In many food products, enzymes are not inactivated during the heating process.

Consequently, enzymatic reactions can take place at even low moisture contents. Water activity also affects the nonenzymatic browning reactions in foods. When water is present, carboxyl and amino compounds are involved as reactants, products, or catalysts in the browning process. Bacterial growth is also affected at fairly high water activity levels. If water activity is maintained at a value below 0.90, most bacteria remain dormant. Most yeasts and molds, however, can grow and multiply at water activity levels as low as 0.80. Physical changes such as texture and aroma can depend greatly on water activity.

Textural changes are most often seen in freeze drying and subsequent storing of foods, particularly meats and fish. The water activity in dried foods can also affect the retention of aroma. Different foods that are stored together will be altered if their individual relative humidities (water activities) are different. Under the action of a driving force created by a difference in water activities changes in the moisture and water activities of these foods will follow their own isotherm curve until an equilibrium water activity is achieved.

Water activity (a_w) is measured as the equilibrium relative humidity (ERH), the percent relative humidity (RH) of an atmosphere in contact with a product at the equilibrium water content. a_w is also ratio of the partial pressure of water in the headspace of a product (P) to the vapour pressure of pure water (P_0) at the same temperature

$a_w = ERH = P/P_0$

The relationship between a_w and the rate of deteriorative reactions is very important to design dehydration systems. Reducing a_w below 0.7 prevents microbiological spoilage.

However, although microbiological spoilage does not occur at $a_w = 0.7$, prevention of other deteriorative reactions needed to preserve a dried food product successfully reduction of a_w to below 0.3.

Mass Transfer

During dehydration, water is vapourized only from the surface. The transfer of water vapour from the wet surface to a stream of moving air is analogous to convection heat transfer, therefore, a mass transfer coefficient is used.

Moisture flux is proportional to the driving force which is the difference in vapour pressure on the surface and the vapour pressure of water in air surrounding the surface. At the same time that water is removed from the surface, water diffuses from the interior of a solid towards the surface.

The later is a general form of diffusion which is analogous to conduction heat transfer. The differential equations for conduction also apply to diffusion, but mass diffusivity is used in place of thermal diffusivity.

Example

A dry food product has been exposed to a 30% RH environment at 15°C for 5 h without a weight change. The moisture content has been measured and is 7.5% (wet basis). The product is moved to a 50% RH environment and a weight increase of 0.1 kg/kg product occurs before equilibrium is reached.

- Determine water activity of the product in the first and second environment.
- Compute moisture contents of the product on dry basis in both environments.

Solution

Equilibrium RH = 30% in first environment.

Product moisture content is 7.5% (wet basis) in first environment.

Weight gained in second environment = 0.1 kg / kg product.

The water activity is equilibrium RH divided by 100. so water activities are 0.3 and 0.5 in first & second environment respectively.

MATERIAL AND ENERGY BALANCES

Material quantities, as they pass through food processing operations, can be described by material balances. Such balances are statements on the conservation of mass. Similarly, energy quantities can be described by energy balances, which are statements on the conservation of energy. If there is no accumulation, what goes into a process must come out. This is true for batch operation. It is equally true for continuous operation over any chosen time interval.

Material and energy balances are very important in the food industry. Material balances are fundamental to the control of processing, particularly in the control of yields of the products. The first material balances are determined in the exploratory stages of a new process, improved during pilot plant experiments when the process is being planned and tested, checked out when the plant is commissioned and then refined and maintained as a control instrument as production continues.

When any changes occur in the process, the material balances need to be determined again.The increasing cost of energy has caused the food industry to examine means of reducing energy consumption in processing. Energy balances are used in the examination of the various stages of a process, over the whole process and even extending over the total food production system from the farm to the consumer's plate.

Material and energy balances can be simple, at times they can be very complicated, but the basic approach is general. Experience in working with the simpler systems such as individual unit operations will develop the facility to extend the methods to the more complicated situations, which do arise. The increasing availability of computers has meant that very complex mass and energy balances can be set up and manipulated quite readily and therefore used in everyday process management to maximise product yields and minimise costs.

Basic Principles

The law of conservation of mass leads to what is called a mass or a material balance.

Mass In = Mass Out + Mass Stored

Raw Materials = Products + Wastes + Stored Materials.

$Sm_R = Sm_P + Sm_W + Sm_S$

(where S (sigma) denotes the sum of all terms).

$Sm_R = m_{R1} + m_{R2} + m_{R3} + \ldots\ldots$ = Total Raw Materials.

$Sm_P = m_{P1} + m_{P2} + m_{P3} + \ldots\ldots$ = Total Products.

$Sm_W = m_{W1} + rn_{W2} + m_{W3} + \ldots\ldots$ = Total Waste Products.

$Sm_S = m_{S1} + m_{S2} + m_{S3} + \ldots$ = Total Stored Materials.

If there are no chemical changes occurring in the plant, the law of conservation of mass will apply also to each component, so that for component A:

m_A in entering materials = m_A in the exit materials + m_A stored in plant.

For example, in a plant that is producing sugar, if the total quantity of sugar going into the plant in sugar cane or sugar beet is not equalled by the total of the purified sugar and the sugar in the waste liquors, then there is something wrong. Sugar is either being burned (chemically changed) or accumulating in the plant or else it is going unnoticed down the drain somewhere. In this case:

$$(m_A) = (m_{AP} + m_{AW} + M_{AS} + m_{AU})$$

where m_{AU} is the unknown loss and needs to be identified. So the material balance is now:

Raw Materials = Products + Waste Products + Stored Products + Losses

where Losses are the unidentified materials.

Just as mass is conserved, so is energy conserved in food processing operations. The energy coming into a unit operation can be balanced with the energy coming out and the energy stored.

Energy In = Energy Out + Energy Stored

$$SE_R = SE_P + SE_W + SE_L + SE_S$$

where:

$SE_R = E_{R1} + E_{R2} + E_{R3} + \ldots\ldots$ = Total Energy Entering

$SE_P = E_{P1} + E_{P2} + E_{P3} + \ldots\ldots$ = Total Energy Leaving with Products

$SE_W = E_{W1} + E_{W2} + E_{W3} + \ldots\ldots$ = Total Energy Leaving with Waste Materials

$SE_L = E_{L1} + E_{L2} + E_{L3} + \ldots\ldots$ = Total Energy Lost to Surroundings

$SE_S = E_{S1} + E_{S2} + E_{S3} + \ldots\ldots$ = Total Energy Stored

Energy balances are often complicated because forms of energy can be interconverted, for example mechanical energy to heat energy, but overall the quantities must balance.

CHAPTER

FLUID FLOW IN FOOD PROCESSING 3

Many raw materials for foods and many finished foods are in the form of fluids. These fluids have to be transported and processed in the factory. Food technologists must be familiar with the principles that govern the flow of fluids, and with the machinery and equipment that is used to handle fluids. In addition, there is an increasing tendency to handle powdered and granular materials in a form in which they behave as fluids. Fluidization, as this is called, has been developed because of the relative simplicity of fluid handling compared with the handling of solids.

The engineering concept of a fluid is a wider one than that in general use, and it covers gases as well as liquids and fluidized solids. This is because liquids and gases obey many of the same laws so that it is convenient to group them together under the general heading of fluids.

The study of fluids can be divided into the study of fluids at rest - fluid statics, and the study of fluids in motion - fluid dynamics. For some purposes, further subdivision into compressible fluids such as gases, and incompressible fluids such as liquids, is necessary. Fluids in the food industry vary considerably in their properties. They include such materials as:

Thin liquids - milk, water, fruit juices,

Thick liquids - syrups, honey, oil, jam,

Gases - air, nitrogen, carbon dioxide,

Fluidized solids - grains, flour, peas.

FLUID STATICS

A very important property of a fluid at rest is the pressure exerted by that fluid on its surroundings, that is the fluid pressure

Pressure is defined as force exerted on an area. Under the influence of gravity, a mass of any material exerts a force on whatever supports it. The magnitude of this force is equal to the mass of the material multiplied by the acceleration due to gravity. The mass of a fluid can be calculated by multiplying its volume by its density, which is defined as its mass per unit volume. Thus the equation can be written:

$$F = mg = Vrg$$

where F is the force exerted, m is the mass, g the acceleration due to gravity, V the volume and ρ (the Greek letter rho) the density. The units of force are Newtons, or kg m s^{-2}. and of pressure Pascals, one Pascal being one Newton m^{-2}, and so one Pascal is also one kg m^{-1} s^{-2}.

For a mass to remain in equilibrium, the force it exerts due to gravity must be resisted by some supporting medium. For a weight resting on a table, the table provides the supporting reaction; for a multi-storey building, the upper floors must be supported by the lower ones so that as you descend the building the burden on the floors increases until the foundations support the whole building. In a fluid, the same situation applies. Lower levels of the fluid must provide the support for the fluid that lies above them. The fluid at any point must support the fluid above. Also, since fluids at rest are not able to sustain shearing forces, which are forces tending to move adjacent layers in the fluid relative to one another, it can be shown that the forces at any point in a fluid at rest are equal in all directions. The force per unit area in a fluid is called the fluid pressure. It is exerted equally in all directions.

Consider a horizontal plane in a fluid at a depth Z below the surface, as illustrated in Fig. 1.

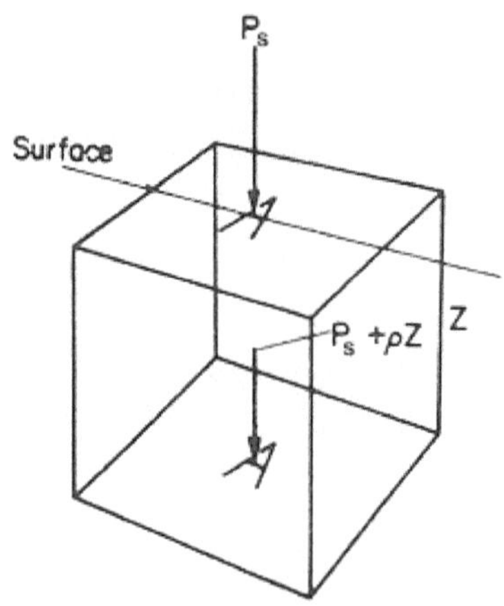

Figure 1 Pressure in a fluid

If the density of the fluid is r, then the volume of fluid lying above an area *A* on the plane is *ZA* and the weight of this volume of fluid, which creates a force exerted by it on the area *A* which supports it, is *ZrAg*. But the total force on the area *A* must also include any additional force on the surface of the liquid. If the force on the surface is P_s per unit area,

$$F = AP_s + Z\rho Ag \quad (1)$$

where F is the total force exerted on the area A and P_s is the pressure above the surface of the fluid (e.g. it might be atmospheric pressure). Further, since total pressure P is the total force per unit area,

$$P = F/A = P_s + Zrg \quad (2)$$

In general, we are interested in pressures above or below atmospheric. If referred to zero pressure as datum, the pressure of the atmosphere must be taken into account. Otherwise the atmospheric pressure represents a datum or reference level from which pressures are measured. In these circumstances we can write:

$$P = Zrg \quad (3)$$

This may be considered as the fundamental equation of fluid pressure. It states that the product of the density of the fluid, acceleration due to gravity and the depth gives the pressure at any depth in a fluid.

EXAMPLE 1. Total pressure in a tank of peanut oil

Calculate the greatest pressure in a spherical tank, of 2 m diameter, filled with peanut oil of specific gravity 0.92, if the pressure measured at the highest point in the tank is 70 kPa.

Density of water $= 1000 \text{ kg m}^{-3}$

Density of oil $= 0.92 \times 1000 \text{ kg m}^{-3} = 920 \text{ kg m}^{-3}$

Z = greatest depth = 2 m

and $g = 9.81 \text{ m s}^{-2}$

Now $P = Zrg$

$= 2 \times 920 \times 9.81 \text{ kg m}^{-1} \text{ s}^{-2}$

= 18,050 Pa = 18.1 kPa.

To this must be added the pressure at the surface of 70 kPa.

Total pressure = **70 + 18.1 = 88.1 kPa.**

Note in **Example 1**, the pressure depends upon the pressure at the top of the tank added to the pressure due to the depth of the liquid; the fact that the tank is spherical (or any other shape) makes no difference to the pressure at the bottom of the tank.

In the previous paragraph, we established that the pressure at a point in a liquid of a given density is solely dependent on the density of the liquid and on the height of the liquid above the point, plus any pressure which may exist at the surface of the liquid. When the depths of the fluid are substantial, fluid pressures can be considerable.

For example, the pressure on a plate 1 m^2 lying at a depth of 30 m will be the weight of 1 m^3 of water multiplied by the depth of 30 m and this will amount to × 30 294.3 = 9.81 × 1000 kPa. As 1 tonne exerts a force on 1 m^2 of 9810 = 9.81 × 1000 Pa = 9.81 kPa the pressure on the plate is equal to that of a weight of 294.3 / 9.81 = 30 tonnes of water.

Pressures are sometimes quoted as **absolute pressures** and this means the total pressure including atmospheric pressure. More usually, pressures are given as **gauge pressures**, which implies the pressure above atmospheric pressure as datum. For example, if the absolute pressure is given as 350 kPa, the gauge pressure is (350 - 100) = 250 kPa assuming that the atmospheric pressure is 100 kPa.

Standard atmospheric pressure is actually 101.3 kPa but for our practical purposes 100 kPa is sufficiently close and most convenient to use. Any necessary adjustment can easily be made.

Another commonly used method of expressing pressures is in terms of "**head**" of a particular fluid. From eqn. (3) it can be seen that there is a definite relationship between pressure and depth in a fluid of given density. Thus pressures can be expressed in terms of depths, or heads as they are usually called, of a given fluid.

The two fluids most commonly used, when expressing pressures in this way, are water and mercury. The main reason for this method of expressing pressures is that the pressures themselves are often measured by observing the height of the column of liquid that the pressure can support. It is straightforward to convert pressures expressed in terms of liquid heads to equivalent values in kPa by the use of eqn. (3).

EXAMPLE 3.2. Head of Water

Calculate the head of water equivalent to standard atmospheric pressure of 100 kPa.

Density of water = 1000 kg m^{-3},

g = 9.81 m s^{-2}

and pressure = 100 kPa

= 100 × 10^3 Pa = 100 × 10^3 kg $m^{-1}s^{-2}$.

but from eqn. (3):

$Z = P/rg$

= (100 × 10^3)/ (1000 × 9.81)

= 10.2 m

EXAMPLE 3.3. Head of mercury

Calculate the head of mercury equivalent to a pressure of two atmospheres.

Density of mercury = 13,600 kg m^{-3}

$$Z = (2 \times 100 \times 10^3)/(13{,}600 \times 9.81)$$
$$= \mathbf{1.5\ m}$$

Liquid Transport Systems

In most processes fluids have to be moved so that the study of fluids in motion is important. Problems on the flow of fluids are solved by applying the principles of conservation of mass and energy. In any system, or in any part of any system, it must always be possible to write a mass balance and an energy balance. The motion of fluids can be described by writing appropriate mass and energy balances and these are the bases for the design of fluid handling equipment.

MASS BALANCE

Consider part of a flow system, such for example as that shown in Fig. 2.

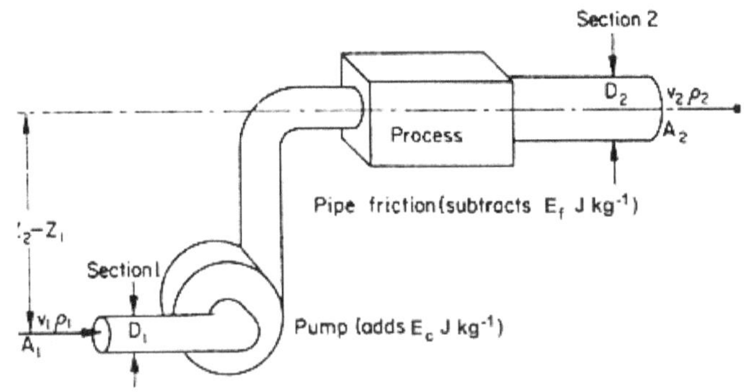

Fig. 2. Mass and energy balance in fluid flow

This consists of a continuous pipe that changes its diameter, passing into and out of a unit of processing plant, which is represented by a tank. The processing equipment might be, for example, a pasteurizing heat exchanger. Also in the system is a pump to provide the energy to move the fluid.

In the flow system of Fig. 2 we can apply the law of conservation of mass to obtain a mass balance. Once the system is working steadily, and if there is no accumulation of fluid in any part the system, the quantity of fluid that goes in at section 1 must come out at section 2. If the area of the pipe at section 1 is A_1, the velocity at this section, v_1 and the fluid density r_1, and if the corresponding values at section 2 are A_2, v_2,

r_2, the mass balance can be expressed as

$$r_1 A_1 v_1 = r_2 A_2 v_2 \quad (4)$$

If the fluid is incompressible $r_1 = r_2$ so in this case

$$A_1 v_1 = A_2 v_2 \quad (5)$$

Equation (5) is known as the **continuity equation** for liquids and is frequently used in solving flow problems. It can also be used in many cases of gas flow in which the change in pressure is very small compared with the system pressure, such as in many air-ducting systems, without any serious error.

EXAMPLE 3.4. Velocities of flow

Whole milk is flowing into a centrifuge through a full 5 cm diameter pipe at a velocity of 0.22 m s^{-1}, and in the centrifuge it is separated into cream of specific gravity 1.01 and skim milk of specific gravity 1.04. Calculate the velocities of flow of milk and of the cream if they are discharged through 2 cm diameter pipes. The specific gravity of whole milk of 1.035.

From eqn. (4):

$r_1 A_1 v_1 = r_2 A_2 v_2 + r_3 A_3 v_3$

where suffixes 1, 2, 3 denote respectively raw milk, skim milk and cream. Also, since volumes will be conserved, the total leaving volumes will equal the total entering volume and so

$A_1 v_1 = A_2 v_2 + A_3 v_3$ and from this equation

$v_2 = (A_1 v_1 - A_3 v_3)/A_2$ (a)

This expression can be substituted for v_2 in the mass balance equation to give:

$r_1 A_1 v_1 = r_2 A_2 (A_1 v_1 - A_3 v_3)/A_2 + r_3 A_3 v_3$

$r_1 A_1 v_1 = r_2 A_1 v_1 - r_2 A_3 v_3 + r_3 A_3 v_3.$

So $A_1 v_1 (r_1 - r_2) = A_3 v_3 (r_3 - r_2)$ (b)

From the known facts of the problem we have:

$A_1 = (p/0.05)\ \ (4)^2 = {}^{3\text{-}}10\ \ 1.96\ m^2$

$A_2 = A_3 = (p/0.02)\ \ (4)^2 = {}^{4\text{-}}10\ \ 3.14\ m^2$

$v_1 = 0.22\ m\ s^{-1}$

$r_1 = 1.035 \times r_w,\ r_2 = 1.04 \times r_w,\ r_3 = 1.01 \times r_w$

where r_w is the density of water.

Substituting these values in eqn. (b) above we obtain:

$-1.96\ \ 10^{-3}\ \ {}^{4\text{-}}10\ \ 3.14- = (0.005)\ 0.22 \times v_3\ \ (0.03)$

so $v_3 = 0.23\ m\ s^{-1}$

Also from eqn. (a) we then have, substituting 0.23 m s^{-1} for v_3,

$v_2 = [({}^{4\text{-}}10\ \ 3.14\ /\ [(0.23\ \ {}^{4\text{-}}10\ \ 3.14) - (0.22\ \ {}^{3\text{-}}10\ \ 1.96$

$= \mathbf{1.1 m\ s^{-1}}$

Energy Balance

In addition to the mass balance, the other important quantity we must consider in the analysis of fluid flow, is the energy balance. Referring again to **Fig. 2**, we shall consider the changes in the total energy of unit mass of fluid, one kilogram, between Section 1 and Section 2.

Firstly, there are the changes in the intrinsic energy of the fluid itself which include changes in:

(1) Potential energy.

(2) Kinetic energy.

(3) Pressure energy.

Secondly, there may be energy interchange with the surroundings including:

(4) Energy lost to the surroundings due to friction.

(5) Mechanical energy added by pumps.

(6) Heat energy in heating or cooling the fluid.

In the analysis of the energy balance, it must be remembered that energies are normally measured from a datum or reference level. Datum levels may be selected arbitrarily, but in most cases the choice of a convenient datum can be made readily with regard to the circumstances.

Potential energy

Fluid maintained above the datum level can perform work in returning to the datum level. The quantity of work it can perform is calculated from the product of the distance moved and the force resisting movement; in this case the force of gravity. This quantity of work is called the potential energy of the fluid.

Thus the potential energy of one kilogram of fluid at a height of Z (m) above its datum is given by E_p, where

$$E_p = Zg \quad (\mathrm{J})$$

Kinetic energy

Fluid that is in motion can perform work in coming to rest. This is equal to the work required to bring a body from rest up to the same velocity, which can be calculated from the basic equation

$v^2 = 2as$, therefore $s = v^2/2a$,

where v (m s^{-1}) is the final velocity of the body, a (m s^{-2}) is the acceleration and s (m) is the distance the body has moved.

Also work done = $W = F \times s$, and from Newton's Second Law, for m kg of fluid

$F = ma$

and so

$E_k = W = mas = mav^2/2a = mv^2/2$

The energy of motion, or kinetic energy, for 1 kg of fluid is therefore given by E_k where

$E_k = v^2/2$ (J).

Pressure energy

Fluids exert a pressure on their surroundings. If the volume of a fluid is decreased, the pressure exerts a force that must be overcome and so work must be done in compressing the fluid. Conversely, fluids under pressure can do work as the pressure is released. If the fluid is considered as being in a cylinder of cross-sectional area A (m^2) and a piston is moved a distance L (m) by the fluid against the pressure P (Pa) the work done is PAL joules. The quantity of the fluid performing this work is ALr (kg). Therefore the pressure energy that can be obtained from one kg of fluid (that is the work that can be done by this kg of fluid) is given by E_r where

$$E_r = PAL / AL\mathrm{r}$$
$$= P/\mathrm{r} \quad (J)$$

Friction loss

When a fluid moves through a pipe or through fittings, it encounters frictional resistance and energy can only come from energy contained in the fluid and so frictional losses provide a drain on the energy resources of the fluid. The actual magnitude of the losses depends upon the nature of the flow and of the system through which the flow takes place. In the system of Fig. 2, let the energy lost by 1 kg fluid between section 1 and section 2, due to friction, be equal to E_f (J).

Mechanical energy

If there is a machine putting energy into the fluid stream, such as a pump as in the system of Fig. 2, the mechanical energy added by the pump per kg of fluid must be taken into account. Let the pump energy added to 1 kg fluid be E_c (J). In some cases a machine may extract energy from the fluid, such as in the case of a water turbine.

OVERVIEW OF FOOD PROCESSING EQUIPMENT

Food processing equipment is an umbrella term referring to the components, processing machines, and systems used to handle, prepare, cook, store, and package food and food products. Although this equipment is primarily aimed toward the transformation—i.e., increasing the palatability, consumability, and digestibility—or preservation—i.e., extending the *shelf life*—of food, some pieces of equipment are also employed to perform preliminary or auxiliary functions, such as handling, preparation, and packaging.

Employed for food and food product applications ranging from bakery goods to beverages and dairy to produce, a wide range of food processing equipment is available to execute the various unit operations necessary during a complete production cycle, such as washing, separating, mixing, baking, freezing, and sealing.

Depending on the demands of the operation (and the overarching food processing application), this equipment can be designed and constructed to handle solid, semi-solid, or liquid food products by batch or continuously. Some of the other design considerations include the food grade material used for construction, hygienic and governmental standards, sizing, cost, and integration of automation or analytical components.

Each of these characteristics can influence the performance and efficiency of the equipment, but choosing the optimal design and construction is dependent on the specifications and requirements of the particular food processing application.

This section focuses on food processing equipment, exploring the types available, and providing examples of each. Additionally, it outlines some of the considerations for designing and choosing equipment for a food processing application.

Types of Food Processing Equipment

While the wide range of food processing equipment available can be classified and categorized in several different ways—e.g., end product form, mode of operation, application, etc.—this article follows industry standards and groups them by their respective functions. The food processing production cycle can be broken into several stages, characterized by a specific function and during which individual *unit operations* are performed. For example, within the preparation stage, the primary function is to prepare the food material for further processing, and some of the unit operations performed include washing and separating. Other secondary functions provide support to the primary functions of the production cycle, such as material handling and system control operations which convey food materials between process stations or maintain the required processing conditions and standards, respectively.

Some of the most common functions by which food processing equipment are grouped include:

- Preparation
- Mechanical processing
- Heat processing
- Preservation

- Packaging

PREPARATION EQUIPMENT

Each time equipment is scaled to a different size (kitchen equipment, lab equipment, pilot plant equipment, plant equipment), the equipment will function slightly differently, which can have small or large effects on the finished product. These effects can be due to small differences in how the equipment works, how full the equipment is, differences in volume to surface area ratios, and more. It is best to be prepared for changes and adjustments when scaling up equipment. Food industry equipment used for initial food preparation processes for products.

Some of the unit operations which manufacturers employ during the raw material preparation stage include cleaning, sorting, grading, or peeling (or skinning). Table 1 below describes some of the preparatory unit operations and outlines the equipment used to execute them.

Mechanical Processing Equipment

Mechanical processing operations are employed (without the application of heat or chemicals) to reduce, enlarge, homogenize, or otherwise change the physical form of solid, semi-solid, and liquid food matter. By altering the form and size of the food matter, manufacturers can facilitate and increase the efficiency and effectiveness of subsequent processes, improve the overall quality and edibility, and allow for a greater range of food products to be produced.

Fig. 3. A close-up of an electrical dough mixer

Table 1 – Food Preparation Equipment by Unit Operation

Unit Operation	Description	Equipment Employed*
Cleaning	Removes foreign matter and contaminants-e.g., soil, oil, insects,	Wet Processes
	skins, chemicals, etc.-from the surface of raw food material via wet	Soak/floatation tanks (soaking)
	and dry cleaning processes	Spray washers (spray washing)
		Washing systems (washing)
		Sterilizers (sterilizing)
		Ultrasonic cleaners
		Dry Processes
		Air classifiers
		Magnetic separators
		Screening separators
Grading	Closely related to, and often precluding, sorting processes	Tungsten lights (candling)
	Assesses several characteristics of food matter (e.g., flavor,	Image processors
	damage, skin color, aroma, etc.) to determine the overall quality	Laboratory equipment
Peeling/Skinning	Removes inedible or undesirable material to increase the overall	Pressure vessels (flash steam peeling)
	quality and/or appearance of the final food product	Stationary/rotating blades (knife peeling)
		Carborundum abrasive rollers/ bowls (abrasion peeling)
		Conveyors and furnaces (flame peeling)
Sorting	Operates similarly to and overlaps with dry cleaning processes	See Dry Processes
	Classifies and separates foreign matter and contaminants from	Sorting machinery
	raw food material based on a measurable physical characteristic	Disc separators (shape sorting)
	(typically size, shape, weight, or color)	Sieves/screens (size sorting)

Unit Operation	Description	Equipment Employed*
		Machine vision sorting systems
		Sorting conveyors
Note: *If applicable, the specific process related to the piece of equipment is indicated in parentheses.		

Within the general mechanical processing classifications—i.e., size reduction, size enlargement, homogenization—there are numerous unit operations, such as cutting, forming, and grinding/crushing, which fall below them. Table 2 below describes the overarching classifications and outlines some of their more specific unit operations and the equipment used to execute them.

Fig. 4. Bread dough bakes in an industrial oven.

Table 2 – Mechanical Food Processing Equipment by Unit Operation

Unit Operation	Description	Equipment Employed
Size Reduction*	Reduces the average particle size of solid food matter through	Grinding/Crushing
	mechanical processes involving compression, shear, or impact force	Impact mills
		Pressure mills
		Attrition mills
		Jaw crushers
		Roll crushers
		Strainers/pulpers

Unit Operation	Description	Equipment Employed
		Cutting/Chopping
		Knives/blades
		Band saws
		Slicing machines
		Meat grinders
SizeEnlargement*	Increases the average particle size of solid food matter through mechanical processes, such as extrusion, agglomeration, or forming	Extrusion
		Non-thermal extruders
		Single-screw extruders
		Twin-screw extruders
		Refrigerated extruders
		Agglomeration
		Rotating pans
		Rotating drums
		High-speed agitators
		Tableting equipment
		Pelletizing equipment
		Forming
		Bread molders
		Pie and biscuit formers
		Confectionary molders
		Enrobing machines
Homogenization*	Also referred to as emulsification Reduces the average particle size and increases the consistency of semi-solid and liquid food matter	Homogenizers Colloid mills High shear mixers
Mixing**	Also referred to as blending	
	Combines and disperses two or more components into one another to achieve and maintain a uniform mixture and/or an alteration to the functional or aesthetic qualities of the food product (e.g., texture)	Fluid Mixers Agitated tanks Paddle mixers
	Type of equipment depends on the form of the food components-gas/ liquid, liquid/ liquid, liquid/solid, solid/solid	Anchor mixers

Unit Operation	Description	Equipment Employed
		Turbine mixers
		Dough/Paste Mixers
		Horizontal dough mixers
		Sigma-blade mixers
		Cutter mixers
		Solids Mixers
		Diffusive (passive) mixers
		Convective (active) mixers
		Drum blenders
Note: *In these sections, the specific unit operation related to the piece of equipment is indicated immediately prior.**In this section, the form of the food material suitable for the piece of equipment is indicated immediately prior.		

Table 3 - Heat Processing Equipment by Unit Operation

Unit Operation	Description	Equipment Employed
Baking	Similar to, and often referred interchangeably with, roasting Employs heated air (heated by convection, conduction, and radiation)-and, in some cases, water vapor-to heat and produce physical and chemical changes in food material, such as texture or flavor Assists in the preservation of food matter by destroying microorganisms and reducing the amount of moisture at the food surface Suitable for producing bread, crackers, biscuits, and other flour- based or dough-based products	**Baking ovens** Direct heating ovens Indirect heating ovens Batch ovens Continuous and semi-continuous ovens
Blanching	Employs heated water or steam to reduce the number of microorganisms and inactivate undesirable enzymes which can cause spoilage Also cleans, removes excess air from, softens, and improves the overall quality Typically follows preparation operations and precedes preservation operations, such as packaging, dehydrating, or freezing Suitable for fruits and vegetables	**Blanchers** Steam blanchers Hot water blanchers
Dehydration	Employs heat to remove (i.e., evaporate) water from solid, semi-solid, or liquid food material with the intention of producing a solid food product with sufficiently low water content Increases the shelf life of food products due to the reduced water content which inhibits microbial growth and enzyme activity Reduces weight and volume and/or transforms the form of the final food product	**Dryers** Convective dryers Contact (conductive) dryers Vacuum dryers Freeze dryers

Unit Operation	Description	Equipment Employed
Evaporation	Removes volatile solvents (typically water) from food material by boiling to increase the concentration of solid contents Increases the shelf life of food products due to the reduced water content, but also increases the rate of chemical deterioration Reduces the weight and volume of the final food product Typically precedes operations, such as crystallization, precipitation, and coagulation Suitable for liquid-based food products	**Heat exchangers** Evaporators Condensers
Frying	Employs heated (~160-180 °C) fat or oil to transfer heat directly to food material Reduces moisture content, forms a surface crust (changes texture and structure), and inactivates microorganisms which improves shelf life and overall quality	**Fryers** Batch fryers Continuous fryers
Pasteurization	Processes food material under medium temperatures (70-100 °C) to inactivate most enzymes and microorganisms (but not spores) which cause spoilage Produces food products with limited shelf lives (short-term preservation method)	**Pasteurizers** In-container pasteurizers Continuous flow pasteurizers
	Little to no impact to quality and characteristics beyond the shelf life Suitable for dairy, fruit/ vegetable-based, wine, beer, and egg products	**Heat exchangers]** Plate heat exchangers Concentric tube heat exchangers
Roasting	Similar to, and often referred interchangeably with, baking Employs heated air (heated by convection, conduction, and radiation)-and, in some cases, water vapor-to heat and produce physical and chemical changes in food material, such as texture or flavor Assists in the preservation of food matter by destroying microorganisms and reducing the amount of water at the food surface Suitable for meats, nuts, vegetables, etc.	**Roasting machinery** Roasting ovens
Sterilization	Processes food material under high temperatures (100+ °C) to inactivate all micro- organisms and enzymes (including microbial spores) Can be heated by steam, hot water, or direct flames May result in a significant impact on quality and characteristics	**Sterilizers/ sterilizing retorts** In-container sterilizers 'Continuous flow sterilizers

Heat Processing Equipment (Preservation by the Application of Heat)

Depending on whether the application (and the specific unit operation) is aimed towards heating or cooling the food material, heat transfer equipment can be used to direct heat towards or away from the material, respectively. This section of the article will focus primarily on the applications and equipment aimed towards heating food products, while the following section—Preservation Equipment—will touch on the applications and equipment aimed towards cooling food products, as well as those intended to preserve and extend the shelf life of food products.

Heat processing equipment—i.e., equipment which heats food—can cause not only physical changes in the food material, but chemical, biochemical, and biological changes as well. These changes can transform and affect the overall quality of the resulting food products—such as by altering the chemical structure or enhancing the flavor—and serve as a preservation method by inhibiting or destroying the microorganisms or enzymes which cause spoilage.

There are many unit operations employed during the heat processing stage, including blanching, baking, roasting, and frying, and Table 3 below describes some of them and outlines the equipment used to execute them.

Preservation Equipment

There is significant overlap between heat processing equipment and preservation equipment, the previous section already covers the former category—i.e., preservation methods (and their respective equipment) which apply heat. Therefore, this section will focus on other preservation methods, processes, and equipment.

Fig. 5. Meat and meat products waiting in refrigerated storage

The preservation stage of the food processing production cycle ultimately aims to prevent or inhibit the spoilage and increase the shelf life of food products. There is a wide range of preservation methods available ranging from refrigeration to irradiation, each of which acts to destroy microorganisms and enzymes within the food material or, at the very least, limit and depress their activity.

DESIGN OF FOOD FREEZING

Freezing is one of the common processes for preservation of foods. Preservation of food by freezing occurs by several mechanisms. The reduction of temperature to levels below 0°C causes a significant reduction in growth rates for microorganisms and the corresponding deterioration of product due to microbial activity. In addition to this enzymatic and oxidation reactions will be slow downed or arrested. In addition, formation of ice crystals with in the product changes the availability of water to participate in reactions. The engineering aspects of food freezing include:

- Design of food freezing system
- Design of refrigeration system
- Prediction of the rate at which freezing progresses.

Freezing Process

Actual freezing process in food products is somewhat more complex than freezing of pure water. In water the temperature decreases as heat is removed from the system until freezing point is reached. After small amount of sub cooling / super cooling, the temperature remains constant as latent heat is removed from the water system. Following this latent heat removal, the temperature deceases again.

In a food product removal of heat energy results in a temperature decrease until initial freezing point is reached. Initial freezing results in crystallization of a portion of the water, resulting in a concentration of the remaining solution and further reduction of the freezing point of that unfrozen portion. This results in additional decrease in temperature. This process continues till *eutectic point* of the solute present in the food product. In single solute system, the removal of heat energy beyond this point does result in temperature decrease, but with crystallization of solute as well as ice crystal formation.

In actual food product system, it is very probable that more than one solute will be present. If the temperature is monitored at *thermal center* of a food (the point that cools most slowly) as heat is removed, the freezing curve obtained will be like

Ice Crystal Formation

The freezing point of a food is the temperature at which a minute crystal of ice exists in equilibrium with surrounding water. However before an ice crystal can form, a nucleus of water molecules must be present.

There are two types of nucleation: homogeneous nucleation (the chance orientation and combination of water molecules) and heterogeneous nucleation (the formation of a nucleus around suspended particles or at cell wall). Heterogeneous nucleation is more likely to occur in foods and takes place during super cooling.

The length of super cooling period depends on the type of food and the rate at which heat is removed. High rates of heat transfer therefore produce a large number of small ice crystals.

The rate of ice crystals growth is controlled by rate of heat transfer. The rate of mass transfer does not control the rate of crystal growth except towards the end of freezing period. The time taken for the temperature of a food to pass through the *critical zone* determines both the number and the size of ice crystals.

Enthalpy Change During Freezing

One of the basic considerations in the design of a system for the freezing process is the refrigeration requirement for reducing the food product temperature to the desired level. The enthalpy changes required will reduce the product from some temperature above the freezing point to some temperature below the freezing point and can be represented by

$$\Delta H = \Delta Hs + \Delta Hu + \Delta H_L + \Delta H_F$$

where the terms on right hand side represent the sensible heat required to reduce the product solids temperature from initial to storage temperature (ΔHs), the sensible heat removed to reduce the unfrozen portion of the product to the storage temperature(ΔH_u), the latent heat removed (ΔH_L), and the sensible heat removed to reduce the frozen water portion of the product to the storage temperature(ΔH_F).

Sensible heat ΔHs is given by,

$\Delta Hs = M\ Cp\ (Ti - T_F)$ where T_F = freezing point temperature

Evaluation of other components is somewhat complex because of changing state of product below initial freezing point. Mass of unfrozen product and frozen product are changing and are temperature dependant. Enthalpy change required to reduce the unfrozen portion of the product to various temperatures below initial freezing point, T_F is given by:

$$\Delta H_u = M_u(T)\ C_{p',u}\ (T)\ (T_F - T)$$

Similarly,

$$\Delta H_F = M_F\ (T)\ C_{p',F}\ (T_F - T)$$

these equations can be writen in differential form as:

$$dH_u = M_u\ (T)\ C_{p',u}(T)\ dT \text{ and}$$

$$dH_F = M_F\ (T)\ C_{p',F}\ dT$$

Latent heat portion is given by:

$$\Delta H_L = M_F\ (T)\ L$$

Unfrozen and frozen portions of product at any temperature below the initial freezing point can be calculated by

$\ln X_A = l'/R_g\,(1/T_{Ao} - 1/T_A)$

Where, T_{Ao} - freezing point of pure liquid, K

X_A - is mole fraction of water in solution

R_g - gas constant , 8.314 kJ/ kg mol K.

T_A - absolute temperature of aqueous solution, K

l' - Latent heat of fusion, J/ mol (6003 J/ mol for dilute liquids)

After obtaining information on the frozen and unfrozen fractions as function of temperature and specific heat of unfrozen fraction the above equations can be evaluated by integration.

Volume changes:

The volume of ice is 9 % greater than that of pure water hence expansion of food after freezing is expected. The degree of expansion depends on:

- Moisture content (higher moisture content produce greater changes)
- Cell arrangement: Plant material with intra cellular air spaces will absorb internal increases in volume without large changes in their overall size.
- The concentrations of solute: High concentrations reduce freezing point and do not freeze or expand.
- Freezer temperature: This determines the amount of unfrozen water and hence the degree of expansion.

Prediction of Freezing Rates/ Time

Most important consideration in food freezing problems is the prediction of time required to accomplish a given freezing process. The concept of "*thermal arrest time*", which may be defined as the time required to reduce the temperature of the product to some stated temperature below the initial freezing point.

During freezing, heat is conducted from the interiors of a food to the surface and is removed by the freezing medium. The factors which influence rate of heat transfer are:

- The thermal conductivity of food
- The area of food available for heat transfer
- The distance that the heat must travel through the food
- The temperature difference between the food and freezing medium
- The insulating effect of boundary film of air surrounding the food.

Prediction of Freezing Time is Complicated because of the following reasons:

- Differences in initial temperature of the food
- Differences in initial size and shape of pieces of food
- Differences in the freezing point and the rate of ice crystal formation with in different regions of pieces of food.
- Changes in density, thermal conductivity, specific heat, and thermal diffusivity with reduction in temperature of food.

Factors Influencing Freezing Time

There are several parameters that influence freezing time and that will influence the design of equipment used for food freezing.

- Freezing medium temperature, where lower magnitude will decrease freezing time.
- Product size will influence freezing time.
- Convective heat transfer coefficient, h_c will influence freezing time significantly.
- The initial freezing point.
- Product properties (T_F, ρ, k, C_P) will influence freezing time predictions.

Design Criteria For Selection Food Freezing Equipment

- It should be of sanitary design.
- All product contact surfaces should be made of Stainless Steel.
- The equipment should be adaptable for different food products.
- It should have high efficiency.
- Quick freezing & continuous operation.
- Material of construction should be non-toxic, non-corrosive and odour less & taintless.
- The equipment should be of simple design and easy to operate.
- Type of product processed.
- Reasonable freezing rates.
- Airflow rates should be uniform and air should make intimate contact with the product.

Changes in Food during Freezing

In any food freezing operation the amount of water getting frozen depends on

- Initial water content
- The way in which it is bound.

During freezing the water inside food is converted into the ice and freezing progresses i.e. the constituents like salts, acids, sugar etc., which are dissolved in the water, become more and more concentrated. This concentrated solution may adversely affect the properties of the product. Hence fast freezing is recommended. During rapid freezing the size of ice crystals will be small and the constituents are trapped inside them and almost in unchanged state. It is important to maintain lower temperature as frequent changes in temperature may lead to recrystallization of water in which the water molecules will migrate from the smaller ice crystals which have higher osmotic pressure, to larger ones which have lower osmotic pressure. Apart from rapid freezing, rapid thawing is also important to preserve biological value of foods. Rapid thawing by use of microwaves will prevent recrystallization.

IMPORTANT CHANGES DURING FREEZING

Freezer burns

Due to irreversible drying of the surfaces leading to brown spots on the surface. This occurs due to changes in water vapour pressure in the product. This occurs mainly when the freezer is switched off-and-on frequently. Suppose during a particular interval of time the product has higher temperature which is the case when the freezer is switched on, the water from the product may come out and condense on colder surfaces. However when the temperature changes are reversed, the water does not return to original sites and this gives rise to freezer burns. This can be prevented by wrapping the food products in polyethylene films which are impervious to water vapor before freezing.

Destruction of cells

This takes place during freezing. The cause is not clear yet but reasons may be manifold.

- Purely mechanical action of growing ice crystals may destroy cells.
- Osmotic drying of cells due to increasing concentration of solution.
- Concentrated solution of acid/sugar/salt may damage the cells.

To prevent this, addition of sugars like sucrose at 10%, gelatin etc has been recommended especially in freeze drying of eggs.

- **Change in proteins:** Precipitation of calcium caseinate commonly called as denaturation by cold may take place.
- **Enzymes:** Enzymes are not destroyed completely during freezing. Peroxides are found to be active even at – 20°C . Hence vegetables are given blanching treatment prior to freezing (either by steaming or boiling water). To prevent browning of fruits 0.1 % ascorbic acid or 0.1% citric acid solution is used. SO_2 gas is also used to prevent browning.

- **Vitamins:** No loss of A, B, D, E during freezing and some loss of C especially at below –20°C may occur.
- **Microorganisms:** They are inhibited – no growth. But psychotropic organisms may grow. Destruction of microorganisms by denaturation of protein is possible.
- **Metabolism:** It is slowed down.

Example 1

A milk chilling unit can remove heat from the milk at the rate of 41.87 MJ/hr. Heat leaks into the milk from the surroundings at an average rate of 4.187 MJ/hr. Find the time required for cooling a batch of 500 kg of milk from 45°C to 5°C . Take C_p of milk as 4.187 kJ/kg°C .

Solution

Heat to be removed from milk Q_1 = 4.187)(5-45) 500) kJ

= 8.374 x 10^4 or 83.74 x 10^3 kJ = 83.74 MJ

Heat leakage to milk Q_2 = 4.187 MJ/hr

41.87 MJ of heat is removed in 60 min (1 hr)

Therefore, 83.74 MJ of heat wil be removed in (83.74 x 60)/41.87 = 120 min

Heat added in two hours = 4.187 x 2 = 8.374 MJ

Now additional time required is determined as follows –

83.74 MJ of heat is removed in 120 min,

Then 8.374 MJ of heat is removed in 12 min

Therefore, total time required is 120 min + 12 min = **132 minutes**

FOOD PACKAGING

Packaging equipment and machineries

Packaging is the science, art, and technology of enclosing or protecting products for distribution, storage, sale, and use. When we consider machines for packaging, there is an endless variety of packaging and processing machineries for food products, and many of them are remarkably similar. One of the major problem in selection of machinery is distinguishing which, if any, offer a specific advantage over the other. Also, the degree of versatility may have to be considered; it is pointless to purchase a do-everything machine when there is a single task at hand, similarly machine with too little adaptability may be a blockage in variable production process.

Intermittent and continuous machinery

Machinery can operate on an intermittent or continuous basis depending on the requirements of the production and the amount of capitalization. Normally, intermittent machinery will take a single package or small number of packages, execute an operation on them all at once, and then pass them to the next process step. A continuous operation machine receives a stream of packages and executes the operation without stopping or reducing the speed of the overall flow of material. However, a four station manual-feed filling machine might be suitable for a small winery, that machine would be useless for a soft-drink bottler who needs several thousand fillings per minute to remain competitive. Equally, while continuous machinery is more economical at high throughput rates, it might make less sense to have a high-capacity operation that is in operation only for several hours a week because of low demand, although for the smallest of markets this may happen because of the employment of used machinery and the lack of properly scaled equipment.

Characteristics and classification of machinery

Choice of packaging machinery depends on

- Technical capabilities, labour requirements
- Worker safety
- Maintainability
- Serviceability
- Reliability
- Ability to integrate into the packaging line
- Capital cost
- Floorspace
- Flexibility (change-over, materials, etc.)
- Energy requirement
- Quality of outgoing packages
- Qualifications (for food, pharmaceuticals, etc.)
- Throughput
- Efficiency
- Productivity
- Ergonomics, etc.

Packaging machines may be classified into following general types:

- Blister packs, skin packs and Vacuum Packaging Machines
- Bottle capping equipment, Over-Capping, Lidding, Closing, Seaming and Sealing Machines
- Cartoning Machines
- Box, Case and Tray Forming, Packing, Unpacking, Closing and Sealing Machines
- Cleaning, Sterilizing, Cooling and Drying Machines
- Conveyors, Accumulating and Related Machines
- Feeding, Orienting, Placing and Related Machines
- Filling Machines: handling liquid and powdered products
- Package Filling and Closing Machines
- Form, Fill and Seal Machines
- Inspecting, Detecting and Check weighing Machines
- Falletizing, Depalletizing, Unit load assembly
- Product Identification: labeling, marking, etc.
- Wrapping Machines
- Converting Machines

Types of Packaging Machinery

Fillers, Feeders, Metering and measuring machines

The job of fillers and feeders is to take product in a determined number or amount to fill the assembled package. There is a wide spectrum of different types of equipment to for this purpose, many of which can handle difficult materials such as thick pastes and creams, or very fine objects.

The simplest of them will fill and dump baskets, cups, or trays of material, are well suited to loose and granular products such as cereal and animal food, and can be used for single objects if the objects are properly guided in feed chutes going to the packaging machine. These may be fitted with weighing sensors to permit a net-weight fill to be directly applied to the product, and may use a combination of small fill trays to estimate the net weight more accurately. It is possible to use auger metering where the product is dispensed by based on timing or counting the rotations of an auger that delivers product into the package.

Many types of powders, particulates, and granular products are subjected to bridging and arching phenomena, where the powder will jam feed chutes if they are not suitably designed for such product. Fine powders practise cohesive arches using inter-particulate bonding whereas larger particles will make interlocking arches where the mechanical locking of irregularly shaped particles provides the cohesiveness to form arches. Both of these phenomena depends on the same principles that traditional arch structures in bridges and buildings use – transferring circular compression into vertical and horizontal force to maintain the arch structure. Proper design of the hoppers for the product as well as adaptive equipment such as agitators or bridge breakers may be necessary to keep the device operating properly.

On the contrary bridging phenomena can be used to stop product from flowing in certain types of measuring devices, such as screw-auger fillers. They are designed with a small disk at the end of the auger that creates blockages when the auger is not rotating and depends on bridging to stop the flow between package fillings.

Direct counting machines can use trays, simple photocells, or complex vision-system pick and-place manipulators to acquire the exact number of products required in a package. Similarly, it is possible to use the package or an inner tray as a measuring device, filling it with the appropriate number of product units, and then enclosing it in a larger secondary package.

Filling by Gravitation

Filling by gravity is used for filling thin liquids like milk into glass bottles/plastic bottles. In gravitational filling, the filling process is stopped when the pre-calibrated filling height has been reached. This system is suitable for filling milk in glass bottles. However, packaging milk in bottles is outdated in India. In volumetric filling process, fixed volume of liquid is filled.

Mechanical Filling

In this type milk powders are metered filled by using screw conveyors. The conveyors are used as metering and dosing devices. However, due to variation in bulk densities of milk powder, care must be exercised to ensure even delivery of the product and which shall be equal in weight from package to package.

Cup Filling Machines

Plastic cups either preformed, from a film in the machine itself or readymade cups when used are placed in the stacker of the machine. Automatic packaging machines which work aseptically are in use for the production of long life products. In these machines the film is passed through a bath of Hydrogen peroxide and then goes to a sterile tunnel in which it is sterilized by either exposure to high temperature or exposure to excess pressure of air. The remaining stage like cup moulding, filling and sealing with lid takes place in the sterile tunnel.

Form-Fill-Seal Systems

Form-Fill-Seal (FFS) systems are preferred methods for many food items, and as the name suggests, it forms packages from roll around a central mandrel that incorporates a product fill tube and allows the forming, filling, and sealing of packages of almost all types of products from solids to liquids like milk. Mostly these may be multiple-head machines to produce small packets of sauces for fast-food restaurant use. With many types of food products, the product handling must be done accurately as the filled pouches are filled to avoid infecting the seal surfaces. Additional types of machinery peculiar to the food industry will insert slices of meat products into folded or vacuum formed film and seal it, or will even extrude a cylinder of cheese food into a tube of plastic film and then flatten and seal the product to create the wrapped processed cheese slices.

Specialty machinery may be necessary for particularly sensitive materials such as fruit juices and meats that may discolour with the slightest gain of dissolved metal. In the most extreme cases like processing of juices and pharmaceutical or cosmetic products, special surfaces must be constructed for contact with the product using materials such as borosilicate glass, Teflon, or other materials, as appropriate.

For some products the film may first be fed through a sterilizing chemical bath and dryer prior to use in the packaging system. The film approaches the back of a long hollow conical tube, and when the centre of the plastic is near the tube, the outer edges of the film form flaps that wrap around the conical tube. The film is pulled downward around the outside of the tube and a vertical heatsealing bar clamps onto the edges of the film, bonding the film by melting the seam edges together.

Packaging in Cans and Tubes

Cans made of tin plate are used for sweetened condensed milk, evaporated milk, canned fruits, canned beans etc. The cans open at the back are loaded into the machine which fills the product in to the cans. Then the lids are applied and closed by either simple or double seaming depending on the product. Aerosol cans are used for the packaging of whipped cream. In these, the product is filled with a propellant which is under pressure.

Flat bottomed collapsible tubes made from aluminium are used for packing cream. After the product is filled from the back of the tube, the head space in the tube is filled with inert gas and the tube is then closed by folding. Polyethylene collapsible tubes are presently used for packing cheese spread/cheese food and ketchups.

Shrink Wrapping Machines

Shrink wrapping includes packing of one or several articles with a thermoplastic film which when subjected to heat shrinks and form a tight wrap around the object. Shrink wrap or shrink film, is a material made up of polymer plastic film. When heat is applied to this material it shrinks tightly over whatever it is covering. Heat can be applied with a hand held heat gun or the package can pass through a heat tunnel on a conveyor. Shrink wrap is usually used as an overwrap on many types of packaging, including cartons, boxes, beverage cans and pallet loads. A variety of products may be enclosed in shrink wrap to stabilize the products, unitize them, keep them clean, add a degree of tamper resistance, etc. It can be primary covering for foods like cheese and Paneer. Shrink wrap film is available in a variety of thicknesses, clarities, strengths and shrink ratios. The two primary films are either crosslinked, or non crosslinked. Major shrink films include PVC and several other compositions like LDPE, LLDPE, PP, EVA etc. Coextrusions and laminations are available for specific mechanical and barrier properties for shrink wrapping food.

In shrink-wrap machine a loose plastic film pouch is made. The product is placed in this pack which passes through a heated tunnel in which the film shrinks and adheres closely to the product. The film is commonly heated by hot air, infrared rays or hot water. Shrink wrapping is also used to hold together several singly wrapped products or small packages in a multiple unit package.

Advantages of shrink wrap packaging are:

- All types of items of regular / irregular shapes and sizes can be shrink wrapped.
- Small items can be utilized and stacked one on top of the other
- Requires minimum packaging material and operation.
- Simple operation
- Easy stacking
- Enable unit packaging or packaging in groups.

Speciality Packaging Machines

Inert gas packing

Inert gas packing using nitrogen, carbon dioxide or a mixture of the two is done by passing the gasses around the product prior to sealing for example, Cheese. Whole milk powder packed in tins with a pin hole is evacuated under vacuum. The vacuum is then broken by the inert gas (usually N2) and the pin hole is sealed immediately.

Vacuum Packaging

Vacuum packaging is done for products like cheese blocks, meat products etc, where there is problem of microorganisms growing on the surface. The product is placed in a plastic pouch and placed in the vacuum packaging machine for the creation of vacuum in the pack and subsequent sealing takes place in the machine itself.

Benefits and Capabilities of Food Packaging

Food packaging machines are used to seal and protect food products so that they are ready storage, handling, distribution, and marketing purposes. Compared to manual food packaging operations, machine-assisted food packaging operations allow for better quality control, greater productivity, safer working conditions, broader production flexibility, and easier scalability. However, the machine must be selected, installed, and maintained properly to ensure it operates as expected. The type of food packaging machine employed depends on the requirements and restrictions of the food packaging operation. Below we provide an overview of the types of machines we have available, including what they are, key benefits, and typical applications.

Tray/Cup Sealers

Tray/cup sealers—also referred to as tray lidding machines, heat sealers, and top sealers—are used to apply and adhere plastic film—such as modified atmosphere packaging (MAP) film—to the openings of product-filled trays and cups. The airtight seal formed helps protect fresh, frozen, and ready-made food against contamination, degradation, and premature spoilage. Compared to manual sealing operations, machine-assisted sealing operations accommodate a greater number of trays/cups in a shorter time period, which allows for faster production speeds and higher productivity.

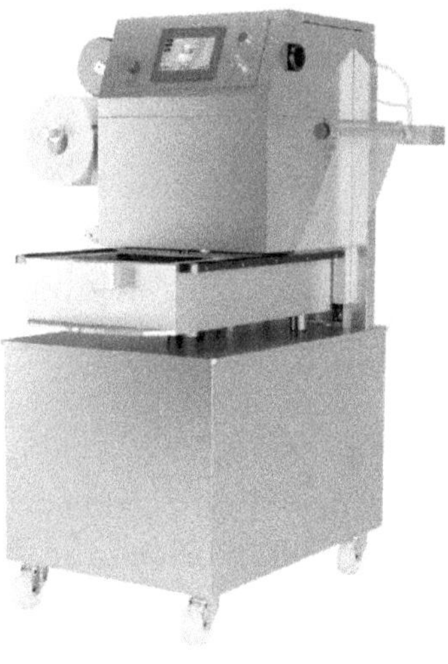

Fig. 6. Tray/Cup Sealers

Fillers/Depositors

Fillers and depositors are machines designed to dose and deposit a specific amount of product into packaging containers. They can accommodate low- to high-viscosity compounds and operate as standalone or integrated units, depending on their design and configuration. Various accessories can also be incorporated to suit different products, such as product mixers and agitators, hoppers, interchangeable nozzles, automatic loading systems, and level detectors. In addition to food and beverage production operations, other typical applications for these machines include bakery, cosmetic manufacturing, household, and industrial processes.

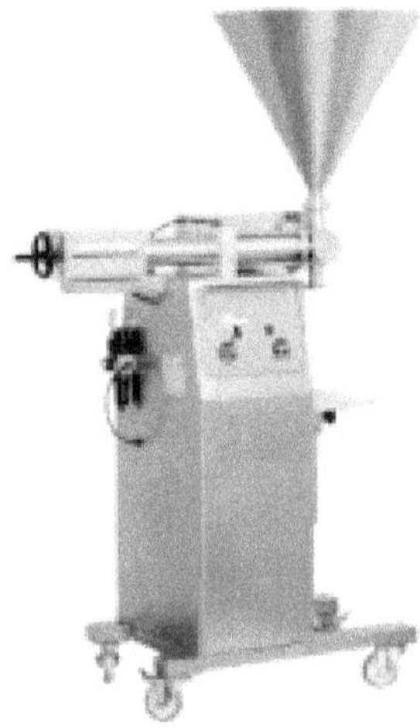

Fig. 7. Fillers/Depositors

Food packaging can serve several functions

- Containment: Holds (i.e., contains) food contents until they are used
- Protection and Preservation: Creates a physical barrier between processed food products and physical, chemical, microbial, and macrobial variables during storage, transportation, and distribution which can cause spoilage, contamination, or loss of quality
- Convenience: Enables food products to be portioned out (by weight or volume) for easier consumer use, and stored, transported, and distributed
- Communication: Helps identify the food contents and indicate handling, storage, and use instructions, as well as allows for an opportunity for branding and marketing

Additionally, food packaging is available in several forms—e.g., boxes, jars, bottles, cans, etc. Depending on the packaging form used to package the food material, packaging operations, and the equipment employed to execute them, can significantly vary. Some of the other factors which may influence the type of packaging equipment employed include the type and form of food product and the storage, handling, distribution, and marketing requirements.

FOOD QUALITY MEASUREMENT INSTRUMENTS

An important aspect of food processing and food preparation is the topic of food quality control. Food quality in a production setting can be degraded and compromised by any number of contributing factors, the most common of which include the presence of :

- Pesticides
- Pathogens
- Heavy metals
- Organic toxins
- Foreign objects

In addition to these factors, food quality needs to be validated in terms of nutritional content and consistency, be assessed for the presence of allergens, verified with respect to adulteration and certification for GMO, and tested for shelf life.

There are a number of instruments used in food laboratories that can be applied to assure food quality which applies to solids as well as beverages. Some of the most common examples of these instruments and their use are described below:

- **Alcoholic beverage analyzers** – Can be used to measure the alcohol content, density, color, and pH of alcoholic beverages.
- **Carbonated beverage analyzers** – measure product density, temperature, current/fresh/inverted sugar concentrations, degree of inversion, and CO_2 levels.
- **Food analysis equipment** – various instruments that can be used to measure the fat, protein, and oil concentrations in food samples and detect the level of gluten in foods.
- **Pesticide detection instruments** – detects the presence of pesticides in food samples.
- **Electron Spin Resonance Spectrometers** – also known as electron paramagnetic resonance (EPR), these instruments can be used to test the purity of products without physically destroying or altering samples.
- **Cell and colony counters** – can be used to measure the colonies of microorganisms that have grown on an agar plate prepared from a sample.
- **Incubators** – are used to provide a controlled environment (i.e. temperature, humidity, CO_2 level) for food safety testing.
- **Chemical imaging (NIR/Raman) systems** – devices that use analysis of samples by detecting and analyzing light that is in the near-infrared, visible, or near-ultraviolet light areas of the electromagnetic spectrum.

- **Magnetic analyzers** – detect low levels of iron in food samples by measuring the imbalance in resonance between two air core coils which can be translated into a signal that reflects the level of iron present in the sample.
- **Moisture analyzers** – also known as moisture balances, these devices are used to establish the percentage of moisture in a food sample, either by weighing the sample before and after an evaporation process or by using an absorption spectrometer to analyze the gas emitted during evaporation to establish its content.
- **Polarimeters** – devices that pass polarized light through a sample and measure the angle at which the emitted light emerges. Optically active substances will cause a change in the polarization angle of the emitted light, which can be used to establish concentrations of sugars such as glucose and sucrose.
- **Refractometers** – are devices that measure the angle of refraction from light that is passed through a liquid, gel, or solid substance and using that to establish parameters such as the salinity and sugar content.
- **Rheometers & Viscometers** – are instruments that can measure the viscosity of a fluid and the behavior of fluids when shear or stress forces are applied to it. Having this information can reveal the properties of the fluid that relate to its structure and elasticity.
- **Saccharimeters** – are instruments that specifically measure the concentration of sugars present in a solution. They do so by measuring the refractive index of the liquid as light as passed through it.
- **Titration Equipment** – can be used to detect and measure the concentration of a substance within a liquid through acid/base titration. The addition of titrant of known concentration to a known volume of solution with unknown concentration can determine that unknown concentration through a reaction neutralization.
- **Other equipment** – additional equipment often employed in food quality applications includes ovens, centrifuges, water baths, and dry baths.

Additional Equipment

Beyond the aforementioned equipment, there are other types of auxiliary equipment that may not directly contribute to the various food processing stages but are still critical to the food and processing industries and their operations. This equipment can serve or support several functions throughout the overall production cycle, including:

- **Utilities:** Utility equipment provides the resources necessary for operating and maintaining a food processing facility, such as pumps, generators, power supplies, and combustion equipment. Typical resources include electricity, heat, water, steam, compressed air, and waste disposal options.
- **Measurement and Control:** Measurement and control equipment is used to ensure that the food processing equipment operates correctly, and the food processing stages proceed as specified. These instruments can be used to analyze ingredients and machines, as well as allow manufacturers to perform and duplicate processing procedures. They can also be used to monitor existing systems and machinery, such as logging data during product testing or quantifying typical performance statistics. Measurement devices are particularly crucial during food production, as minor changes in cooking temperature, ingredient ratios, and operation times can lead to drastic changes in the finished product. Some of the measurement and control devices commonly used for food processing include:
    - Precision controls
    - Pressure gauges
    - Scales and weighing systems
    - Thermometers
    - Timers
- **Material Handling:** Material handling equipment is used to transport food material through and between preparation, processing, and post-processing stations. Types of equipment used for material handling in the food and food processing industries include:
    - Conveyors systems
    - Fluid transport devices (e.g., pumps, hoses, tubes, and piping)
- **Storage:** Storage equipment allows food material to be held between processing stages (or operations), and the final food product to be stowed and warehoused prior to distribution. Food and food products can be stored by a variety of methods depending on the type of food being stored, the length of storage (i.e., short-term or long-term) and the storage requirements, and include:
    - Bags, bins, boxes, and silos (i.e., types of storage containers for solids)
    - Tanks, vats, and vessels (i.e., types of storage containers for liquids)
    - Cold storage rooms (for chilled or frozen products)
- **Product Distribution:** Distribution equipment prepares and packs completed food products for transportation to their final destination (generally a store

or other type of marketplace). Smaller packaged food products are typically grouped and packed together prior to transportation to optimize efficiency, while larger packaged food products can be wrapped and shipped individually on pallets. Some of the equipment employed for food product distribution includes:

- Cartoners
- Palletizers
- Stretch and shrink wrappers

- Equipment Cleaning: Regular cleaning and sanitation operations are essential to maintaining the hygienic environment required for food processing. Food processing equipment is often designed to help optimize these operations with components assembled such that can be cleaned after dismantling or via cleaning in place (CIP) methods. Cleaning and sanitation equipment used for food processing applications includes:
    - Cleaning in Place (CIP) systems
    - Ball spray devices (for process and storage tanks)
    - Dry steam vapor cleaners (for food processors and beverage production equipment)

DESIGN AND SELECTION CONSIDERATIONS FOR FOOD PROCESSING EQUIPMENT

As outlined above, there is a wide range of food processing equipment available to satisfy the various functions which are integral to the food and food processing industries. While individual pieces of equipment may have some distinctive considerations—typically based on the specific functions and unit operations which they perform—to keep in mind when designing and selecting equipment for a particular food processing application, there are also a few factors an industry professional or procurement agent can consider across the board to ensure that their food processing equipment fulfills their needs. These factors include:

- Function
- Form
- Hygienic design
- Sizing
- Construction requirements
- Operational characteristics
- Cost

Function

The particular function for which a piece of equipment is intended largely determines the general type of equipment employed. For example:

- Material handling equipment includes conveyors and tube systems
- Preparation equipment includes sterilizers or wash systems
- Heat processing equipment includes ovens and fryers
- Preservation equipment includes freezers and dehydrators
- Product distribution equipment includes wrapping and palletizing systems

Form

The form in which the food material being processed comes in significantly influences the specific type of equipment employed as some food processing equipment is better suited for one form over another. For example, among material handling equipment, pumps are better suited for transporting liquid food material, while conveyors are better suited for transporting solid food material.

Hygienic Design

As the products produced by the food and food processing industries are intended for consumption, the equipment employed to execute the necessary processes and unit operations is designed with consideration to hygiene and sanitation standards and regulations set by a variety of government agencies and private organizations. These standards and regulations outline the requirements and best practices aimed towards ensuring the manufacture of safe food and protection of public health and safety. For example, industrial professionals must select not only appropriate food-grade materials but design and construct the equipment such that the risk of contamination and food quality degradation is minimized and the methods of cleaning and sanitizing are optimized.

Sizing

There are several factors which help determine the optimal size for food processing equipment, but ultimately, the goal is to balance the material and resources used for each unit operation and the required production output. Typically, processing equipment is oversized between 10–20% to compensate for potential operational issues, such as equipment breakdown or demand fluctuation, or environmental conditions, such as temperature or humidity changes. However, depending on the production requirements of the particular facility, multiple, small-sized equipment can also be employed to allow for greater operational flexibility.

Construction Requirements

As indicated above, food processing equipment must be constructed with consideration to hygienic design. Beyond this requirement, equipment must also be designed and built to withstand the stress from the various mechanical, chemical, thermal, and physical processes and unit operations of the food processing cycle. Other construction requirements to keep in mind include:

- Dimensional and weight restrictions
- Facilitation of cleaning and maintenance operations
- Use of standard vs. custom parts

Operational Characteristics

Food processing equipment is typically designed and built with a particular function or unit operation in mind. However, the method in which these functions and operations are executed can vary depending on the design of the equipment, and additional components can be integrated to facilitate smoother operation. Some characteristics of food processing equipment to consider include:

- Batch vs. continuous processing
- Manual vs. automatic operation
- Integration of analytical or quality control units
- Integration of ergonomic or safety components

Cost

While it is necessary to choose equipment which effectively fulfills the requirements of the food processing application, it is also important to consider the overall costs to better determine whether the particular selection is worth the investment. Some factors to consider when doing a cost-benefit analysis include:

- **Construction Material:** The majority of the cost of processing equipment can be attributed to the construction materials as the raw materials typically used (e.g., carbon steel, stainless steel, aluminum, etc.) can range between a couple hundred to a couple thousand USD per ton. The materials chosen can also influence the cost of the actual construction process, as different materials may have different handling and fabricating requirements such as treatment procedures or higher precision machining.
- **Standard vs. Custom Equipment:** As expected, custom-designed equipment is more expensive than standard, off-the-shelf equipment. Therefore, if possible and suitable, industry professionals and procurement agents should opt for the latter type, especially for processing products such as pumps, heat exchangers, valves, evaporators, distillation columns, and centrifuges.

APPLICATIONS OF FOOD PROCESSING EQUIPMENT

Processing equipment is used throughout the food and food processing industries for various food products. As outlined above there is a wide range of food processing equipment available, not all of which are used in every facility as specific industrial subsections favor using certain types of equipment over others for their particular processes and unit operations.

Some of the subsections into which the food and food processing industries are segmented include:

- Alcoholic beverages
- Bakery and confectionary products
- Dairy
- Fish and seafood
- Fruit
- General food products
- Meat and poultry
- Non-alcoholic beverages
- Vegetables

RHEOLOGICAL PROPERTIES OF FLUID FOODS

Introduction

It is necessary to study properties of fluid food products for designing and lay-outing of transport system (piping and pumping layout). For the fluid food products, the design of transport system mainly depends on the type and description of flow characteristics of the product. Some of the properties are interdependent and some are dependent on the fluid food composition and therefore it is necessary to measure dependant properties and we can predict its rheological properties.

Most important dependant fluid food property is viscosity i.e. resistance against flow, generally indicated by μ i.e. dynamic viscosity / η kinematic viscosity ($\eta = \mu/\rho$). In food industry μ is broadly used to describe a single parameter known as 'consistency'. But this approach may lead to confusion in many cases due to non-Newtonian behaviour of many fluid food products. The stress and rate of shear diagram indicate varieties of food products classified under different categories, which is considered to be non-Newtonian.

Classification

The fluids can be classified into following categories depending on the response to the applied shear force.

Newtonian fluids

Newtonian fluids are fluids which exhibit a linear increase in the shear stress with the rate of shear. These fluids exhibit a linear relationship between the shear stress and the rate of shear. The slope 'μ' is constant therefore, the viscosity of a Newtonian fluid is independent of the rate of shear. These fluids exhibit a pure viscous flow i.e. the product begins to flow with the slightest force and the rate of flow is proportional to the magnitude of force applied. The examples of Newtonian fluids are milk, clear fruit juices, sucrose solution, most types of honey, corn syrup etc. The equation for characterizing Newtonian fluid is

$$T = \mu\,(-dv/dx) \qquad \text{(Eq-1)}$$

Where,

T = shear stress,

μ = dynamic viscosity ($\eta = \mu/\rho$),

$-dv/dx$ = velocity gradient

Non-newtonian fluids

A non-Newtonian fluid is broadly defined as one for which the relationship between shear stress and shear rate is not a constant. When the shear rate is varied, the shear stress doesn't vary in the same proportion. These fluids exibit either shear thinning or shear thickening behaviour and some exhibit a yield stress. The two most commonly used equations for characterizing non-Newtonian fluids are the power law model (Eq 2) and Herschel-Bulkley model (Eq 3) for fluids.

$$T = K\,(\gamma)^n \qquad \text{(Eq-2)}$$

$$T = T_0 + K\,(\gamma)^n \qquad \text{(Eq-3)}$$

Where,

T = shear stress,

K = consistency constant,

γ = shear rate,

n = flow behaviour index,

T_0 = yield stress

There are several types of non-Newtonian flow behaviour, characterized by the way a fluid viscosity changes in response to variation in shear rate. The most common non-Newtonian fluids are:

Time-independent flow of non-Newtonian foods

The fluid foods whose viscosity is not influenced by the shearing time at a constant shear rate show two distinct patterns of stress – shear rate relationship i.e shear-thinning and shear-thickening.

Pseudoplastic/shear-thinning fluids

This type of fluids will display a decreasing consistency with an increasing shear rate. Probably the most common of the non-Newtonian fluids, psedo-plastic include emulsions and dispersions of many types. This type of flow behaviour is some times called shear-thinning. The shear stress ('*T*' *or* 'σ') *versus* shear rate (γ) curve is convex toward the stress axis. The shear thinning behaviour of a fluid or semi-solid food is expressed by the power law model or de Waele's model:

$$T = K(\gamma)^{+n} \quad \text{(Eq-4)}$$

Where,

T = shear stress,

K = consistency constant $(\text{Pa s})^n$,

γ = shear rate,

n = flow behaviour index ('n' has a positive value between zero and unity)

Protein concentrates, skim milk concentrate, milk ultrafiltration retentates, concentrated fruit juices such as unpectinized apple juice (50-65°Brix), orange juice (50-65°Brix) etc., melted chocolates, thawed frozen egg, fruit and vegetable purees and gum solutions are the examples of psedo-plastic fluid food products.

Dilatant/shear-thickening flow behaviour

This type of fluid will display an increasing viscosity with increase in shear rate. Dilatancy is frequently observed in fluids containing high level of deflocculated solids, such as candy compounds, cooked corn starch paste, certain types of honey etc. Dilatancy is also referred to as shear-thickening flow behaviour. The stress shear rate curve is concave toward the stress axis and the value of 'n' in the power law (Eq-4 above) is negative.

Time-dependent flow of non-Newtonian foods

Certain non-Newtonian fluids show a time-dependent stress-shear relationship which can be one of the following types:

- **Thixotropy:-** When at a constant shear rate, the stress decreases over a period of time due to structure breakdown until eventually it reaches a steady value, the product is said to be thixotropic. Aged condensed milk, cream and ice cream mix, egg white etc. revel thixotropy.
- **Rheopectic:-** This essentially the opposite of thixotropic behaviour, in which the fluids viscosity increases with time as it is sheared at a constant rate.

Rheopectic fluids are rarely encountered. Both thixotropic and rheopectic may occur in combination with any of the previously discussed flow behaviour, or only at certain shear rates.

- **Plastic:-** This type of fluid will behave as a solid under static conditions. A certain amount of force must be applied to the fluid before any flow is induced, this force is called *yield* value. Tomato ketchup is a good example of this type fluid, its yield value will often make it refuse to pour from the bottle until the bottle is shaken or struck, allowing the catchup to flow freely. Once the yield value is exceeded and flow begins, plastic fluids may display Newtonian, pseudoplastic, or dilatant flow characteristics.

Important Properties of Fluid Foods

- **Bounce:** The resilience rate at which the sample returns to its original shape after partial compression.
- **Chewiness:** It is the number of chews (@1 chew/sec.) needed to masticate the sample to a state ready for swallowing.
- **Coarseness:** Degree to which mass feels coarse in the mouth.
- **Cohesiveness:** Degree to which the sample deforms before rupturing during biting.
- **Denseness:** It is a compactness of cross-section of the sample after complete biting.
- **Dryness:** Degree to which the sample feels dry in the mouth.
- **Fracturability:** Force with which the sample shatters/breaks. It encompasses crispiness, crumbliness, crunchiness and brittleness.
- **Graininess:** Degree to which a sample contains small grainy particles.
- **Gumminess:** Energy required to disintegrate a semisolid food to a state ready to swallow.
- **Hardness:** Force required to deform the product to a given distortion. Generally the distortion is between tongue and palate.
- **Heaviness:** Weight of product perceived when first placed on the tongue.
- **Moisture absorption** (related to sensory): Amount of saliva absorbed by the product in the mouth.
- **Moisture released:** Amount of wetness and juiciness from the product.
- **Mouth coating:** Type and degree of coating in the mouth after mastication.
- **Roughness:** It is the degree of abrasiveness of product surface which is perceived by the tongue.

- **Slipperiness:** It is the degree to which the product slides over the tongue.
- **Smoothness:** It represents the absence of any particle, lumps, etc. in the product.
- **Springiness/sponginess:** Degree to which the product returns to its original position i.e. shape and size after partial compression (without failure). Here compression is between the palette and teeth.
- **Uniformity:** Degree to which the sample is even throughout.
- **Uniformity of bite :** Evenness of force applied on the product while biting.
- **Uniformity of chew :** The degree to which the chewy characteristics of the product are even throughout the mastication
- **Viscosity :** Force required to draw a liquid from spoon over to the tongue.
- **Wetness :** Amount of moisture perceived on the product surface.

NEWTONIAN AND NON-NEWTONIAN FLUIDS

From the fundamental definition of viscosity, we can write:

$F/A = \mathrm{m}v/Z = \mathrm{m}(\mathrm{d}v/\mathrm{dz}) = \mathrm{t}$

where t (tau) is called the shear stress in the fluid. This is an equation originally proposed by Newton and which is obeyed by fluids such as water. However, for many of the actual fluids encountered in the food industry, measurements show deviations from this simple relationship, and lead towards a more general equation:

$\mathrm{t} = k(\mathrm{d}v/\mathrm{dz})^n$ (eq-6)

which can be called the **power-law equation**, and where k is a constant of proportionality.

Where $n = 1$ the fluids are called Newtonian because they conform to Newton's equation and $k = \mathrm{m}$; and all other fluids may therefore be called non-Newtonian. Non-Newtonian fluids are varied and are studied under the heading of rheology, which is a substantial subject in itself and the subject of many books. Broadly, the non-Newtonian fluids can be divided into:

- *Those in which $n < 1$.* As shown in Fig. 8 these produce a concave downward curve and for them the viscosity is apparently high under low shear forces decreasing as the shear force increases. Such fluids are called pseudoplastic, an example being tomato puree. In more extreme cases where the shear forces are low there may be no flow at all until a yield stress is reached after which flow occurs, and these fluids are called thixotropic.
- *Those in which $n > 1$.* With a low apparent viscosity under low shear stresses, they become more viscous as the shear rate rises. This is called dilatancy and

examples are gritty slurries such as crystallized sugar solutions. Again there is a more extreme condition with a zero apparent viscosity under low shear and such materials are called rheopectic. Bingham fluids have to exceed a particular shear stress level (a yield stress) before they start to move.

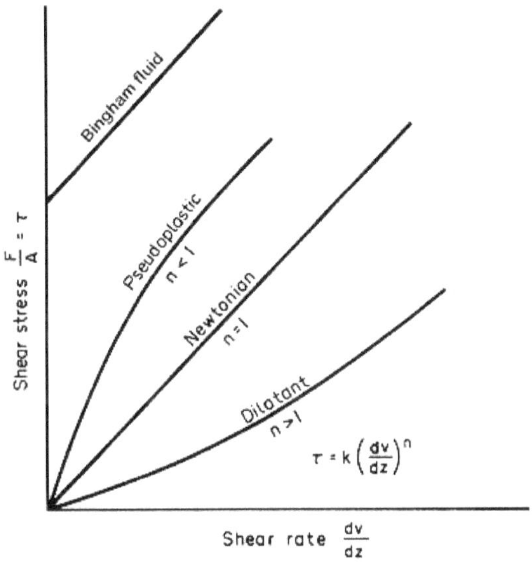

Figure 8. Shear stress/shear rate relationships in liquids.

In many instances in practice non-Newtonian characteristics are important, and they become obvious when materials that it is thought ought to pump quite easily just do not. They get stuck in the pipes, or overload the pumps, or need specially designed fittings before they can be moved. Sometimes it is sufficient just to be aware of the general classes of behaviour of such materials. In other cases it may be necessary to determine experimentally the rheological properties of the material so that equipment and processes can be adequately designed.

REYNOLDS NUMBER

Reynolds number is a dimensionless quantity that is used to determine the type of flow pattern as laminar or turbulent while flowing through a pipe. Reynolds number is defined by the ratio of inertial forces to that of viscous forces. The flow characteristics for laminar flow are influenced by liquid properties, flow rate, and the dimensions of liquid–solid interfaces. As the mass flow rate is increased, the forces of momentum or inertia increase; but these forces are resisted by friction or viscous forces within the flowing liquid. As these opposing forces reach a certain balance, changes in the flow characteristics occur. Based on experiments conducted by Reynolds (1874), the inertial forces are a function of liquid density, ρ, tube diameter, *D*, and average fluid velocity, *u*.

The viscous forces, on the other hand, are a function of liquid viscosity. A dimensionless number, called a Reynolds number, is defined as the ratio of the inertial to the viscous forces:

$$N_{Rc} = \frac{\text{internal forces}}{\text{viscous forces}} \qquad \text{(eq 7)}$$

or

(eq 8)

Key Facts

- Osborn Reynolds discovered that the flow regime depends mainly on the ratio of the inertia forces to viscous forces in the fluid.
- When the **viscous forces** are dominant (slow flow, low Re), they are sufficient to keep all the fluid particles in line, then the flow is laminar.
- When the **inertial forces dominate** over the viscous forces (when the fluid flows faster and Re is larger), the flow is turbulent.

It is defined as:

- in which V is the mean flow velocity, D is a characteristic linear dimension, ρ fluid density, μ dynamic viscosity, and μ kinematic viscosity.
- The Reynolds number can be used to compare a real situation (e.g., airflow around an airfoil and water flow in a pipe) with a small-scale model.

The **Reynolds number** represents the ratio of inertial forces to viscous forces and is a convenient parameter for predicting if a flow condition will be laminar or turbulent. It is defined as a characteristic length multiplied by a characteristic velocity and divided by the kinematic viscosity.

It can be interpreted that when the viscous forces are dominant (slow flow, low Re), they are sufficient enough to keep all the fluid particles in line, then the flow is laminar. Even very low Re indicates viscous creeping motion, where inertia effects are negligible. When the inertial forces dominate over the viscous forces (when the fluid flows faster and Re is larger), the flow is turbulent. The transition from laminar to turbulent flow depends on the surface geometry, surface roughness, free-stream velocity, surface temperature, and type of fluid, among other things.

Application of Reynolds Number

The **Reynolds number** plays an important part in calculations in fluid dynamics and heat transfer problems.

- It is essential to calculate the friction factor in a few of the equations of fluid mechanics, including the Darcy-Weisbach equation.
- It is essential for heat transfer calculations since many other characteristic numbers (e.g.,, Nusselt number) depend on the flow regime.
- It is used when modeling the movement of organisms swimming through water.
- It is used to predict the transition from laminar to turbulent flow and is used to scale similar but different-sized flow situations, such as between an aircraft model in a wind tunnel and the full-size version.
- The predictions of the onset of turbulence and the ability to calculate scaling effects can be used to help predict fluid behavior on a larger scale, such as in local or global air or water movement and thereby the associated meteorological and climatological effects.

CHAPTER

SENSIBLE AND LATENT HEAT CHANGES 4

Heat may be stored in a substance in two forms: sensible heat and latent heat. Sensible heat is readily perceived by the sense of touch and produces a temperature rise or fall as heat is added or removed from a substance. Sensible heat is the heat that changes the temperature of a substance without any phase change.

The heat required to bring about a change of state at constant pressure with no change in temperature is called the latent heat.

It is the heat that necessary to change the physical state of 1 kg of any substance, i.e. from solid to liquid or liquid to vapor or vice versa.

In the case of food materials, latent heat refers to latent heat of fusion during freezing of products and the latent heat of melting during heating and thawing operation.

In foods of high fat content and in the temperature range, where phase transitions in fats occur, it is necessary to take into account the latent heat of crystallization of fats.

The latent heat of fusion of fats is in the range from 80 to 200 J/g. It increases with increasing length of the fatty acid chain and the degree of saturation.

Energy takes many forms such as heat, kinetic energy, chemical energy, potential energy but, because of interconversions, it is not always easy to isolate separate constituents of energy balances. However under some circumstances certain aspects predominate. In many heat balances, other forms of energy are insignificant; in some chemical situations, mechanical energy is insignificant and in some mechanical energy situations, as in the flow of fluids in pipes, the frictional losses appear as heat but the details of the heating need not be considered.

We are seldom concerned with internal energies.Therefore practical applications of energy balances tend to focus on particular dominant aspects and so a heat balance, for example, can be a useful description of important cost and quality aspects of a food process situation. When unfamiliar with the relative magnitudes of the various forms of energy entering into a particular processing situation, it is wise to put them all down. Then after some preliminary calculations, the important ones emerge and other minor ones can be lumped together or even ignored without introducing substantial errors. With experience, the obviously minor ones can perhaps be left out completely though this always raises the possibility of error.

Energy balances can be calculated on the basis of external energy used per kilogram of product, or raw material processed, or on dry solids. or some key component. The energy consumed in food production includes direct energy which is fuel and electricity used on the farm, in transport, in factories, in storage, selling, etc.; and indirect energy which is used to build the machines, to make the packaging, to produce the electricity and the oil and so on.

Food itself is a major energy source, and energy balances can be determined for animal or human feeding; food energy input can be balanced against outputs in heat and mechanical energy and chemical synthesis.In the SI system there is only one energy unit, the joule. However, kilocalories are still used by some nutritionists, and British thermal units (Btu) in some heat-balance work.The two applications used in this book are heat balances, which are the basis for heat transfer, and the energy balances used in analysing fluid flow.

HEAT BALANCES

The most common important energy form is heat energy and the conservation of this can be illustrated by considering operations such as heating and drying. In these, **enthalpy** (total heat) is conserved and as with the mass balances so enthalpy balances can be written round the various items of equipment. or process stages, or round the whole plant, and it is assumed that no appreciable heat is converted to other forms of energy such as work.

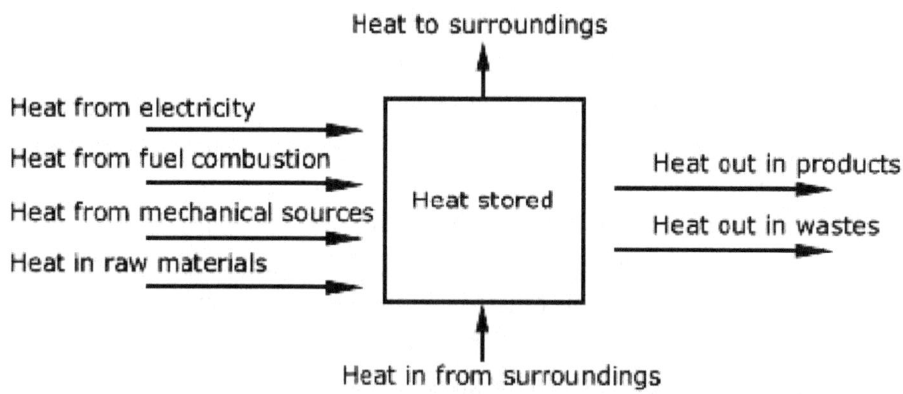

Fig. 1. Heat balance

Enthalpy (H) is always referred to some reference level or datum, so that the quantities are relative to this datum.

Working out energy balances is then just a matter of considering the various quantities of materials involved, their specific heats, and their changes in temperature or state (as quite frequently latent heats arising from phase changes are encountered). Figure 1 illustrates the heat balance.

Heat is absorbed or evolved by some reactions in food processing but usually the quantities are small when compared with the other forms of energy entering into food processing such as sensible heat and latent heat.

Latent heat is the heat required to change, at constant temperature. the physical state of materials from solid to liquid, liquid to gas, or solid to gas.

Sensible heat is that heat which when added or subtracted from food materials changes their temperature and thus can be sensed.

The units of specific heat (c) are J kg^{-1} $°C^{-1}$, and sensible heat change is calculated by multiplying the mass by the specific heat by the change in temperature, $m\ c\ DT$, (J).

The units of latent heat are J kg^{-1} and total latent heat change is calculated by multiplying the mass of the material, which changes its phase, by the latent heat. Having determined those factors that are significant in the overall energy balance, the simplified heat balance can then be used with confidence in industrial energy studies. Such calculations can be quite simple and straightforward but they give a quantitative feeling for the situation and can be of great use in design of equipment and process.

EXAMPLE 1. Heat demand in freezing bread

It is desired to establish freezing of 10,000 loaves of bread each weighing 0.75 kg from an initial room temperature of 18°C to a final store temperature of -18°C. If this is to be carried out in such a way that the maximum heat demand for the freezing is twice the average demand, estimate this maximum demand, if the total freezing time is to be 6 h.

If data on the particular bread are unavailable, in the literature are data on bread constituents, calculation methods, enthalpy/temperature tables

- Tabulated data (Appendix 1) suggests specific heat above freezing 2.93 kJ kg^{-1} $°C^{-1}$, below freezing 1.42 kJ kg^{-1} $°C^{-1}$, latent heat of freezing 115 kJ kg^{-1} and freezing temperature is -2°C.

  Total enthalpy change, $(\Delta H) = [18 - (-2)]\ 2.93 + 115 + [-2 - (-18)]\ 1.42$

  $= \mathbf{196\ kJ\ kg^{-1}}$.

- Formula (Appendix 1) assuming the bread is 36% water gives

-specific heat above freezing

$4.2 \times 0.36 + 0.84 \times 0.64 = 2.05$ kJ kg^{-1} °C^{-1},

specific heat below freezing

$2.1 \times 0.36 + 0.84 \times 0.64 = 1.29$ kJ kg^{-1} °C^{-1},

latent heat

0.36×335

= **121 kJ kg^{-1}.**

Total enthalpy change, (DH) = [18 - (-2)]2.05 + 121 +[-2-(-18)] 1.29

= **183 kJ kg^{-1}.**

- Enthalpy/temperature data for bread of 36% moisture suggest

  $H_{18.3°C}$ = 210.36 kJ kg^{-1},

  $H_{-17.3°C}$ = 65.35 kJ kg^{-1}.

  So from +18°C to -18°C total enthalpy change (DH)

  = 210 - 65 = **145 kJ kg^{-1}.**

- The enthalpy/temperature data in Mannheim et al. can also be used to estimate "apparent" specific heats as

DH /DT = c and so using the data:

T°C	=	-20.6	-17.8	15.6	18.3
H kJ kg^{-1}	=	55.88	65.35	203.4	210.4

$$\text{Giving } c_{-18} = \frac{\Delta H}{DT} = \frac{65.35 - 55.88}{20.6 - 17.8} = 3.4 \text{ kJ kg}^{-1}\,°\text{C}^{-1}$$

$$\text{Giving } c_{18} = \frac{\Delta H}{DT} = \frac{210.4 - 203.4}{18.3 - 15.6} = 2.6 \text{ kJ kg}^{-1}\,°\text{C}^{-1}$$

Note that the "apparent" specific heat at -18°C, 3.4 kJ kg^{-1} °C is higher than the specific heat below freezing in (a), 1.42, and in (b) , 1.29 kJ kg^{-1} °C^{-1}. The reason for the high apparent specific heat at -18°C is due to some freezing still continuing at this temperature. It is suggested that at -18°C only about two-thirds of the water is actually frozen to ice.

This implies only two-thirds of the latent heat has been extracted at this temperature.

Making this adjustment to the latent-heat terms, estimates for the total emthalph change (a) and (b) give 158kJ kg^{-1} and 142kJ kg^{-1} respectively, much improving the agreement with (c) of 145 kJ kg^{-1}.

Taking ΔH = 150 kJ kg^{-1}

Total heat change

$$= 150 \times 10{,}000 \times 0.75 = 1.125 \times 10^6 \text{ kJ}.$$

Total time = 6h = 2.16 x 10^4s.

$$(\Delta H / Dt) = 52 \text{ kJ s}^{-1} = 52 \text{ kW on average.}$$

And if the maximum rate of heat removal is twice the average:

$$(DH / Dt)_{max} = 2 \times 52 = \underline{104 \text{ kW}}.$$

Example 1 illustrates the application of heat balances, and it also illustrates the advisability of checking or obtaining corroborative data unless reliable experimental results are available for the particular system that is being considered. The straightforward application of the tabulated overall data would have produced a result about 30% higher than that finally calculated. On the other hand, for some engineering calculations to be within 30% may be about as close as you can get.

In some cases, it is adequate to make approximations to heat balances by isolating dominant terms and ignoring less important ones. To make approximations with any confidence, it is necessary to be reasonably sure about the relative magnitudes of the quantities involved. Having once determined the factors that dominate the heat balance, simplified balances can then be set up if appropriate to the circumstances and used with confidence in industrial energy studies. This simplification reduces the calculation effort, focuses attention on the most important terms, and helps to inculcate in the engineer a quantitative feeling for the situation.

EXAMPLE 2. Dryer heat balance for casein drying

In **drying** casein the dryer is found to consume 4 m^3/h of natural gas with a calorific value of 800 kJ/mole. If the throughput of the dryer is 60 kg of wet casein per hour, drying it from 55% moisture to 10% moisture, estimate the overall thermal efficiency of the dryer taking into account the latent heat of evaporation only.

Basis: 1 hour of operation

60 kg of wet casein contains:

60 x 0.55 kg water = 33 kg moisture

and 60 x (1 - 0.55) = 27 kg bone dry casein.

As the final product contains 10% moisture, the moisture in the product is 27/9 = 3 kg.

and so moisture removed = (33 - 3) = 30 kg

Latent heat of evaporation = 2257 kJ kg^{-1} (at 100 °C)

so heat necessary to supply = 30 x 2257 = 6.8 x 10^4 kJ

Assuming the natural gas to be at standard temperature and pressure at which 1 mole occupies 22.4 litres

$$\text{Rate of flow of natural gas} = 4\ m^3\ h^{-1} = \frac{4 \times 1000}{22.4} = 179\ \text{moles}\ h^{-1}$$

Heat available from combustion = 179 x 800 = 14.3 x 10^4 kJ/h

$$\text{Approximate thermal efficiency of dryer} = \frac{\text{heat needed}}{\text{heat used}}$$

= 6.8 x 10^4 / 14.3 x 10^4 = %48

To evaluate this efficiency more completely it would be necessary to take into account the sensible heat of the dry casein solids and the moisture, and the changes in temperature and humidity of the combustion air, which would be combined with the natural gas. However, as the latent heat of evaporation is the dominant term the above calculation gives a quick estimate and shows how a simple energy balance can give useful information.

Similarly energy balances can be carried out over thermal processing operations, and indeed any processing operations in which heat or other forms of energy are used.

EXAMPLE 3. Heat balance for cooling pea soup after canning

An autoclave contains 1000 cans of pea soup. It is heated to an overall temperature of 100°C. If the cans are to be cooled to 40°C before leaving the autoclave, how much cooling water is required if it enters at 15°C and leaves at 35°C?

The specific heats of the pea soup and the can metal are respectively 4.1 kJ kg^{-1} $°C^{-1}$ and 0.50 kJ $kg^{-1}°C^{-1}$. The weight of each can is 60 g and it contains 0.45 kg of pea soup. Assume that the heat content of the autoclave walls above 40°C is 1.6 x 10^4 kJ and that there is no heat loss through the walls.

Let *w* = the weight of cooling water required; and the datum temperature be °40C, the temperature of the cans leaving the autoclave.

Heat entering

Heat added to cans = weight of cans x specific heat × temperature above datum

= 1000 x 0.06 x 0.50 x (40 - 100) kJ = 1.8 x 10^3 kJ.

Heat added to can contents = weight pea soup x specific heat x temperature above datum

= 1000 x 0.45 x 4.1 x (100 - 40)

= 1.1×10^5 kJ.

Heat added to water = weight of water x specific heat x temperature above datum

= w x 4.21 x (40 - 15)

= -104.6 w kJ.

Heat leaving

Heat in cans = 1000 x 0.06 x 0.50 x (40 - 40) (cans leave at datum temperature)

= 0.

Heat in can contents = 1000 x 0.45 x 4.1 x (40 – 40)

= 0.

Heat in water = w x 4.186 x (35 - 40)

= -20.9 w.

Heat-energy Balance Of Cooling Process; 40°C As Datum Line

Heat entering	(kJ)	Heat Leaving	(kJ)
Heat in cans	1,800	Heat in cans	0
Heat in can contents	110,000	Heat in can contents	0
Heat in autoclave wall	16,000	Heat in autoclave wall	0
Heat in water	105.3w	Heat in water	-20.9 w
Total heat entering	127,800 -105.3w	Total heat leaving	-20.9 w
Total heat entering	= Total heat leaving		
127,800 - 105.3w		= -20.9 w	
And so w		= 1514 kg	
Amount of cooling water required		= 1514 kg.	

Other Forms of Energy

The most common **mechanical** energy is motor power and it is usually derived, in food factories, from **electrical** energy but it can be produced from steam engines or water power. The electrical energy input can be measured by a suitable wattmeter, and the power used in the drive estimated. There are always losses from the motors due to heating, friction and windage; the motor efficiency, which can normally be obtained from the motor manufacturer, expresses the proportion (usually as a percentage) of the electrical input energy which emerges usefully at the motor shaft and so is available.

When considering movement, whether of fluids in pumping, of solids in solids handling, or of foodstuffs in mixers, the energy input is largely mechanical. The flow situations can be analysed by recognising the conservation of total energy whether as energy of motion, or potential energy such as pressure energy, or energy lost in friction. Similarly, chemical energy released in combustion can be calculated from the heats of combustion of the fuels and their rates of consumption. Eventually energy emerges in the form of heat and its quantity can be estimated by summing the various sources.

THERMAL PROPERTIES OF FROZEN FOODS

Introduction

The food product properties of interest when considering the freezing process include density, specific heat, thermal conductivity, enthalpy, and latent heat. These properties must be considered in the estimation of the refrigeration capacity for the freezing system and the computation of freezing times needed to assure adequate residence times. The approach to prediction of property magnitudes during the freezing process depends directly on the relationship between unfrozen water fraction and temperature.

It is important to study thermal properties of foods because they affect the design of food processing equipment. The food products undergo changes in composition during such process as freezing, evaporation and dehydration. There are different methods available to measure the thermal properties of food, but the available data differ depending on the method used. The important thermal properties of food are as follows:

Density

The density is mass per unit volume. Usually the density is expressed in grams per mL or cc. Mathematically a "per" statement is translated as a division. cc is a cubic centimeter and is equal to a ml.

The influence of freezing on food product density is relatively small but a dramatic change does occur at and just below the initial freezing temperature. This change can be predicted by the following equation, as discussed by Heldman:

ρ = 1/ " (m si /ρ si)

Specific Heat

A measure of the heat required to raise the temperature of a substance. When the heat DQ is added to a body of mass m, raising its temperature by DT, the ratio C given in Eq. (1) is defined as the heat capacity of the body.

$C_p = DQ/ DT$ Eq. (1)

The specific heat capacity of a food product can be predicted, based on product composition and the specific heat capacity of individual product components. The following expression was proposed:

$Cp = \Sigma(C\ psi.\ m\ si\ )$

where each factor on the right-side of the equation is the product of the mass fraction of a product component and the specific heat capacity of that component. The specific heat values for product components were estimated by Choi and Okos (1986). The above equation can be used to predict the specific heat capacity of product solids by removing the term for the water fraction. These specific heat magnitudes for the product solids can be used in the prediction of product enthalpy and apparent specific heat.

$C_p = 4.180\ X_w + 1.711\ X_p + 1.98\ X_f + 1.547\ X_c + 908/0\ X_{a,;}$ kJ/kg^{0}C (Cho's and Oko's Model)

Where,

X_w: water fraction

Xp: Protein fraction

X_f: Fat fraction

$X_{c:}$ Carbohydrate fraction

$X_{A:}$ Ash fraction

Thermal Conductivity

Thermal conductivity (λ) is the intrinsic property of a material which relates its ability to conduct heat. Heat transfer by conduction involves transfer of energy within a material without any motion of the material as a whole. Conduction takes place when a temperature gradient exists in a solid (or stationary fluid) medium. Conductive heat flow occurs in the direction of decreasing temperature because higher temperature equates to higher molecular energy or more molecular movement. Energy is transferred from the more energetic to the less energetic molecules when neighboring molecules collide.

Thermal conductivity is defined as the quantity of heat (Q) transmitted through a unit thickness (L) in a direction normal to a surface of unit area (A) due to a unit temperature gradient (ΔT) under steady state conditions and when the heat transfer is dependent only on the temperature gradient. In equation form this becomes the following:

Thermal Conductivity = heat × distance / (area × temperature gradient)

$\lambda = Q \times L / (A \times \Delta T)$

The thermal conductivity magnitudes of most food products are a function of water content and the physical structure of the product. Many models suggested for prediction of thermal conductivity are based on moisture content and do not consider structural orientation. The Choi's and Oko's Model for prediction of thermal conductivity is as follows.

$K = 0.58\,X_w + 0.155\,X_p + 0.25\,X_c + 0.16\,X_f + 0.135\,X_{a,}$ W/m ºK (Cho's and Oko's Model)

Where,

X_w: water fraction

Xp: Protein fraction

X_f: Fat fraction

$X_{c:}$ Carbohydrate fraction

$X_{a:}$ Ash fraction

Thermal Diffusivity

A measure of the rate at which a temperature disturbance at one point in a body travels to another point. It is expressed by the relationship K/dC_p, where K is the coefficient of thermal conductivity, d is the density, and C_p is the specific heat at constant pressure. Very little thermal diffusivity data are available, but it can be determined using relationship of specific heat, thermal conductivity and mass density of the food product.

Freezing Point Depresssion

Probably one of the more reveling properties of water in food is the freezing point depression. Since all food products contain relatively large amounts of moisture or water in which various solutes are present, the actual or initial freezing point of water in the product will be depressed to some level below that expected for pure water.

The magnitude of this freezing point depression becomes a direct function of the molecular weight and concentration of the solute in the food product and in solution in the water.

R_g is gas constant

$R_g = w \times W_A = 18 \times 0.462$

where

$R_g = 8.316$ Joules/mole °C

W_A = molecular weight of water = 18 moles/g

w = fraction of water

Thermodynamics of Food Freezing

Freezing is one the more common processes for the preservation of foods. It is well known that lowering the temp reduces the activity of microorganisms and enzyme systems, thus preventing deterioration of the food products. In addition to the influence of temp reduction on m.o. and enzymes, crystallization of the water in the product tends to reduce the amount of liquid water in the system and inhibit microbial growth or enzyme activity in the secondary action.

The engineering aspects of food freezing include several interesting areas. In order to design a refrigeration system that will serve a food freezing process, some indication of the refrigeration requirements or enthalpy change which occurs during product freezing is required. This aspect is related to the type of product being frozen. The second aspect of food freezing that is closely related to engineering is the rate at which freezing progresses. This aspect is related to the refrigeration requirement, but the temperature difference existing between the product and freezing medium are also of significance.

The rate of freezing is also closely related to product properties and quality. Product properties resulting from very rapid freezing are significantly different from those obtained by slow freezing. This difference is dependent primarily on the manner in which ice is formed within the product structure. In addition, the rate of freezing will establish the rate of production for a particular food- freezing operation. For this purpose the most rapid rate of freezing is desirable provided that product quality is not sacrifice.

Example 1

A formulated food product contains the following components – water 80%, protein 2%, carbohydrate 17%, fat 0.1% and ash 0.9%. Predict the specific heat in W/kg K using Choi's and Oko's model.

Solution:

$$C_p = 4.180\ X_w + 1.711\ X_p + 1.98\ X_f + 1.547\ X_c + 908/0\ X_a$$

$$= 4.180\ (0.8) + 1.711\ (0.02) + 1.98\ (0.001) + 1.547\ (0.17) + 0.908\ (0.009)$$

$$= 3.651\ kJ/kg^\circ C$$

$$= 0.8726\ kCal/kg^\circ C$$

$$= 1.0147\ W/kg^\circ C$$

Example 2

Calculate the thermal conductivity of milk using choi & OKOS model, if milk contains 87.5% water, 3.7% protein, 3.7% fat, 4.6% lactose and 0.5% ash at 10°C.

Solution

$K = 0.58\ X_w + 0.155\ X_p + 0.25\ X_c + 0.16\ X_f + 0.135\ X_a$

$= 0.58\ (0.875) + 0.155\ (0.037) + 0.25\ (0.046) + 0.16\ (0.037) + 0.135\ (0.005)$

$= 0.49 + 0.005735 + 0.0115 + 0.00592 + 0.000675$

$= 0.51383$ W/m ºK

Example 3

Compute the temperature at which ice formation begins in an ice cream mix with the following composition: 10% butter fat, 12% solids-not-fat, 15% sucrose and 0.22% stabilizer.

The solute accounted for in the ice-cream mix is sucrose (W = 342) and lactose (W = 342), which represents 54.5 % of the SNF in the mix. – Molality is computed as:

Fraction solute = 0.15 + 0.545(0.12) = 0.2154 g/ g product

When expressed in terms of water fraction (62.78%), thus 0.2154/0.6278 = 0.3431 g solute/g solvent or 343.1 g solute / 1000g solvent

And m = 343.1/342 = 1.003

APPENDIX 1

THERMAL DATA FOR SOME FOOD PRODUCTS

Freezing point (°C)	Percent water	Specific heat above freezing (kJ kg-1°C-1)	Latent heat of fusion below freezing (kJ kg-1)		
Fruit					
Apples	-2	84	3.6	1.88	280
Bananas	-2	75	3.35	1.76	255
Grapefruit	-2	89	3.81	1.93	293
Peaches	-2	87	3.78	1.93	289
Pineapples	-2	85	3.68	1.88	285
Watermelons	-2	92	4.06	2.01	306
Vegetables					
Asparagus	-1	93	3.93	2.01	310

Freezing point (°C)	Percent water	Specific heat above freezing (kJ kg-1°C-1)	Latent heat of fusion below freezing (kJ kg-1)		
Beans (green)	-1	89	3.81	1.97	297
Cabbage	-1	92	3.93	1.97	306
Carrots	-1	88	3.6	1.88	293
Corn	-1	76	3.35	1.8	251
Peas	-1	74	3.31	1.76	247
Tomatoes	-1	95	3.98	2.01	310
Meat					
Bacon	-2	20	2.09	1.26	71
Beef	-2	75	3.22	1.67	255
Fish	-2	70	3.18	1.67	276
Lamb	-2	70	3.18	1.67	276
Pork	-2	60	2.85	1.59	197
Veal	-2	63	2.97	1.67	209
Miscellaneous					
Beer	-2	92	4.19	2.01	301
Bread	-2	32-37	2.93	1.42	109-121
Eggs	-3		3.2	1.67	276
Ice cream	-3 to -18	58-66	3.3	1.88	222
Milk	-1	87.5	3.9	2.05	289
Water	0	100	4.19	2.05	335

Specific heats, latent heats of freezing and thermal conductivities of foodstuffs can be estimated if the percentage of water in the foodstuff is known. If this percentage is p then:

- Specific heat = 4.19p/100 + 0.84(100- p)/100 kJ kg-1 °C-1 above freezing
- = 2.1 p/100 + 0.84(100- p)/100 kJ kg-1 °C-1 below freezing.
- Latent heat = 335p/100 kJ kg-1
- Thermal conductivity = 0.55p/100 + 0.26(100 -p)/l00 J m-1 s-1 °C-1 above freezing

= 2.4p/100 + 0.26(100 -p)/l00 J m-1 s-1 °C-1 below freezing.

CHAPTER

ENERGY AND CONTROLS IN FOOD PROCESSES 5

GENERATION OF STEAM

Steam plays an integral role in the food processing industry for various purposes such as cooking, drying, sterilizing, packaging, etc. Since steam is energy efficient and reliable in transferring heat to food products, it is essential to check and maintain the steam quality. Steam quality in food processing industry directly affects the food and causes contamination as it is in direct contact during process operations. Steam quality in food processing industry varies as per its utilization purpose and different levels of contamination risk.

Steam is a convenient means to convey energy in food processing operations. It is produced from inexpensive and abundant water. Pressure control valves can be used to precisely regulate and maintain the temperature of steam. Large amounts of energy are contained in a relatively small mass of steam, so heat-transfer equipment can be compact. Steam is easily and inexpensively conveyed over fairly long distances and into remote locations of the process.

Steam used by food processors commonly falls into two broad categories. The first is the so-called "culinary," "sanitary" or "clean" steam. This type of steam is used for direct injection into the product or to clean or sterilize product contact surfaces. We will refer to steam in this category as "culinary" steam. Any additives in culinary steam must meet all applicable FDA and USDA requirements for human consumption.

The second category of steam used by food processors is often referred to as "plant steam," "utility steam" or just simply "steam." This document will call steam in this category "plant" steam. Ordinary plant steam can be used in most applications that do not involve contact with food products or with surfaces that contact food products. An example of a non-contact use of plant steam is indirect heating. Most people are familiar with radiant heaters (a type of indirect steam heater) used in residential and commercial applications.

Definitions

- **Condensation** – conversion of water from the vapor state to a liquid phase.
- **Culinary steam** – steam that is suitable for direct injection into food products or direct contact with food products or surfaces that contact foods. Culinary steam must meet all applicable codes of the appropriate regulatory agency (e.g. FDA and USDA) for the application.
- **Heat transfer** – energy in transit as the result of a temperature difference.
- **Indirect heating** – heat transfer from a warmer body to a cooler body through a physical barrier such as steel.
- **Latent heat** – heat given off or absorbed in a process (as fusion or vaporization) other than a change of temperature.
- **Latent heat of vaporization** – the heat added to water to make it vaporize without a change in temperature.
- **Plant steam** – steam that is not or cannot be classified as "culinary steam." In general, steam is water vapor kept under pressure to supply energy for heating, cooking or mechanical work.
- **Quality** – property of steam determined by the amount of free moisture carried with the steam.
- **Sensible heat** – heat given off or absorbed in a process characterized by a change in temperature.
- **Specific heat** – a ratio of the amount of heat transferred to raise a unit mass of material one degree in temperature to that required to raise a unit mass of water one degree in temperature at some specified, initial temperature.
- **Steam trap** – mechanical device that automatically traps steam in an enclosed space and simultaneously allows condensate (and sometimes air) to be drained away.

EFFECTIVE HEAT TRANSFER

Steam is most effective at transferring heat when it condenses on a surface at a lower temperature. The latent heat of vaporization is given up by the steam as it condenses. Maximum heat transfer occurs when steam is allowed to condense on a heat transfer surface without restriction or obstruction. Realistically, there are obstacles that prevent perfect condensation by insulating heat transfer surfaces. Figure 1 shows examples of layers of films that can build up on the product and steam side of a heat transfer surface.

Some of the most common examples of these films in food processing operations are: stagnant layers of product, burned-on product, rust, scale, water (condensate), layers of single-celled organisms, and air. These films must be eliminated or reduced to optimize heat transfer.

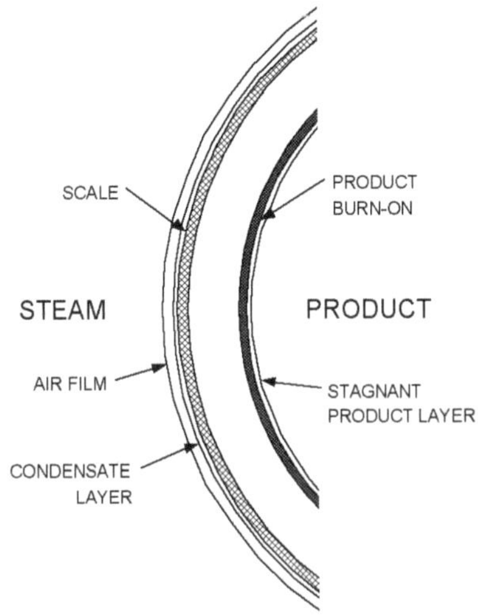

Fig. 1. Films on heat transfer surfaces that slow heating

Stagnant layers of product can be removed (or reduced) by agitation, especially surface-scraping agitation. Product burn-on can be minimized by surface-scraping agitation and regulation of the heat-transfer surface temperature. Rust, scale and organic films may be controlled by methods such as frequent cleaning of heat-transfer surfaces, specification of corrosion-resistant materials, and selection and maintenance of boiler-water additives.

Condensed water must be quickly and completely drained from the steam side of the process. This is accomplished by providing dry steam to the process and utilizing correctly specified and installed steam traps. Air can be removed from the steam side of the process by proper design and installation of equipment, following startup procedures designed to remove air, and installation of air vents.

TYPES OF STEAM AND ITS QUALITY

Plant Steam

Plant steam is the lowest quality of steam and consists of a high amount of impurities and contaminants. The impurity level in plant steam is determined by the factors such as,

- The quality of feed water entering the steam boiler
- The amount of chemicals used in the steam boiler for the water treatment management
- TDS level, operating pressure, and other components of steam boiler operation

Filtered Steam

Filtered steam, also known as culinary steam, is the most utilized type of steam in food processing sector. It passes through a fine stainless steel filter, usually of five-micron pore size.

The impurities from feed water treatment chemicals and cross-contamination are present in filtered steam. The amount of contaminants passing through the filter depends on the nature of the boiler problem, frequency of boiler maintenance, and the velocity of the steam.

Clean Steam

Clean steam is the highest level of steam quality in food processing and beverage industry. It is generated from purified water in a specific clean steam generator. Clean steam is mainly used for process operations that rely on the superior quality of products.

Pure Steam

Pure steam is slightly higher than clean steam in terms of purity and is largely used in pharmaceutical industries. With the increasing requirement of steam quality in food processing industry, pure steam is also used in the food and beverage processing sector.

Determining Steam Quality

Steam quality refers to the amount of water present in the generated steam. It is also defined as dryness fraction that is measured by the following formula,

Dryness Fraction = Mass of Steam/Mass of steam + Entrained water

On the other hand, steam purity is a quantitative measure of the dissolved solids, volatiles, or other particles in the vapor remaining in the steam after the separation from the steam boiler.

CAUSES OF STEAM CONTAMINATION

Various factors contribute to the contamination of the steam quality. Following are some of the factors that should be controlled to minimize impurity and contamination of steam.

Chemical Usage

The feed water quality in steam boilers has a significant impact on the efficiency of the boiler system and steam generation.

Along with the components present in the feed water, further chemicals are added in boiler water treatment to reduce the possibility of corrosion, erosion, and scaling. Excess use of chemicals may affect the steam quality.

Boiler Carryover

Boiler carryover is not entirely steam and consists of harmful boiler treatment chemicals in the form of entrained water and foam. The two factors that mainly cause carryover in the boiler are:

Priming

During steam generation, some water particles are carried along with the steam. The process of formation of wet steam is known as priming. It is caused due to,

- Negligence in regular maintenance and proper water treatment
- High water level
- Operating the steam boiler below the design pressure
- Sudden steam demand

Foaming:

Foaming is the constant formation of bubbles or foam in the steam boiler caused by the oily substances in water that reduce the surface tension in the boiler water. The greater amount of foaming leads to severe issues in steam boilers. The major contributors of foaming in steam boilers are,

- High levels of Total Dissolved Solids (TDS) in the boiler system
- Excess amount of water treatment chemicals
- Boiler water contamination from other areas of the process
- High alkalinity (>1 000 ppm)
- **Cross Contamination:** Several food processing plants return condensate for reducing water, energy, and chemical utilization. As the condensate passes through the steam boiler system, it may be subjected to cross-contamination from other sources.
- **Corrosion and Scaling:** Carbon dioxide and oxygen cause severe corrosion of steam/condensate pipework and boilers. As a result, corrosion products can precipitate and form deposits that contaminate steam supplies and any area using steam.
- **Non-condensable gases:** Dissolved gases in feed water such as oxygen, ammonia, and carbon dioxide may lead to corrosion, reduced heat transfer, etc. in a steam boiler. It is essential to control the amount of gases in the boiler system.

Preventive Measures to Ensure Steam Quality for Food Processing Industry

- **Efficient Operation:** It is necessary to ensure hassle-free steam boiler operation. With a constant load of steam boiler and design parameters, the entrained moisture carried over within steam should be less than 2%
- **Chemical Usage:** The amount of chemicals in the steam boiler should be as per the requirement and not be in excess.
- **TDS Control:** TDS level must be controlled as per the water treatment guideline and at levels that minimize the effect of foaming
- **Condensate assessment:** Condensate and steam sampling should be performed regularly to ensure appropriate water treatment procedures. Since steam boiler carryover is dependent on various factors, infrequent testing may not always identify the occurrence of carryover.
- **Monitoring of Cross Contamination:** Cross-contamination of the steam boiler system from other sources can occur at any time and, therefore, should be monitored regularly.

FOOD PROCESSING EQUIPMENT

Shell and Tube Heater

Shell and tube heaters are commonly used to heat a flowing liquid by condensing plant steam or a pumped heat transfer media. A thin-tube wall separates the heating media from the product being heated. In the case of a pumped heat-transfer media (such as hot water), steam is often used to heat the media in a separate heat exchanger. Figure 2 shows a typical shell and tube heater.

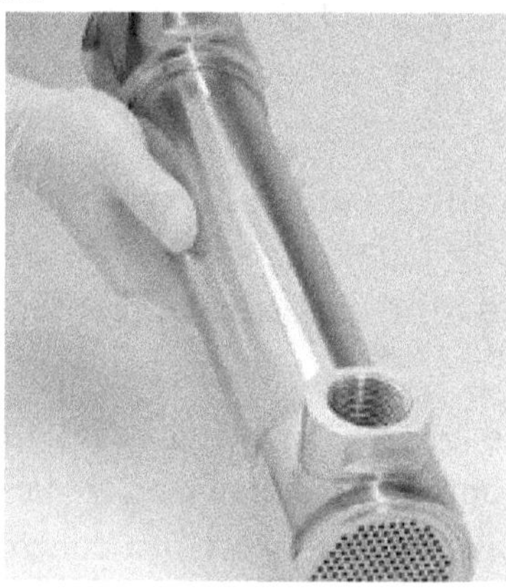

Fig. 2. Shell and tube heater

Steam Injection Heater

Steam injection heating for food products is a direct-contact process in which culinary steam is mixed with a pumpable food (or ingredient). Heating occurs when the steam transfers some of its internal energy to the food product. Steam gives up all of its latent heat of vaporization while condensing and, depending upon the system pressure, some of its sensible heat. Since the steam directly contacts the food product and the condensate becomes incorporated into it, the steam source must be culinary. Typical steam injection units are compact, inexpensive and simple to control. An example is provided in figure 3.

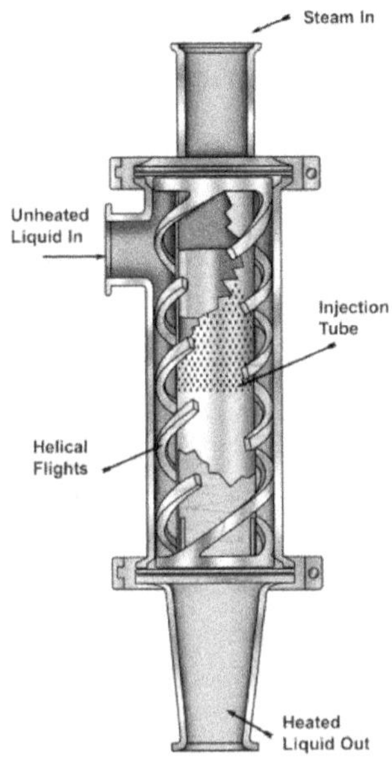

Fig. 3. Steam injection heater

Steam Infusion

Steam infusion is another method of steam heating where the food product is heated by relatively low-pressure, culinary steam. A steam infusion device usually incorporates a chamber where the product is cascaded through a steam environment.

Steam Sparger

Steam sparging is common in open tanks or kettles containing liquid products or water. Culinary steam is bubbled into the liquid from nozzles located near the bottom of the tank.

Steam Jacketed Kettle

A jacket, used to distribute steam over a wide surface area, consists of a thin space formed between two, parallel, metallic surfaces. Steam jackets are typically used to heat bulk products held in tanks and kettles. An example of a steam-jacketed kettle is shown in Fig. 4. Condensing steam, held captive within the jacket, transfers heat to the product in the kettle. A layer of insulation over the jacket protects operators and conserves heat.

Steam Generation Systems

The systems for generation of steam can be divided into two major classifications: firetube and water-tube. Both systems are used in the food industry, but water-tube systems are designed for the more modern applications. The steam generation system or boiler is a vessel designed to bring water into contact with a hot surface, as required to convert liquid to vapor. The hot surface is maintained by using hot gases, usually combustion gases from natural gas or other petroleum products. The boiler vessel is designed to contain the steam and to withstand the pressures resulting from the change of state for water.

Fig. 4. Steam Jacketed Kettle

Fire-tube steam generators (Fig. 5) utilize hot gases within tubes surrounded by water to convert the water from liquid to vapor state. The resulting heat transfer causes the desired change of state, with the vapors generated contained within the vessel holding the water. A water-tube steam generator (Fig. 2) utilizes heat transfer from hot gas surrounding the tubes to the water flowing through the tubes to produce steam. The heat transfer in the water-tube system tends to be somewhat more rapid because of the ability to maintain turbulent flow within the liquid flow tube.

Water-tube boilers generally operate with larger capacities and at higher pressures. These systems have greater flexibility and are considered safer to operate than the counterpart fire-tube systems. The safety feature is associated most closely with the change-of-phase occurring within small tubes in a water-tube system rather than in a large vessel in a fire-tube system. The latter system does have an advantage when the load on the system varies considerably with time. Nearly all modern installations in the food industry are of the water-tube design. One of the more recent developments is the utilization of alternate fuels as a source of energy for steam generation. In particular, combustible waste materials from processing operations have become a viable alternative. In many situations, these materials are available in large quantities and may present a disposal problem.

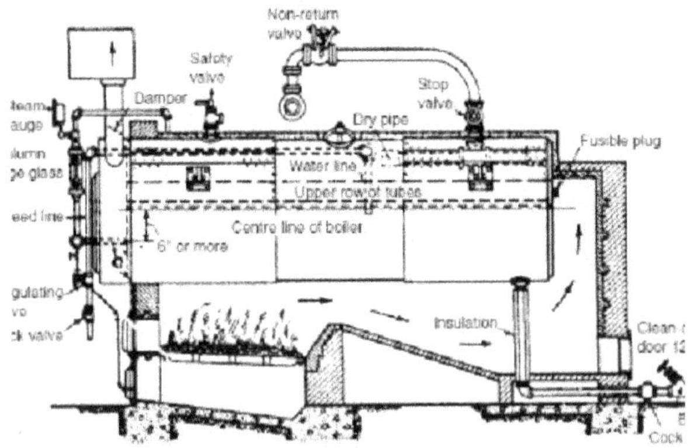

Fig. 5. The horizontal return tubular (HRT) fire-tube boiler

STEAM UTILIZATION

The capacity of the steam generation system in a food processing plant is established by requirements of the individual operations using steam. The requirements are expressed in two ways: (1) the temperature of steam needed as a heating medium, and (2) the quantity of steam required to supply the demands of the operation. Since the temperature requirement is a function of pressure, this establishes one of the operating conditions of the system. In addition, the steam properties are a function of pressure (and temperature), which in turn influences the quantity of steam utilized.

The steps involved in determining the capacity of a steam generation system include the following. The thermal energy requirements of all operations utilizing steam from a given system are determined. In most situations, those requirements will establish maximum temperature required and therefore the pressure at which the steam generation system must operate.

After the operating pressure of the system is established, the properties of steam are known and the thermal energy available from each unit of steam can be determined. This information can then be used to compute quantities of steam required for the process. An important consideration in sizing the pipe connecting the process to the steam generation system is the volume of steam required. Using the quantity of steam required as expressed in mass units, and the specific volume of the steam being used, the volumetric flow rate for steam leading to the process is computed.

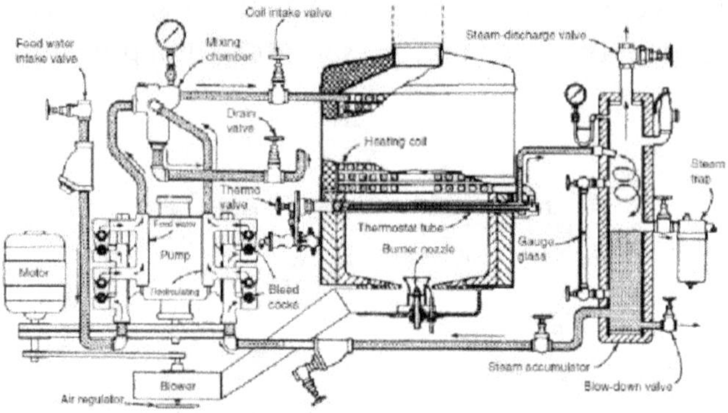

Fig. 6. Water-tube steam generator.

FUEL UTILIZATION

The energy requirements for food processing are met in a variety of ways. In general, the traditional energy sources are utilized to generate steam as well as to provide for other utilities used in the processing plant. As illustrated in Table 1, the energy types include natural gas, electricity, petroleum products, and coal. Although the information presented was collected in 1973 and percentages of natural gas utilization have declined somewhat, it seems evident that food processing has a definite dependence on petroleum products and natural gas.

To release the energy available from natural gas and petroleum products, they are exposed to a combustion process. This is a rapid chemical reaction involving fuel components and oxygen. The primary fuel components involved in the reaction include carbon, hydrogen, and sulfur, with the last being an undesirable component. The oxygen for the reaction is provided by air, which must be mixed with fuel in the most efficient manner.

ELECTRIC POWER UTILIZATION

Electric power has become so commonplace in the food industry that modern plants could not operate without this power source. In fact, most plants of significant size have acquired "back-up" electrical power generators to use in case disruptions occur in the primary supply. It is quite evident that electric power represents the most versatile and flexible power source available. In addition, the cost of electric power is very attractive when compared with other sources.

Electrical Terms and Units

As in most physical systems, electricity has its own set of terms and units. These terms and units are entirely different from most physical systems, and it requires careful analysis to relate the terms to applications. This presentation is elementary and is intended to be a brief introduction to the subject. The following terms are essential.

- Electricity can be defined as the flow of electrons from atom to atom in an electrical conductor. Most materials can be considered conductors, but will vary in the ability to conduct electricity.
- Ampere is the unit used to describe the magnitude of electrical current flowing in a conductor. By definition, 1 ampere (A) is 6.06×10^{18} electrons flowing past a given point per second.
- Voltage is defined as the force causing current flow in an electrical circuit. The unit of voltage is the volt (V).
- Resistance is the term used to describe the degree to which a conductor resists current flow. The ohm (Ù) is the unit of electrical resistance.
- Direct current is the type of electrical current flow in a simple electrical circuit. By convention, current is considered to flow from a positive to a negative terminal of a voltage generator.
- Alternating current describes the type of voltage generated by an AC (alternating current) generator. Measurement of the actual voltage generated would indicate that the magnitude varies with time and a uniform frequency. The voltage ranges from positive to negative values of equal magnitudes. Most electrical service in the United States operates at 60 cycles per second (60 Hz).
- Single-phase is the type of electrical current generated by a single set of windings in a generator designed to convert mechanical power to electrical voltage. The rotor in the generator is a magnet that produces magnetic lines as it rotates. These magnetic lines produce a voltage in the iron frame (stator) that holds the windings. The voltage produced becomes the source of alternating current.
- Three-phase is the type of electrical current generated by a stator with three sets of windings. Since three AC voltages are generated simultaneously, the voltage can be relatively constant. This type of system has several advantages compared with single-phase electricity.
- Watt is the unit used to express electrical power or the rate of work. In a direct current (DC) system, power is the product of voltage and current, whereas computation of power from an alternating current (AC) system requires use of a power factor.
- Power factors are ratios of actual power to apparent power from an alternating current system. These factors should be as large as possible to ensure that excessive current is not carried through motors and conductors to achieve power ratings.

- Conductors are materials used to transmit electrical energy from source to use. Ratings of conductors are on the basis of resistance to electrical flow.

Electric Motors

The basic component of an electric energy utilization system is the electric motor. This component converts electrical energy into mechanical energy to be used in operation of processing systems with moving parts.

The majority of the motors used in food processing operations operate with alternating current (AC), and their operation depends on three basic electrical principles. These principles include the electromagnet, formed by winding insulated wire around a soft iron core. Current flow through the wire produces a magnetic field in the iron core; orientation of the field is dependent on the direction of current flow.

The second electrical principle involved in the operation of a motor is electromagnetic induction. This phenomenon occurs when an electric current is induced in a circuit as it moves through a magnetic force field. The induced electric current produces a voltage within the circuit, with magnitude that is a function of the strength of the magnetic field, the speed at which the current moves through the field, and the number of conductor circuits in the magnetic field.

The third electrical principle is alternating current. As indicated earlier, this term refers to a current that changes direction of flow in a consistent manner. Normal electric service is of 60 Hz, indicating that the change in current flow direction occurs 60 times per second.

A second component of an electric motor is the rotor: a rotating drum of iron with copper bars. The rotor is placed between the two poles or windings of the stator. The current flow to the stator and the resulting electromagnetic field produces current flow within the copper bars of the rotor. The current flow within the rotor creates magnetic poles, which in turn react with the magnetic field of the stator to cause rotation of the rotor. Due to the 60 Hz alternating current to the stator, the rotation of the rotor should be 3600 revolutions per minute (rpm), but it typically operates at 3450 rpm.

Although there are numerous types of electric motors, they operate on these same basic principles. The most popular motor in the food processing plant is the single-phase, alternating current motor. There are different types of single-phase motors; the differences are related primarily to the starting of the motor.

The selection of the proper motor for a given application is of importance when ensuring that efficient conversion of electrical to mechanical energy occurs. The selection process takes into account the type of power supply available, as well as the use of the motor. The type and size of load must be considered, along with the environmental conditions of operation and the available space.

Electric Lighting

Another primary use of electric power in food processing plants is to provide illumination of work spaces. Often the work productivity of workers within the plant will be dependent on the availability of proper lighting. The design of a lighting system for a workspace will depend on several factors. The light must be distributed properly within the space, and the light source must be of sufficient size and efficiency. The light source must be supported properly and easily replaced or serviced. Finally, the cost of the entire system will be a factor to consider.

Light can be defined as visually evaluated radiant energy. Light is a small portion of the electromagnetic spectrum and varies in color depending on the wavelength. The intensity of a light at a point location is measured in the unit lux: the magnitude of illumination at a distance of one meter from a standard candle. A light source can be expressed in lumens: the amount of light on one square meter of surface when the intensity is one lux.

Two types of light sources are used in food processing plants: the incandescent lamp and the fluorescent lamp. The incandescent lamp uses a tungsten filament through which current flows. Due to the high electrical resistance of the filament, the flow of current through it causes it to glow white-hot. These types of lamps will provide efficiencies of approximately 20 lumens per watt.

A fluorescent lamp uses an inductance coil to create a current discharge within the tube. The heat from the discharge causes electrons to be removed from mercury vapor within the tube. The return of the electrons to the shell of mercury vapor causes emission of ultraviolet rays. These rays react with phosphor crystals at the tube surface to produce light. Fluorescent lamps are two to three times more efficient than comparable incandescent lamps. Although there are other factors to consider when comparing incandescent and fluorescent lamps, the efficiency and the longer life of fluorescent lamps are the most important.

OVERALL VIEW OF AN ENGINEERING PROCESS

Using a material balance and an energy balance, a food engineering process can be viewed overall or as a series of units. Each unit is a unit operation. The unit operation can be represented as shown in Fig. 7.

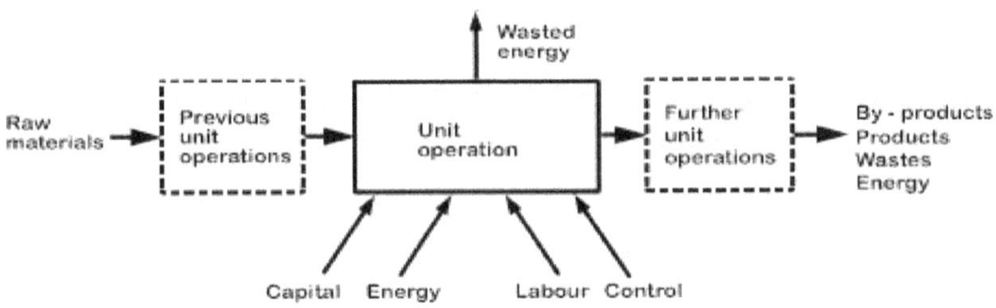

Fig. 7. Unit operation

Into the box go the raw materials and energy, out of the box come the desired products, by-products, wastes and energy. The equipment within the box will enable the required changes to be made with as little waste of materials and energy as possible. In other words, the desired products are required to be maximized and the undesired by-products and wastes minimized. Control over the process is exercised by regulating the flow of energy, or of materials, or of both.

CONSERVATION OF MASS AND ENERGY

The law of conservation of mass states that mass can neither be created nor destroyed. Thus in a processing plant, the total mass of material entering the plant must equal the total mass of material leaving the plant, less any accumulation left in the plant. If there is no accumulation, then the simple rule holds that "what goes in must come out". Similarly all material entering a unit operation must in due course leave.

For example, if milk is being fed into a centrifuge to separate it into skim milk and cream, under the law of conservation of mass the total number of kilograms of material (milk) entering the centrifuge per minute must equal the total number of kilograms of material (skim milk and cream) that leave the centrifuge per minute.

Similarly, the law of conservation of mass applies to each component in the entering materials. For example, considering the butter fat in the milk entering the centrifuge, the weight of butter fat entering the centrifuge per minute must be equal to the weight of butter fat leaving the centrifuge per minute. A similar relationship will hold for the other components, proteins, milk sugars and so on.

The law of conservation of energy states that energy can neither be created nor destroyed. The total energy in the materials entering the processing plant, plus the energy added in the plant, must equal the total energy leaving the plant.

This is a more complex concept than the conservation of mass, as energy can take various forms such as kinetic energy, potential energy, heat energy, chemical energy, electrical energy and so on.

During processing, some of these forms of energy can be converted from one to another. Mechanical energy in a fluid can be converted through friction into heat energy. Chemical energy in food is converted by the human body into mechanical energy.

For example, consider the pasteurizing process for milk, in which milk is pumped through a heat exchanger and is first heated and then cooled. The energy can be considered either over the whole plant or only as it affects the milk. For total plant energy, the balance must include: the conversion in the pump of electrical energy to kinetic and heat energy, the kinetic and potential energies of the milk entering and leaving the plant and the various kinds of energy in the heating and cooling sections,as well as the exiting heat, kinetic and potential energies.

To the food technologist, the energies affecting the product are the most important. In the case of the pasteurizer, the energy affecting the product is the heat energy in the milk. Heat energy is added to the milk by thc pump and by the hot water passing through the heat exchanger. Cooling water then removes part of the heat energy and some of the heat energy is also lost to the surroundings.

The heat energy leaving in the milk must equal the heat energy in the milk entering the pasteurizer plus or minus any heat added or taken away in the plant.

Heat energy leaving in milk = initial heat energy
+ heat energy added by pump
+ heat energy added in heating section
- heat energy taken out in cooling section
- heat energy lost to surroundings.

The law of conservation of energy can also apply to part of a process. For example, considering the heating section of the heat exchanger in the pasteurizer, the heat lost by the hot water must be equal to the sum of the heat gained by the milk and the heat lost from the heat exchanger to its surroundings.

From these laws of conservation of mass and energy, a balance sheet for materials and for energy can be drawn up at all times for a unit operation. These are called material balances and energy balances.

Process Control In Food Processing

Process control is an integral part of modern processing industries; and the food processing industry is no exception. The fundamental justification for adopting process control is to improve the economics of the process by achieving, amongst others, the following objectives:

- reduce variation in the product quality, achieve more consistent production and maximise yield,
- ensure process and product safety,
- reduce manpower and enhance operator productivity,
- reduce waste and
- optimist energy efficiency.

Processes are operated under either steady state, i.e. process conditions do not change, or unsteady state conditions, process conditions depend on time. The latter occurs in most real situations and requires control action in order to keep the product within specifications. Although there are many types of control actions and many different reasons for controlling a process, the following two steps form the basis of any control action:

- accurate measurement of process parameters;
- manipulation of one or more process parameters using control systems in order to alter or correct the process behaviour.

It is essential to note that a well designed process ought to be easy to control. More importantly, it is best to consider the controllability of a process at the very outset, rather than attempt to design a control system after the process plant has been developed.

Control Systems

Control systems can be of two types: manual control and automatic control.

Manual Control

In manual control, an operator periodically reads the process parameter which requires to be controlled and, when its value changes from the set value, initiates the control action necessary to drive the parameter towards the set value.

Figure 8 shows a simple example of manual control where a steam valve is adjusted to regulate the temperature of water flowing through the pipe.

An operator is constantly monitoring the temperature in the pipe. As the temperature changes from the set point on the thermometer, the operator adjusts the steam valve, i.e. either increases or decreases its flow, in order to get the temperature in the pipe back to the set point. Further action may be required if the temperature does not return to the set point within a reasonable time.

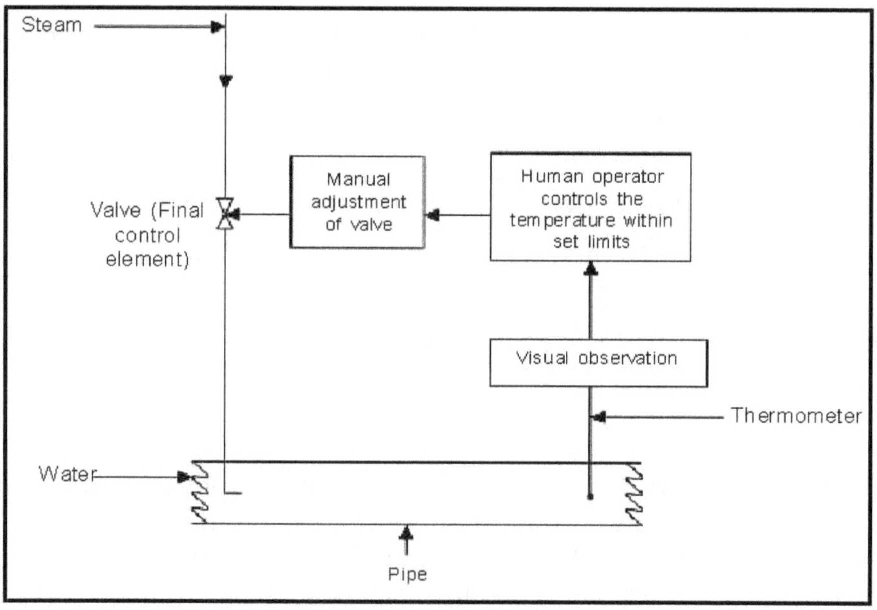

Fig. 8 A simple example of manual control

It is clear that the success of manual control operation depends on the skill of individual operators in knowing when and how much adjustment to make. Therefore, manual control may be used in those applications where changes in the manipulated parameter cause the process to change slowly and by a small amount. This is possible in plants where there are few processing steps with infrequent process upsets and the operator has sufficient time to correct before the process parameter overshoots acceptable tolerance. Otherwise, this approach can prove to be very costly in terms of labour, product inconsistencies and product loss.

Automatic Control

In automatic control, the process parameters measured by various sensors and instrumentation may be controlled by using control loops. A typical control loop consists of three basic components:

- Sensor: the sensor senses or measures process parameters and generates a measurement signal acceptable to the controller.
- Controller: the controller compares the measurement signal with the set value and produces a control signal to counteract any difference between the two signals.
- Final control element: the final control element receives the control signal produced by the controller and adjusts or alters the process by bringing the measured process property to return to the set point, e.g. liquid flow can be controlled by changing the valve setting or the pump speed.

An automatic control system can be classified into four main types:

- on/off (two position) controller
- proportional controller (P-controller)
- proportional integral controller (PI controller)
- proportional integral derivative controller (PID controller).

On/Off (Two Position) Controller

This is the simplest automatic controller for which the final control element, e.g. valve, is either completely open or at maximum, or completely closed or at minimum. There are no intermediate values or positions for the final control element.

Thus, final control elements often experience significant wear, as they are continually and rapidly switched from open to closed positions and back again. To protect the final control element from such wear, on/off controllers are provided with a dead band. The dead band is a zone bounded by an upper and a lower set point. As long as the measured process parameter remains between these set points, no changes in the control action are made.

On/off controllers with a dead band are found in many instances in our daily lives: home heating system, oven, refrigerator and air conditioner. All these appliances oscillate periodically between an upper and lower limit around a set point. Figure 9 illustrates the action of the control system. It is interesting to note from the figure that the use of dead band reduces the wear and tear on the final control element, but amplifies the oscillations in the measured process parameter.

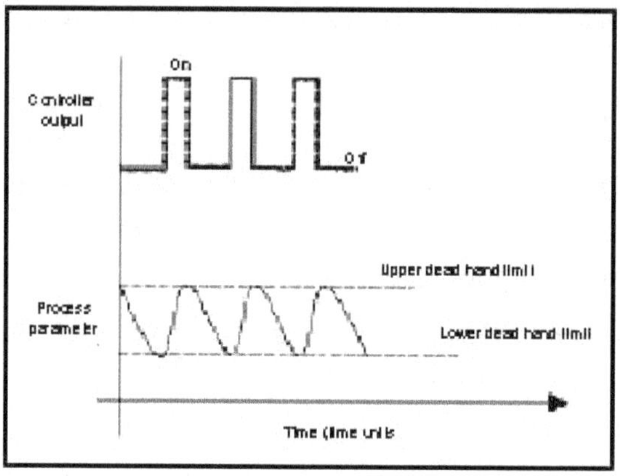

Fig. 9 Action of an on/off controller with dead bands

Such controllers have three main advantages:

- low cost,
- instant response and
- ease of operation.

However, it is important to ascertain that the upper and lower limit values are acceptable for a specific process. The main disadvantages of this type of control action are:

- it is not suitable for controlling any process parameter likely to suffer large sudden deviations from the set point and
- the quality of control is inferior to the continuous controller.

PROCESS CONTROL IN MODERN FOOD PROCESSING

Control applications in food processing, can be discussed in the context of three categories of products: (1) bulk commodity processing, e.g. grain milling, milk, edible oil, sugar and starch production, where control is arguably most advanced, (2) manufactured products, e.g. pasta, cheese, in-container and aseptically processed products, and (3) products which have been subjected to processing methods essentially designed to retain their original structure, e.g. meat, fish, fruits and vegetables. Regardless of the nature of the products, process control in food processing has moved on from just attempting to control single variables, e.g. level, temperature, flow, etc., to systems which ensure smooth plant operation with timely signalling of alarms. The systems are also geared to provide vital data at shop floor level right through to vertically structured systems which encompass supervisory control and data acquisition (SCADA), manufacturing execution systems (MES) and interfacing with complex enterprise resource planning systems (ERP), which may be connected across multiple production sites.

Programmable Logic Controller

The programmable logic controller (PLC) is the most common choice in modern control. It is a microprocessor-based system which can communicate with other process control components through data links. PLCs commonly use the so called ladder logic which was originally developed for electrical controls using relay switches. Programs can be written using a variety of languages. Once the program sequence has been entered into the PLC, the keyboard may be locked or removed altogether for security. When a process is controlled by a PLC, it uses inputs from sensors to take decisions and update outputs to drive actuators.

Thus, a control loop is a continuous cycle of the PLC reading inputs, solving the ladder logic and then changing the outputs. A real process will inevitably change over time and the actuators will drive the system to new states (or modes of operation). This implies that the control performance relies on the sensors available and its performance is limited by their accuracy.

Supervisory Control and Data Acquisition

The supervisory control and data acquisition (SCADA) system is not a full control system, but is a software package that is positioned at a supervisory level on top of hardware to which it is interfaced, generally via PLCs, or other hardware modules. SCADA systems are designed to run on common operating systems.

Two basic layers can be distinguished in a SCADA system: (1) the client layer which serves as the human-machine interface and (2) the data server layer which handles process data and control activities, by communicating with devices such as PLCs and other data servers. Such communication may be established by using common computing networks.

Modern data servers and client stations often run on Windows NT or Linux platforms. SCADA-based control systems also lend themselves to being scalable by adding more process variables, more specialised servers, e.g. for alarm handling, or more clients. This is normally done by providing multiple data servers connected to multiple controllers. Each data server has its own configuration database and realtime database (RTDB) and is responsible for handling a section of the process variables, e.g. data acquisition, alarm handling or archiving. Reports can also be produced when needed, or be automatically generated, printed and archived. SCADA systems are generally reliable and robust and, more often than not, technical support and maintenance are provided by the vendor.

Manufacturing Execution Systems

Manufacturing execution systems (MES) are software packages which have been used for a number of years in process industries to support key operations and management functions ranging from data acquisition to maintenance management, quality control and performance analysis. However, it is only in recent years that there has been a concerted attempt to integrate factory floor information with enterprise resource planning systems (ERP).

Modern MES include supply chain management, combine it with information from the factory floor and deliver results to plant managers in real time, thus integrating supply chain and production systems with the rest of the enterprise. This gives a holistic view of the business that is needed to support a 'manufacture to order' model.

MES can manage production orders and can track material use and material status information. The software collects data and puts it into context, so that the data can be used for both realtime decision making and performance measurement and for historical analysis. A typical control system currently available for dairy plants is shown in Fig. 10.

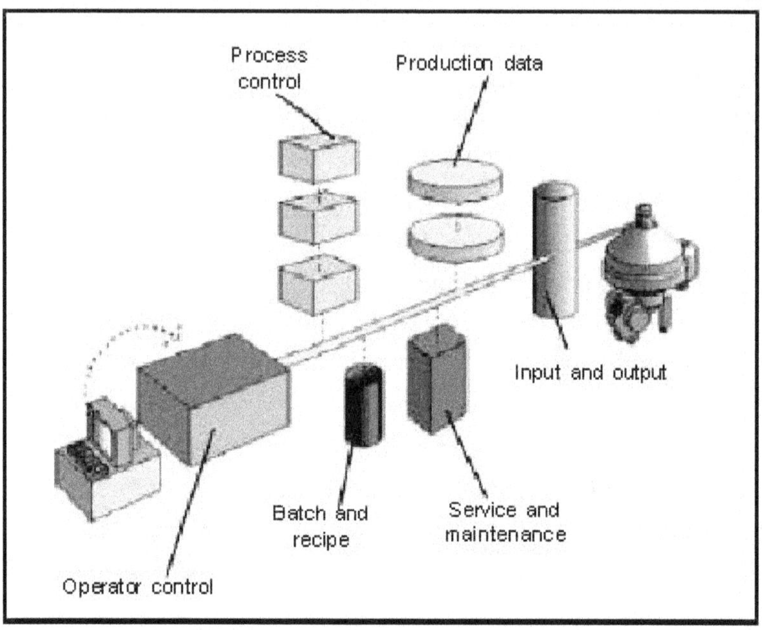

Fig. 10 A typical control system for dairy plant

The system combines supervision and control into a single concept, but its architecture is essentially open and modular, thereby delivering the operational flexibility needed. Operator control is the link between the plant operator and the process control modules, which enables actions such as routing and storage tank selection to be initiated at the click of a mouse.

Using a variety of software tools, it is possible to incorporate a graphical presentation of the plant and other written information. Using the zoom facility, an operator can quickly access more detail on any given section of the plant. This feature can also be used to train new operators. Process data are also stored in this section, which can be easily accessed, for instance via 'pop up' control windows. The process control model contains information on the process and the parameters which have to be controlled. It can also include flow routing and control, storage information and sequences of the cleaning cycles. These modules can provide realtime control capability.

The batch and recipe section contain information on recipe, ingredients and product specifications. This ensures use of ingredients in the right proportion and consistent product quality. All critical parameters can be monitored in realtime graphs and displayed on screen as well as logged for reports. The production data modules form a logbook of the activities of each processing unit – giving the origin of the product, how it has been treated and its final destination. The logbook can also be used for process evaluation and establishing traceability.

The service and maintenance modules log information on equipment run times, number of valve strokes, alarm limits on equipment, etc., which identify poor functioning or worn units, well ahead of a possible breakdown. This enables the implementation of a preventative maintenance programme. Finally, the input and output modules handle the actual connections from the process control and supervision to the physical parts (like valves, pumps, etc.) to give a complete and detailed inventory.

SENSORS

utomation systems in the food manufacturing and processing sector depend on data from huge numbers of sensors. The most important parameters monitored by these sensors include pressure, temperature, flow and level. Many of the specifications for these sensors are the same as those in any other process industry but, in the food sector, hygienic design is often an additional and critical requirement. And in some cases, the sensors must also be suitable for use in hazardous areas.

Sensors with added value

Many sensor products are available in hygienic versions designed specifically for food industry applications, and leading manufacturers of such sensors are increasingly focusing their development efforts on applying innovative technology to provide added value for users of their products.

A good example of these innovations is a new generation of level sensors that operate on the guided wave radar (GWR) principle. GWR sensors are installed vertically at the top of a tank and emit high frequency electromagnetic pulses that are guided downwards along a stainless-steel rod. The pulses are reflected from the surface of the medium in the tank back towards the sensor. The transit time of the pulses is used to evaluate the distance to the surface.

This technology provides important advantages over other types of level sensor. GWR sensors are unaffected by changing media (provided that the dielectric constant is greater than 5), changing temperature, gas blankets or vapours, or build-up of product on the probe itself. In addition, they can accurately and reliably measure liquid levels in many foaming applications.

The probes for the food industry versions of these sensors are made of high-grade stainless steel, and all other components follow hygienic design principles to meet the requirements of the European Hygienic Engineering and Design Group, EHEDG. Probes with different lengths are available and can be easily interchanged by the user to meet the needs of the application, or can be cut to length if required. The best of these sensors also has an IO-Link interface.

Another example of sensors that offer added value is provided by temperature sensors that incorporate two sensor elements with opposing temperature characteristics. The elements – Pt1000 and NTC thermistor – operate independently and are continuously monitored by a microprocessor in the sensor. This verifies that the measured temperature value is valid, based on a user defined temperature differential or "drift" between the two sensors. If the specified drift is exceeded, it is signalled immediately via a separate diagnostic output. Both measured temperature values are available by IO-Link.

The use of these sensors in temperature-critical processes provides assurance of measurement, and therefore process performance, between calibration intervals and may therefore reduce requirements for costly calibration.

Types of sensors used in food processing industries

Because of their versatility and high level of functionality or application different sensor are used in the food processing industries are explained in this below.

- Proximity sensors (inductive, capacitive, and ultrasonic)
- Temperature sensor (resistance temperature detector, infrared sensor, thermistor and thermocouple)
- Humidity sensors (optical, gravimetric, capacitive, resistive, piezoresistive, and magnetoelastic sensors)

- Bio-sensor (amperometric, conductometer, thermometric biosensor and potentiometric biosensor)
- Chemosensory
- Pressure sensors
- E-tongue taste sensors
- Torque sensors,
- Freshness sensor
- pH-sensor and gas sensor

Intelligent Sensor Methodology

Many different approaches have been developed to standardise the data modelling process including CRISP-DM and Analytics Solutions Unified Method for Data Mining/predictive analytics (ASUM-DM) from Microsoft.

However, many of these methods are focussed on general data-driven modelling projects and not specific to ML modelling of sensor data in food and drink production environments. The transition to Industry 4.0 means that there is a growing variety of online sensors available to monitor production environments. Developing these sensors has considerations around precision of measurements, cost, positioning, and deployment in industrial environments, which will impact the volume and granularity of data available to develop a ML model.

While undertaking the projects in the reported case studies, the need for a unifying methodology between the two areas became apparent. Figure 11 presents a methodology for combining sensor measurements and ML to create intelligent sensors to address the specific challenge of efficient resource use within food and drink manufacturing. The methodology has been devised as a synthesis of the methodologies applied in the reported case studies and hence is grounded in the practical application of the problem. While this methodology has been developed for the food and drink industry, it may be adapted for other industries and applications.

Resource Efficiency Problem

The intelligent sensor developer should first define the specific resource problem to be addressed, similar to the business need step in other data modelling methodologies. Specific resources challenges may include minimising the consumption of resources utilised during the process in addition to emissions and waste generated.

The problem should be well-defined, take into consideration the scope of influence of a company (e.g., if they alter the composition of their waste streams, what impact might this have on available treatments) and also ensure that any related economic and social implications are considered.

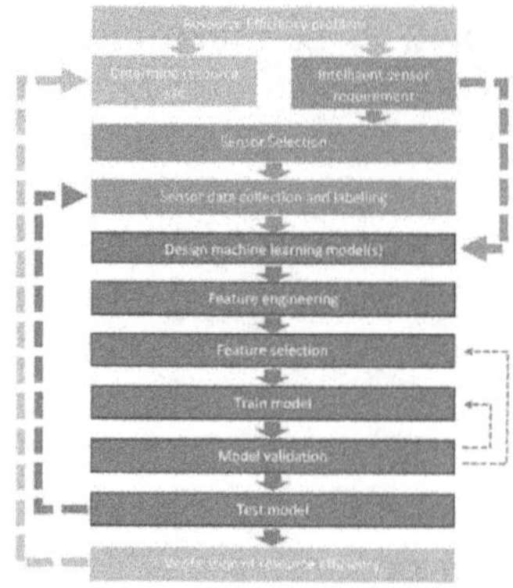

Fig. 11. Intelligent sensing process for resource efficiency. The steps in the green boxes are directly related to the resource efficiency problem, the orange boxes to the sensor aspects and the blue to the machine learning process. Solid arrows indicate the main direction of methodology, thick dashed lines indicate scenarios that require a return to a previous step and thin dashed arrows indicate stages that will undergo iterative improvement.

Determine Resource Key Performance Indicator (KPI)

Decide the metrics that will be used to monitor the resource and how it will be measured. Metrics may include the amount of resources utilised or waste generated.

INTELLIGENT SENSOR REQUIREMENT

The developer must specify the required output from the predictive model. This could be a value related to a key quality parameter of a product predicted through a regression model (e.g., moisture content) or to determine if the product contains damage or is of acceptable quality or not through a classification model. Alternatively, the purpose of the model could be to predict something related to a unit operation (process). This could include predicting whether the process had reached its end-point or not (classification) or the predicted time remaining (regression) until optimal end-point. It could also include identifying if a fault has occurred in the process (anomaly detection) to determine if an intervention is required.

Sensor Selection

Although obvious, the developer must ensure that when selecting a sensor its sensing modality can record data sufficient to achieve the intelligent sensor requirement. For example, if the requirement is to determine the grade of a fruit or vegetable-based on size, the sensor must produce information on size and an imaging system would be appropriate. If the requirement is to determine a property such a moisture content, then a sensor sensitive to changes in moisture, such as NIR or dielectric, should be used.

A sensor may be required to provide predictions on the internal aspect of food, so sensors capable of measuring internally such as X-ray, US, and electrical methods would be required. For certain applications, choice of sensing technologies may be limited, whereas for others there may be many. The developer should select from the appropriate technologies based on technical specifications, such as accuracy, precision, and resolution in addition to other factors such as size, cost and ease of installation and use. Food safety is an essential aspect of food and drink production, therefore any sensor should be easy to clean and not present any safety or contamination risks.

Sensor Data Collection and Labelling

At this stage, the developer must collect and label the sensor data required to train the ML models. Considerations need to be made in terms of the volume of sensor data and how representative it is of the system under investigation. Regarding volume, the developer must decide between the trade-off that always exists: ML models generally perform better with more data, but this comes with time costs associated with collecting and labelling the data.

With ML it is important that the data set is appropriate for the modelling approach to be used. For example, when developing classification models, it is important to collect enough data for each output class to ensure effective performance.

If only a small number of samples is expected in one class (e.g., rejections based on a rare quality defect), an anomaly detection model may be more suitable. If supervised ML methods are to be used, the recorded data requires a label to determine its class for classification models or value for regression models. There are different ways to label the data that the developer should consider.

Often labelling is completed by humans, which can be extremely costly in terms of time and disruptions to production processes. Labelling of data is one of the primary challenges with utilising ML methods within production environments. If labelling of all collected data is not possible, semi-supervised or unsupervised methods should be explored in addition to domain adaption and transfer learning. The latter two are methods that apply a model trained in one or more "source" domains to a different, but related, "target" domain. However, they rely on a trained model already existing.

Once the data is collected the developer needs to partition it into training, validation, and test sets. Training data is used to train the models and validation is used to tune the model hyperparameters and the model input variables. A hyperparameter is an adjustable algorithm parameter (e.g., number of layers and nodes in an ANN) that must be either manually or automatically tuned in order to obtain a model with the optimal performance. The test data is finally used to evaluate the performance of the model with data that was not used for any of the training and validation processes. Test data provides an unbiased evaluation of the final model fit on data outside of the training data set.

Partitioning of the data may vary depending on the volume of data available and how representative the data is to the system being modelled with splitting their dataset into 70% training, 15% validation and 15% test.

Design Machine Learning Model(s)

In the first step of the modelling stage the developer must determine the most suitable ML algorithm(s) to use. A range of different ML algorithms exist, and it is often difficult to determine the most suitable one for a specific application. Often ML practitioners will assess a range of different algorithms to determine which results in the best performance based on the validation set. Once the algorithm has been selected, the developer must also determine model hyperparameters. Developers often initially set hyperparameters based on their own past experience, similar work available in the literature or initial prototype models.

Feature Engineering

The developer will be required to use domain knowledge to extract variables from raw recorded sensor signals and process them so that they are in a suitable modelling format. Feature extraction methods tend to be unique to each different sensor and could be based on the physical interpretation of the recorded signal or an appropriate signal transformation. Feature engineering is not always necessary, as certain ML techniques, such as CNNs, automate this step. However, these techniques often require significantly larger volumes of data.

Feature Selection

Of the engineered features, the developer must determine if there are redundant or useless features which harm the learning process. Feature selection is important as a high degree of dimensionality within input variables can cause overfitting in ML models. Overfitting is the generation of a model that corresponds too closely or exactly to the training dataset (and sometimes noise), which negatively impacts future predictions.

The developer may use one or both of two categories of feature selection techniques. Firstly, supervised selection involves examining input variables in conjunction with a trained model where the effect of adding or removing variables can be assessed against model performance at predicting the target variable.

The tuning of model input variables is incorporated into the model validation stage. The second approach, called unsupervised selection, performs statistical tests on the input variables (e.g., the correlation between variables) to determine which are similar or do not convey significant information. An example of this is to use PCA. This creates a projection of the data resulting in entirely new input features, or principal components.

Train Models

At this stage, the developer will train the ML model. Initially model parameters (such as node weights on ANN) should be assigned values using methods such as random initialisation, random seeds or fixed values (e.g., all zeros). The training dataset is used to determine the values for these parameters that most accurately fit the data through an optimisation algorithm.

Optimisation algorithms are used to find the model parameters that minimise the cost function (measure of model prediction error). A range of different optimisation algorithms exists and are generally selected as the one that is most appropriate for a specific ML algorithm. For example, stochastic gradient descent and adaptive moment estimation are two common optimisation algorithms used when training a neural networks. Similar to the primarily ML algorithm, the optimisation algorithms have several hyperparameters, including batch size and number of epochs, which will affect the success of training stage.

Verification of Resource Efficiency

At the final stage, the developer should apply the model to monitor the resource problem and deploy it to achieve more efficient resource utilisation. If the model does not deliver this, different steps of the process may need to be reapplied. For example, choosing alternative key quality parameters for the intelligent sensor to predict.

The remainder of this article will focus on four case studies which combine sensors and ML for a variety of applications within food and drink manufacturing. The case studies use optical and/or US sensors and will demonstrate how the intelligent sensing methodology can improve sustainability in the manufacturing process. In addition, the case studies highlight some of the challenges of using ML methods.

ULTRASONIC SENSING

Ultrasonic data was collected from three test sections. A rectangular rig with a 1.2 mm thick SS340 bottom plate (300 mm by 40 mm) and clear PMMA sides (40 mm height) and two circular pipes constructed of PMMA and SS316 were used. The pipes had approximate dimensions of length 300 mm, internal diameter 20 mm and wall thickness 2 mm. Three materials (tomato paste, gravy, and concentrated malt) were used to foul the test sections. The materials were chosen because of their different compositions, which is known to affect surface adhesion and cleaning behaviour.

For both experimental rigs, 15 g of fouling material was placed in the centre of the plate for the flat rig, 30 mm from the exit of the pipes. Cleaning was performed by water, with a fluid temperature of either 12 or 45°C and flowrate of 6 m/s, until all of the surface fouling was removed. Seven repeats were performed for all combinations of fouling rig, temperature and fouling material. The US and temperature data were recorded using the same equipment as CS1. For the circular pipes, different US transducers were used (2 MHz Yushi) and these were glued to the bottom of the pipes. In both configurations, the US transducers were located on the external bottom surfaces in the location the fouling was placed. A camera (Logitech® C270 3MP) was also used to record images of the cleaning processes.

The camera was placed above the rig in the same location as the fouling for the flat rig and slightly above the exit of the pipe rigs. For all configurations, the camera location was adjusted to optimise the view of the fouling and cleaning process.

Ultrasonic and temperature data were recorded every 4 s and camera images were recorded every 20 s. The camera images were used to label the recorded US data as either dirty (fouling present) or clean (no fouling present).

Supervised classification ML techniques were developed to predict if the pipe section was dirty or clean from the recorded US measurements. The intelligent sensor ML model was developed by evaluating several ML models: KNN, SVM, Random Forests (RF), and Adaboost (using decisions trees as base learning). The performance of the algorithms was assessed by comparing the predicted class (dirty or clean) to the actual condition.

Separate models were developed for the flat rig, PMMA pipe and SS316 pipe. For these models, the majority of the experimental data was used in the training data set. However, for each fouling material and temperature combination, two experimental runs were excluded from the training dataset and used for testing the models.

Input features were engineered from the US waveforms. Feature selection was performed using a K-best predictors method and the number of input features varied for experimental geometry and classification method.

DYNAMICS SENSOR CHARACTERISTICS

The static accuracy of a sensor indicates how much the sensor signal correctly represents the measured quantity after it stabilizes (i.e. beyond the transient period.) Important **static characteristics** of sensors include sensitivity, resolution, linearity, zero drift and full-scale drift, range, repeatability and reproducibility.

Sensitivity is a measure of the change in output of the sensor relative to a unit change in the input (the measured quantity.) Example: The speakers you purchase for your home entertainment may have a rated sensitivity of 89 dB Signal Pressure Level per Watt per meter.

Resolution is the smallest amount of change in the input that can be detected and accurately indicated by the sensor. Example: What is the resolution of an ordinary ruler? Of a Vernier Calipers?

Linearity is determined by the calibration curve. The static calibration curve plots the output amplitude versus the input amplitude under static conditions. Its degree of resemblance to a straight line describes the linearity.

Drift is the deviation from a specific reading of the sensor when the sensor is kept at that value for a prolonged period of time. The *zero drift* refers to the change in sensor output if the input is kept steady at a level that (initially) yields a zero reading. Similarly, the *full -scale drift* is the drift if the input is maintained at a value which originally yields a full scale deflection. Reasons for drift may be extraneous, such as changes in ambient pressure, humidity, temperature etc., or due to changes in the constituents of the sensor itself, such as aging, wear etc.

The *range* of a sensor is determined by the allowed lower and upper limits of its input or output. Usually the range is determined by the accuracy required. Example:

Sometimes the range may just be determined by physical limitations. Example: a pocket ruler.

Repeatability is defined as the deviation between measurements in a sequence when the object under test is the same and approaches its value from the same direction each time. The measurements have to be made under a short enough time duration so as not to allow significant long term drift. Repeatability is usually specified as a percentage of the sensor range. Example:

Reproducibility is the same as repeatability, except it also incorporates long time lapses between subsequent measurements. The sensor has to be operation between measurements, but must be calibrated. Reproducibility is specified as a percentage of the sensor range per unit of time. Example:

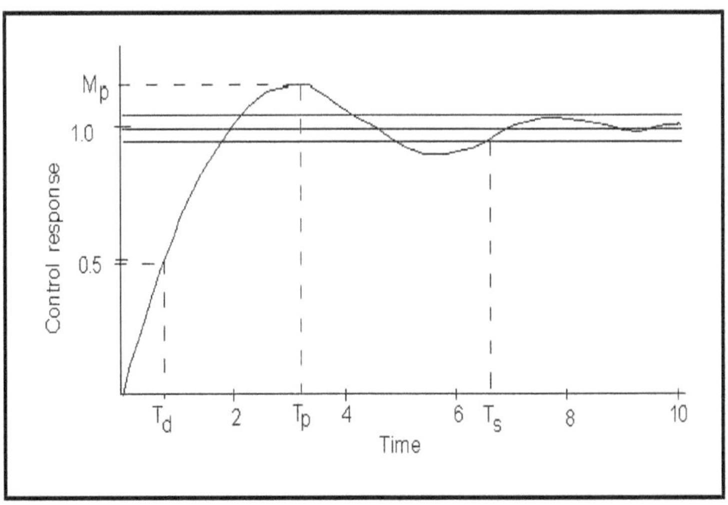

Fig. 12. Sensor Response Characteristics

The **dynamic characteristics** of a sensor represent the time response of the sensor system. Knowledge of these is essential to fruitfully use a sensor. Important common dynamic responses of sensors include rise time, delay time, peak time, settling time percentage error and steady-state error. These are defined following Fig. 12 which shall be used to represent them graphically.

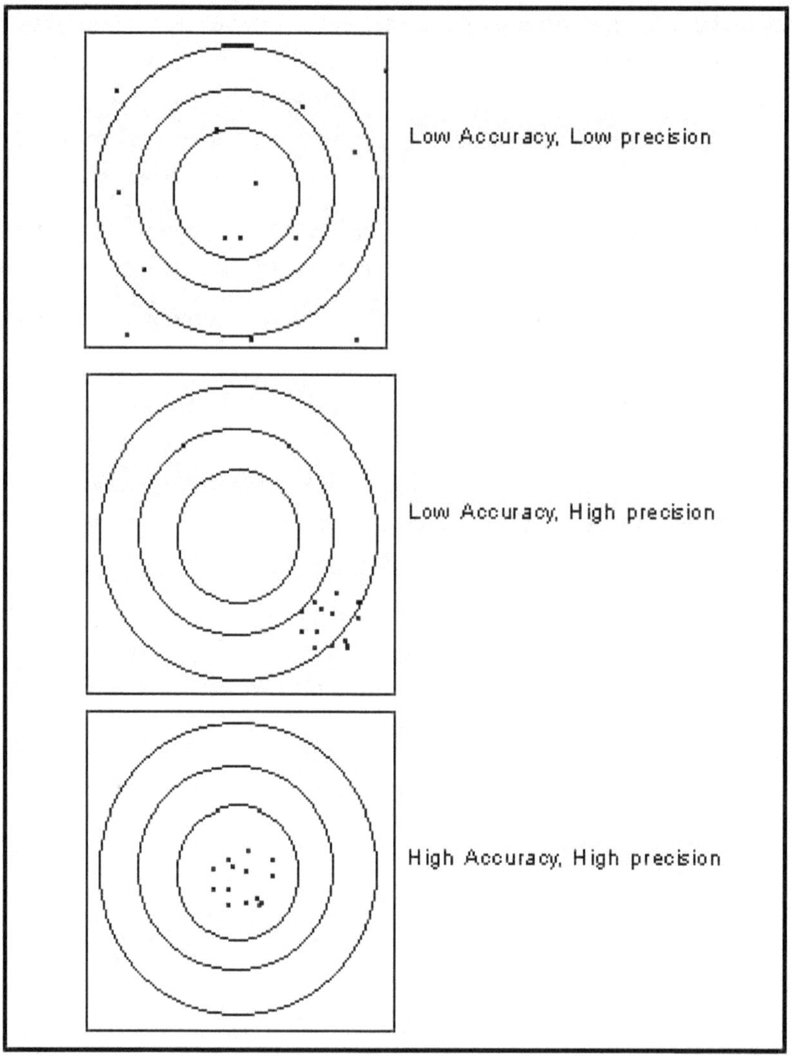

Fig. 13. Illustration of Accuracy and Precision. The center of the target represents the desired 'true' value.

Rise time is the time it takes a sensor to pass between 10% to 90% of the steady state response.

Delay time is the time it takes to reach %50 of the steady state value for the first time.

Peak time is the time it takes to reach the maximum reading for the first time for a given excitement.

Settling time is the time it takes the sensor to settle down to within a certain percentage of the steady state value (say ýÿ 1%).

The *percentage overshoot* is a measure of the peak minus the steady state value, expressed as a percentage of the ssv.

Steady-state error is the deviation of the actual steady-state value from the desired value. It can be corrected by calibration.

Two other terms are used to describe sensor behavior, and shall be described here. These are **accuracy** and **precision**.

The *accuracy* of a system is the difference between the actual and the measured value. The distinction between accuracy and resolution is important. While resolution is the smallest change in measurement value that can be indicated, the accuracy is determined by the actual indication of the sensor and the real value of the measured entity. Thus while resolution depends solely on the sensor hardware, the accuracy is a function of the sensor hardware as well as the measurement environment and other extraneous factors that modify the sensor behavior.

The precision is a statistical measure. It is usually indicated by the standard deviation (or variance) of a set of readings of the sensor for the same input. The distinction between accuracy and precision is clarified by Fig. 13.

CHAPTER

HEAT TRANSFER IN FOOD PROCESSING 6

Heat transfer is an important operation in the food industry. Whether it is called cooking, baking, drying, sterilizing or freezing, heat transfer is part of processing of almost every food. Heat transfer is a dynamic process in which heat is transferred spontaneously from one body to another cooler body. The rate of heat transfer depends upon the differences in temperature between the bodies, the greater the difference in temperature, the greater will be the rate of heat transfer.

Heat transfer is an operation that occurs repeatedly in the food industry. Whether it is called cooking, baking, drying, sterilizing or freezing, heat transfer is part of the processing of almost every food. An understanding of the principles that govern heat transfer is essential to an understanding of food processing.

Heat transfer is a dynamic process in which heat is transferred spontaneously from one body to another cooler body. The rate of heat transfer depends upon the differences in temperature between the bodies, the greater the difference in temperature, the greater the rate of heat transfer.

Temperature difference between the source of heat and the receiver of heat is therefore the driving force in heat transfer. An increase in the temperature difference, increases the driving force and therefore increases the rate of heat transfer. The heat passing from one body to another travels through some medium which in general offers resistance to the heat flow. Both these factors, the temperature difference and the resistance to heat flow, affect the rate of heat transfer. As with other rate processes, these factors are connected by the general equation:

rate of transfer = driving force / resistance

For heat transfer:

rate of heat transfer = temperature difference/ heat flow resistance of medium

During processing, temperatures may change and therefore the rate of heat transfer will change. This is called unsteady state heat transfer, in contrast to steady state heat transfer when the temperatures do not change. An example of unsteady state heat transfer is the heating and cooling of cans in a retort to sterilize the contents. Unsteady state heat transfer is more complex since an additional variable, time, enters into the rate equations.

Heat can be transferred in three ways: by conduction, by radiation and by convection.

In **conduction,** the molecular energy is directly exchanged, from the hotter to the cooler regions, the molecules with greater energy communicating some of this energy to neighbouring molecules with less energy. An example of conduction is the heat transfer through the solid walls of a refrigerated store.

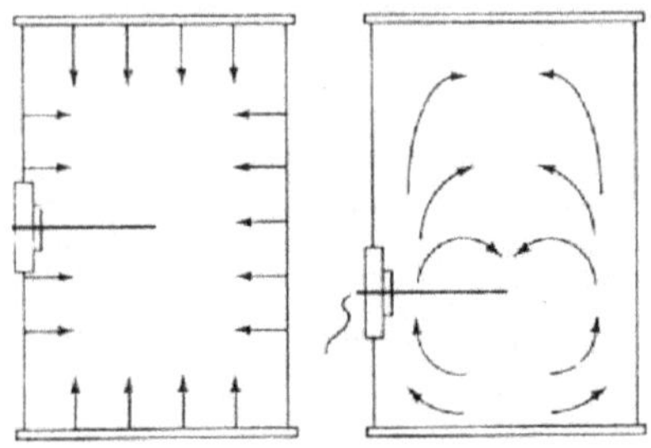

Fig. 1: Heat transfer in containers by (a) conduction and (b) convection

Radiation is the transfer of heat energy by electromagnetic waves, which transfer heat from one body to another, in the same way as electromagnetic light waves transfer light energy. An example of radiant heat transfer is when a foodstuff is passed below a bank of electric resistance heaters that are red-hot.

Convection is the transfer of heat by the movement of groups of molecules in a fluid. The groups of molecules may be moved by either density changes or by forced motion of the fluid. An example of convection heating is cooking in a jacketed pan: without a stirrer, density changes cause heat transfer by natural convection; with a stirrer, the convection is forced.

In general, heat is transferred in solids by conduction, in fluids by conduction and convection. Heat transfer by radiation occurs through open space, can often be neglected, and is most significant when temperature differences are substantial. In practice, the three types of heat transfer may occur together. For calculations it is often best to consider the mechanisms separately, and then to combine them where necessary.

HEAT TRANSFER SYSTEM

Heat can be transferred from one object to another in three ways: by conduction, by convection and by radiation.

Conduction is the movement of heat by direct transfer of molecular energy within solids. The molecules with greater energy communicating some of this energy to neighbouring molecules with less energy. An example of conduction is the heat transfer through the solid walls of a refrigerated store.

Convection is the transfer of heat by the movement of groups of molecules in a fluid. The groups of molecules may be moved by either density changes or by forced motion of the fluid. An example of convection heating is cooking in a jacketed pan: without a stirrer, density changes cause heat transfer by natural convection; while with a stirrer, the convection is forced.

Radiation is the transfer of heat energy by electromagnetic waves, which transfer heat from one body to another, in the same way as electromagnetic light waves transfer light energy. An example of radiant heat transfer is when a foodstuff is passed below a bank of electric resistance heaters that are redhot (electric grill).

In general, heat is transferred in solids by conduction and in fluids by conduction and convection (Fig. 1). Heat transfer by radiation occurs through open space, can often be neglected, and is most significant when temperature differences are substantial. In practice, the three types of heat transfer may occur simultaneously.

For calculations it is often best to consider the mechanisms separately, and then to combine them where necessary.

Conduction

In the case of heat conduction, the equation, rate = driving force/resistance, can be applied directly. The driving force is the temperature difference per unit length of heat-transfer path, also known as the temperature gradient.

Instead of resistance to heat flow, its reciprocal called the conductance is used. This changes the form of the general equation to:

rate of heat transfer = driving force x conductance,

that is:

$dQ/dt = kA\, dT/dx$ (1)

where dQ/dt is the rate of heat transfer, the quantity of heat energy transferred per unit of time, A is the area of cross-section of the heat flow path, dT/dx is the temperature gradient, that is the rate of change of temperature per unit length of path, and k is the thermal conductivity of the medium.

Notice the distinction between thermal conductance, which relates to the actual thickness of a given material (k/x) and thermal conductivity, which relates only to unit thickness.

The units of k, the thermal conductivity, can be found from eqn. (2) by transposing the terms

$$k = dQ/dt \times 1/A \times 1/(dT/dx) \quad (2)$$

$= J\ s^{-1} \times m^{-2} \times 1/(°C\ m^{-1})$

$= J\ m^{-1}\ s^{-1}\ °C^{-1}$

Equation (2) is known as the Fourier equation for heat conduction.

Note: Heat flows from a hotter to a colder body, that is in the direction of the negative temperature gradient. Thus a minus sign should appear in the Fourier equation. However, in simple problems the direction of heat flow is obvious and the minus sign is considered to be confusing rather than helpful, so it has not been used.

THERMAL CONDUCTIVITY

On the basis of eqn. (2) thermal conductivities of materials can be measured. Thermal conductivity does change slightly with temperature, but in many applications it can be regarded as a constant for a given material.

In general, metals have a high thermal conductivity, in the region 50-400 J $m^{-1}\ s^{-1}\ °C^{-1}$. Most foodstuffs contain a high proportion of water and as the thermal conductivity of water is about 0.7 J $m^{-1}\ s^{-1}°C^{-1}$ above °0C, thermal conductivities of foods are in the range 0.7 - 0.6 J $m^{-1}\ s^{-1}°C^{-1}$. Ice has a substantially higher thermal conductivity than water, about 2.3 J $m^{-1}\ s^{-1}°C^{-1}$. The thermal conductivity of frozen foods is, therefore, higher than foods at normal temperatures.

Most dense non-metallic materials have thermal conductivities of 0.5-2 J $m^{-1}\ s^{-1}°C^{-1}$. Insulating materials, such as those used in walls of cold stores, approximate closely to the conductivity of gases as they are made from non-metallic materials enclosing small bubbles of gas or air. The conductivity of air is 0.024 J $m^{-1}\ s^{-1}\ °C^{-1}$ at °0C, and insulating materials such as foamed plastics, cork and expanded rubber are in the range 0.06 -0.03 J $m^{-1}\ s^{-1}\ °C^{-1}$. Some of the new foamed plastic insulating materials have thermal conductivities as low as 0.026 J $m^{-1}\ s^{-1}\ °C^{-1}$. When using published tables of data, the units should be carefully checked. Mixed units, convenient for particular applications, are sometimes used and they may need to be converted.

Conduction through a Slab

If a slab of material, as shown in Fig. 2, has two faces at different temperatures T_1 and T_2 heat will flow from the face at the higher temperature T_1 to the other face at the lower temperature T_2.

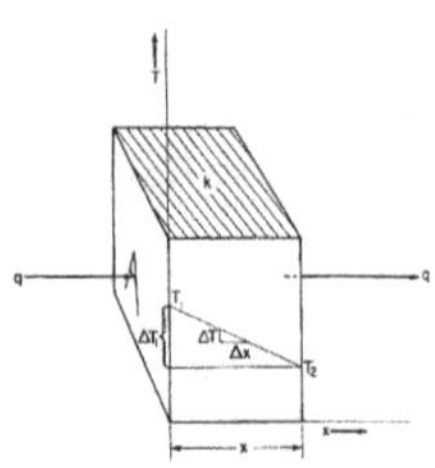

Fig. 2. Heat conduction through a slab

The rate of heat transfer is given by Fourier's equation:

$dQ/dt = kA\ DT/Dx = kA\ dT/dx$

Under steady temperature conditions dQ/dt = constant, which may be called q:

and so $q = kA\ dT/dx$

but dT/dx, the rate of change of temperature per unit length of path, is given by $(T_1 - T_2)/x$ where x is the thickness of the slab,

so $q = kA(T_1 - T_2)/x$

or $q = kA\ DT/x = (k/x)\ A\ DT$ (3)

This may be regarded as the basic equation for simple heat conduction. It can be used to calculate the rate of heat transfer through a uniform wall if the temperature difference across it and the thermal conductivity of the wall material are known.

EXAMPLE 1. Rate of heat transfer in cork

A cork slab 10 cm thick has one face at -12°C and the other face at 21°C. If the mean thermal conductivity of cork in this temperature range is 0.042 J m^{-1} s^{-1} $°C^{-1}$, what is the rate of heat transfer through 1 m^2 of wall?

T_1 = °21C T_2 = °12-C DT = °33C

$A = 1\ m^2$ $k = 0.042\ J\ m^{-1}\ s^{-1}\ °C^{-1}$ $x = 0.1$ m

q = **0.042**__ x 1 x 33

0.1 = **13.9 J s^{-1}**

Heat Conductances

In tables of properties of insulating materials, heat conductances are sometimes used instead of thermal conductivities. The heat conductance is the quantity of heat that will pass in unit time, through unit area of a specified thickness of material, under unit temperature difference, For a thickness x of material with a thermal conductivity of k in J m^{-1} s^{-1} $°C^{-1}$, the conductance is $k/x = C$ and the units of conductance are

J m^{-2} s^{-1} $°C^{-1}$.

Heat conductance = $C = k/x$.

Heat Conductances in Series

Frequently in heat conduction, heat passes through several consecutive layers of different materials. For example, in a cold store wall, heat might pass through brick, plaster, wood and cork.

In this case, eqn. (3) can be applied to each layer. This is illustrated in Fig. 3.

SYSTEMS FOR HEATING AND COOLING FOOD PRODUCTS

In a food processing plant, heating and cooling of foods is conducted in equipment called heat exchangers. Heat exchangers can be broadly classified into noncontact and contact types. As the name implies, in noncontact-type heat exchangers, the product and heating or cooling medium are kept physically separated, usually by a thin wall. On the other hand, in contact-type heat exchangers, there is direct physical contact between the product and the heating or cooling streams.

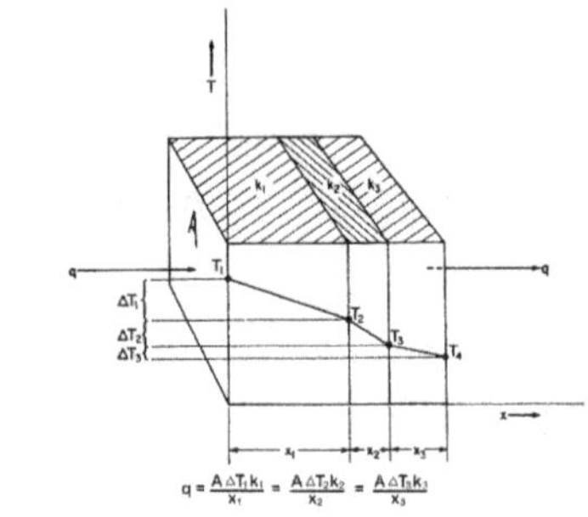

Fig. 3 Heat conductances in series

For example, in a steam-injection system, steam is directly injected into the product to be heated. In a plate heat exchanger, a thin metal plate separates the product stream from the heating or cooling stream while allowing heat transfer to take place without mixing.

Plate Heat Exchanger

The plate heat exchanger invented more than 70 years ago has found wide application in the dairy and food beverage industry. This heat exchanger consists of a series of parallel, closely spaced stainless-steel plates pressed in a frame. Gaskets, made of natural or synthetic rubber, seal the plate edges and ports to prevent intermixing of liquids. These gaskets help to direct the heating or cooling and the product streams into the respective alternate gaps. The direction of the product stream versus the heating/cooling stream can be either parallel flow (same direction) or counter flow (opposite direction) to each other. The plates used in the plate heat exchanger are constructed from stainless steel: Special patterns are pressed on the plates to cause increased turbulence in the product stream, thus achieving better heat transfer.

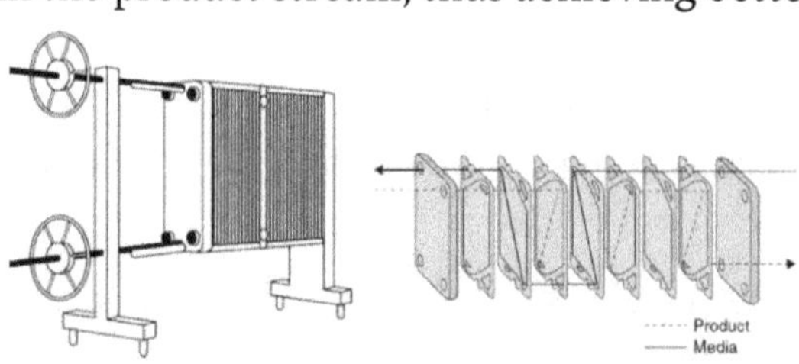

Fig. 4. (a) Plate heat exchanger (b) Schematic view of fliuid flow between plates.

Plate heat exchangers are suitable for low-viscosity liquid foods. If suspended solids are present, the equivalent diameter of the particulates should be less than 0.3 cm larger particulates can bridge across the plate contact points and "burn on" in the heating section.

In industrial-size plate heat exchangers, product flow rates from 5000 to 20,000 kg/h often are obtained. When using plate heat exchangers, care should be taken to minimize the deposition of solid food material such as milk proteins on the surface of the plates. This deposition, also called fouling, will decrease the heat transfer rate from the heating medium to the product; in addition, the pressure drop will increase over a period of time. Eventually, the process is stopped and the plates are cleaned. For dairy products, which require ultra-high temperature applications, the process time is often limited to 3 to 4 h.

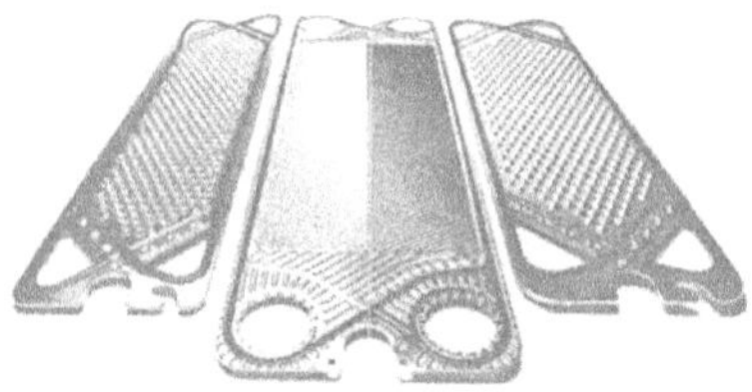

Fig. 5. Patterns pressed on plates used on a plate heat exchanger.

Plate heat exchangers offer the following advantages:

- The maintenance of these heat exchangers is simple, and they can be easily and quickly dismantled for product surface inspection.
- The plate heat exchangers have a sanitary design for food applications.
- Their capacity can easily be increased by adding more plates to the frame.
- With plate heat exchangers, we can heat or cool product to within 1°C of the adjacent media temperature, with less capital investment than other noncontact-type heat exchangers.
- Plate heat exchangers offer opportunities for energy conservation by regeneration.

A liquid food is heated to pasteurization or other desired temperature in the heating section; the heated fluid then surrenders part of its heat to the incoming raw fluid in the regeneration section. The cold stream is heated to a temperature where it requires little additional energy to bring it up to the desired temperature. For regeneration, additional plates are required; however, the additional capital cost may be recovered quickly by lowered operating costs.

The plates used in the plate heat exchanger are constructed from stainless steel: Special patterns are pressed on the plates to cause increased turbulence in the product stream, thus achieving better heat transfer. An example of such a pattern is a shallow herringbone-ribbed design, as shown in Fig. 5.

Plate heat exchangers are suitable for low-viscosity (<5 Pa s) liquid foods. If suspended solids are present, the equivalent diameter of the particulates should be less than 0.3 cm. Larger particulates can bridge across the plate contact points and "burn on" in the heating section.

In industrial-size plate heat exchangers, product flow rates from 5000 to 20,000 kg/h often are obtained. When using plate heat exchangers, care should be taken to minimize the deposition of solid food material such as milk proteins on the surface of the plates. This deposition, also called fouling, will decrease the heat transfer rate from the heating medium to the product; in addition, the pressure drop will increase over a period of time. Eventually, the process is stopped and the plates are cleaned. For dairy products, which require ultra-high temperature applications, the process time is often limited to 3 to 4 h.

Plate heat exchangers offer the following advantages:

- The maintenance of these heat exchangers is simple, and they can be easily and quickly dismantled for product surface inspection.
- The plate heat exchangers have a sanitary design for food applications.
- Their capacity can easily be increased by adding more plates to the frame.
- With plate heat exchangers, we can heat or cool product to within 1°C of the adjacent media temperature, with less capital investment than other noncontact-type heat exchangers.
- Plate heat exchangers offer opportunities for energy conservation by regeneration.

As shown in a simple schematic in Figure 6, a liquid food is heated to pasteurization or other desired temperature in the heating section; the heated fluid then surrenders part of its heat to the incoming raw fluid in the regeneration section. The cold stream is heated to a temperature where it requires little additional energy to bring it up to the desired temperature. For regeneration, additional plates are required; however, the additional capital cost may be recovered quickly by lowered operating costs.

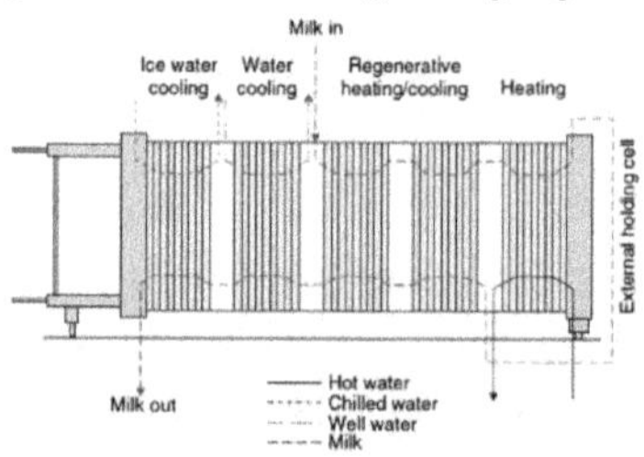

Fig. 6. A five-stage plate pasteurizer for processing milk.

TUBULAR HEAT EXCHANGER

The simplest noncontact-type heat exchanger is a double-pipe heat exchanger, consisting of a pipe located concentrically inside another pipe. The two fluid streams flow in the annular space and in the inner pipe, respectively. The streams may flow in the same direction (parallel flow) or in the opposite direction (counterflow). Figure 7 is a schematic diagram of a counterflow double-pipe heat exchanger.

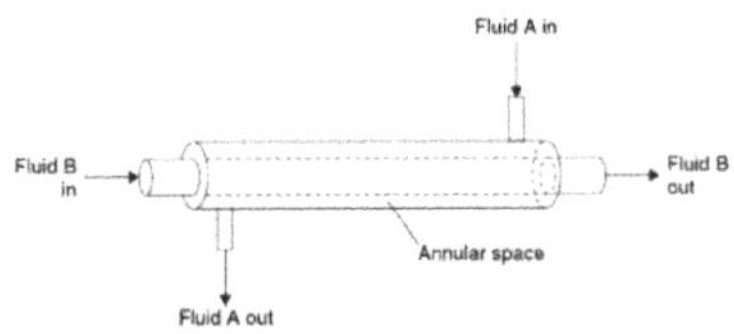

Fig. 7. Schematic illustration of a tubular heat exchanger

A slight variation of a double-pipe heat exchanger is a triple-tube heat exchanger, shown in Fig. 8. In this type of heat exchanger, product flows in the inner annular space, whereas the heating/cooling medium flows in the inner tube and outer annular space. The innermost tube may contain specially designed obstructions to create turbulence and better heat transfer. Some specific industrial applications of triple-tube heat exchangers include heating single-strength orange juice from 4 to 93°C and then cooling to 4°C; cooling cottage cheese wash water from 46 to 18°C with chilled water; and cooling ice cream mix from 12 to 0.5°C with ammonia.

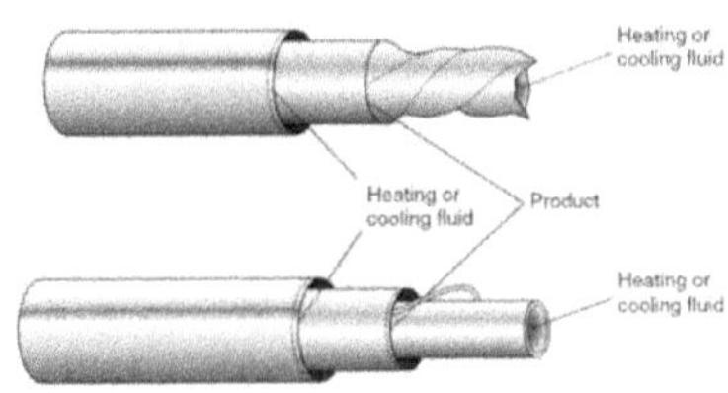

Fig. 8. Schematic illustration of a triple-tube heat exchanger

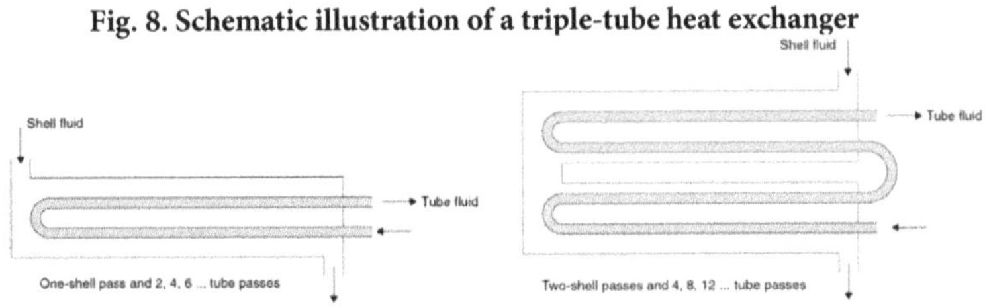

Fig. 9. A shell-and-tube heat exchanger

Another common type of heat exchanger used in the food industry is a shell-and-tube heat exchanger for such applications as heating liquid foods in evaporation systems. As shown in Fig. 9, one of the fluid streams flows inside the tube while the other fluid stream is pumped over the tubes through the shell.

By maintaining the fluid stream in the shell side to flow over the tubes, rather than parallel to the tubes, we can achieve higher rates of heat transfer. Baffles located in the shell side allow the cross-flow pattern. One or more tube passes can be accomplished, depending on the design. The shell-and-tube heat exchangers shown in Figure 9 are one shell pass with two tube passes, and two shell passes with four tube passes.

SCRAPED-SURFACE HEAT EXCHANGER

In conventional types of tubular heat exchangers, heat transfer to a fluid stream is affected by hydraulic drag and heat resistance due to film buildup or fouling on the tube wall. This heat resistance can be minimized if the inside surface of the tube wall is scraped continuously by some mechanical means. The scraping action allows rapid heat transfer to a relatively small product volume. A scraped-surface heat exchanger, used in food processing, is shown schematically in Figure 10.

The food contact areas of a scraped-surface cylinder are fabricated from stainless steel (type 316), pure nickel, hard chromium-plated nickel, or other corrosion-resistant material. The inside rotor contains blades that are covered with plastic laminate or molded plastic (Fig. 9). The rotor speed varies between 150 and 500 rpm. Although higher rotation speed allows better heat transfer, it may affect the quality of the processed product by possible maceration. Thus, we must carefully select the rotor speed and the annular space between the rotor and the cylinder for the product being processed.

As seen in Figure 10, the cylinder containing the product and the rotor is enclosed in an outside jacket. The heating/cooling medium is supplied to this outside jacket. Commonly used media include steam, hot water, brine, or a refrigerant. Typical temperatures used for processing products in scraped-surface heat exchangers range from -35 to 190°C.

The constant blending action accomplished in the scraped-surface heat exchanger is often desirable to enhance the uniformity of product flavor, color, aroma, and textural characteristics.

In the food processing industry, the applications of scraped-surface heat exchangers include heating, pasteurizing, sterilizing, whipping, gelling, emulsifying, plasticizing, and crystallizing. Liquids with a wide range of viscosities that can be pumped are processed in these heat exchangers; examples include fruit juices, soups, citrus concentrate, peanut butter, baked beans, tomato paste, and pie fillings.

STEAM-INFUSION HEAT EXCHANGER

A steam-infusion heat exchanger provides a direct contact between steam and the product. As shown in Fig. 11, product in liquid state is pumped to the top of the heat exchanger and then allowed to flow in thin sheets in the heating chamber.

The viscosity of the liquid determines the size of the spreaders. Products containing particulates, such as diced vegetables, meat chunks, and rice, can be handled by specially designed spreaders. High rates of heat transfer are achieved when steam contacts tiny droplets of the food.

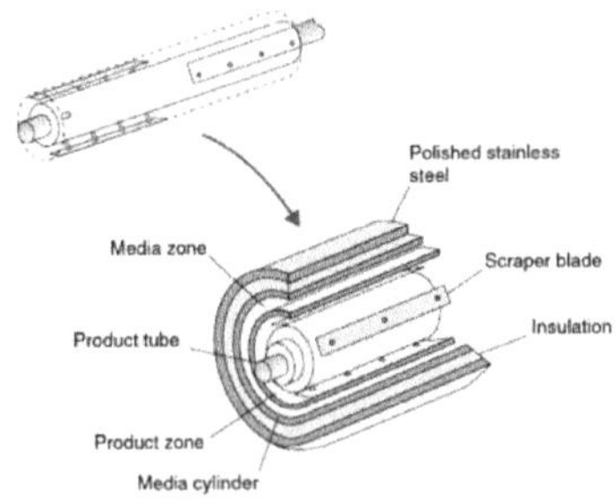

Figure 10. A scraped-surface heat exchanger with a cutaway section illustrating various components

The temperature of the product rises very rapidly due to steam condensation. The heated products with condensed steam are released from the chamber at the bottom. A specific amount of liquid is retained in the bottom of the chamber to achieve desired cooking.

The temperature difference of the product between the inlet and the outlet to the heating chamber may be as low as 5.5°C, such as for deodorizing milk (76.7 to 82.2°C), or as high as 96.7°C, such as for sterilizing puddings for aseptic packaging (48.9 to 145.6°C).

The water added to the product due to steam condensation is sometimes desirable, particularly if the overall process requires addition of water. Otherwise, the added water of condensation can be "flashed off" by pumping the heated liquid into a vacuum cooling system.

The amount of water added due to condensation can be computed by measuring the temperature of the product fed to the heat exchanger and the temperature of the product discharged from the vacuum cooler.

This type of heat exchanger has applications in cooking and/or sterilizing a wide variety of products, such as concentrated soups, chocolate, processed cheese, ice cream mixes, puddings, fruit pie fillings, and milk.

THERMAL PROPERTIES OF FOODS

Most processed foods and many freshly consumed foodstuffs receive some type of heating or cooling during handling or manufacturing. Design and operation of processes involving heat transfer requires special attention, due to the heat-sensitivity of foods. Both theoretical and empirical relationships used when designing or operating heat processes require knowledge of the thermal properties of the foods under consideration. Food thermal properties can be defined as those properties controlling the transfer of heat in a specified food.

They are usually grouped as thermodynamical properties (e.g., specific volume, specific heat, enthalpy, and entropy) and heat transport properties (thermal conductivity and thermal diffusivity).

However, when considering the heating or cooling of foods, it is well known that other physical properties must be considered because of their intrinsic relationship with the mentioned "pure" thermal properties.

Therefore, a well-known group of thermal and related properties, thermophysical properties, provides a powerful tool for the design and prediction of heat transfer operation during handling, processing, canning, and distribution of foods. In spite of the profuse generation and compilation of thermophysical properties of food, they are not always available to the design engineer.

Finding relevant data is usually the controlling step in the design of a given operation, and as foods are such a broad set of materials, the best solution is frequently the experimental determination. This section provides data and information for calculating the thermal processes of foodstuffs, including a brief description of the most widely used methods for the measurement and determination of thermophysical properties.

Thermophysical properties of foods include, in a broad sense, different types of parameters associated with the heat transfer operations of food processing. Heat transfer involves the transfer of heat into or out of a food. There are three ways that heat can be transferred: by radiation, conduction, or convection. Radiation is the transfer of heat by electromagnetic waves (e.g., in a microwave oven).

Conduction is the movement of heat by direct transfer of molecular energy within solids (e.g., heating of a food by direct fire through metal containers). Convection is the transfer of heat by groups of molecules that move because of a gradient of density or agitation (e.g., the stirring of liquid foods).

Heat transfer can take place in 1) a steady-state manner, by keeping the temperature difference constant between two materials; or 2) in an unsteady-state manner, when temperature is constantly changing. Calculation of heat transfer under these conditions is extremely complicated, but can be simplified by making a number of assumptions and, in some cases, using prepared graphic or tabulated information to give approximate solutions.

Food Properties during Freezing

The thermophysical properties of foods change dramatically during the freezing process and must be considered. One characteristic of food freezing is that the temperature changes gradually with the phase change, which implies that the fraction of water frozen always changes continuously with temperature below the freezing point.

Water Content

Water content (Xw) is not a thermophysical property but significantly influences all thermophysical properties. If the food is a living commodity, such as fruits and vegetables, its water content will change with maturity, cultivars, stage of growth, and harvest and storage conditions. Values of most thermophysical properties can be calculated directly from the water content. X_w is usually expressed as water mass fraction (Kg water; Kg^{-1} food; wet basis).

Experimental Data and Prediction Models

Foods show extended variability in composition and structure. This characteristic must be kept in mind when modeling their thermal properties. Foods are generally nonhomogeneous, varying in composition and structure not only between products, but also within a single product. Consider for example the thermal conductivity of onions, a typical "laminated" vegetable. The composition and structure (soluble solids, fibers, porosity, etc.) will not only vary between onions with cultivars and growing conditions, but within a given onion with harvest and post harvest conditions (e.g., side of bulb exposed to sun in the case of solar curing). The same situations can also be found in the case of fruit and meat production.

Some thermophysical property models for food systems are only functions of water content. However, the contents of proteins, fats, and carbohydrates differ greatly from one food to another and therefore have variable effects on the properties of a composite food material. Some models include different components of the food, expressing for example a linear combination of the properties of water, fat, protein, carbohydrate, and/or ash.

A large volume of experimental data can be found that has been reviewed by different authors. Thermophysical properties are dependent on the temperature, as well as the materials chemical composition and physical structure. Since foodstuffs are composite materials, it is apparent that the relevant information is the average or effective value. This value is clearly some function of those of the components. Food materials are complex substances. Therefore, the generation of predictive models requires a physical representation of the food under study.

Firstly, foodstuffs are composed of different chemical species, and any model proposed for the prediction of a given property must consider the individual contribution of the "species", namely ash, proteins, carbohydrates, fats, fiber, water, and minor components. This is done by means of some weighing factor accounting for the proportions in which they are present.

Secondly, some food materials can be considered as the described solid continuously dispersed in a gas or liquid phase, usually air or water. This includes porous foods, suspensions, flours, and powders. In this case, both the volumetric fraction and the spatial distribution of each phase are considered.

The value of the thermophysical property will be a function of the temperature, and porosity or water content. Since water can be either liquid or solid, particular attention is paid to frozen foods. This article is oriented, whenever possible, to the presentation of correlating expressions more than to the tabulation of punctual data, in order to provide predictive tools of a more general nature, together with the description of the accepted measuring techniques. Efforts have been made to include the most recent data and correlations whenever possible.

Density

Definition and Units

Density (ρ) is the unit mass per unit volume. The SI unit for density is [Kg m^{-3}]. In particular, when the foodstuff is a porous solid, density plays an important role in heat transfers intrinsically or through the definition of porosity. A few definitions are necessary:

Substance density, r_s, (also true density) is the density measured when the substance has been broken, milled, or mashed to guarantee that no pores remain. Particle density, r_p, is the density of a sample that has not been structurally modified. In the case of pores not externally connected to the surrounding atmosphere, particle density will include these closed pores.

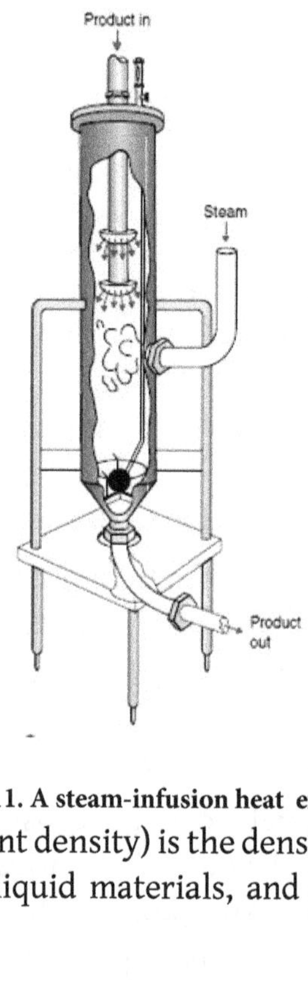

Fig. 11. A steam-infusion heat exchanger

Bulk density, r_b, (or apparent density) is the density measured, in order to include the volume of the solid and liquid materials, and all pores closed or open to the surrounding atmosphere.

HEAT TRANSFER CHARACTERISTICS OF FOOD

It is necessary to have information on both the heat resistance of microorganisms or enzymes and the rate of heat penetration into the food for determination of process time of a food.

Heat is transferred from steam or pressurized water through the container and into the food. Generally the surface heat transfer coefficient is very high and is not a limiting factor in heat transfer. The following factors influence the rate of heat penetration into a food:

- Type of product
- Size of the container
- Agitation of the container
- Temperature of the retort
- Shape of the container. Tall containers promote convection currents in convective heating foods.
- Type of container. Heat penetration is faster through metal than through glass or plastics owing to differences in thermal conductivity.

In this section, the objective is to learn as to how the thermal properties of food products affect the heat penetration and the quantity of heat. Two thermal properties of importance are thermal conductivity and thermal diffusivity in determining heat penetration. Specific heat and latent heat are important in determining the quantity of heat required for the process.

- **Thermal conductivity** is the property indicating the rate at which heat flows through a food product. A product with high thermal conductivity lets the heat flow easily, whereas a material with low thermal conductivity, also known as an insulator, puts resistance to the flow of heat. Fourier's heat conduction equation could be used to derive the units of thermal conductivity, i.e., W/(m°C). It does change slightly with temperature, but in many applications it can be regarded as a constant for a given material. Most foodstuffs contain a high proportion of water and as the thermal conductivity of water is about 0.7 J m^{-1} s^{-1} $°C^{-1}$ above 0°C, thermal conductivities of foods are in the range of 0.6-0.7 J m^{-1} s^{-1} $°C^{-1}$. Ice has a substantially higher thermal conductivity than water, about 2.3 J m^{-1} s^{-1} $°C^{-1}$.
- The thermal conductivity of frozen foods is, therefore, higher than foods at normal temperatures.

  Typical thermal conductivities

  Metals: k = 50-400 W/m°C

Water: k = 0.597 W/m°C

Air: k = 0.0251 W/m°C

Insulating materials: k = 0.035 - 0.173 W/m°C

For foods it is represented as

$k = 0.25\ m_c + 0.155\ m_p + 0.16\ m_f + 0.135\ m_a + 0.58\ m_m$

Where m is mass fraction and subscripts c: carbohydrate, p: protein, f: fat, a: ash and m: moisture. or

$k = 0.55p/100 + 0.26(100"p)/100\ J\ m^{-1}s^{-1}\ °C^{-1}$ above freezing

$= 2.4p/100 + 0.26(100 - p)/100\ J\ m^{-1}\ s^{-1}\ °C^{-1}$ below freezing.

Where p is the percentage of water in the foodstuff.

- **Thermal diffusivity** is the actual ability of a food to conduct heat to adjacent molecules. Thermal diffusivity is a derived property that is the ratio of thermal conductivity and the product of density and specific heat. The units of thermal diffusivity, therefore, work out to be m2 /s. Higher value of thermal diffusivity means faster heat penetration and vice versa.
- **Specific heat**: The specific heat is an important quantity that determines the amount of energy that must be supplied or withdrawn from a unit mass of material in order to increase or decrease its temperature by one degree.
- Knowledge of the specific heat of a material is, therefore, important in the design of processes such as chilling, freezing, warming, sterilization and cooking. Specific heat has the units of kJ/(kg.K) in SI system of units.

  Specific heat = $4.19p/100 + 0.84(100 " p)/100\ kJ\ kg^{-1}\ °C^{-1}$ above freezing

  $= 2.1\ p/100 + 0.84(100 " p)/100\ kJ\ kg^{-1}\ °C^{-1}$ below freezing.

  p is percentage of water in food stuff.
- **Phase transitions**: It is important to determine the temperature at which transitions occur, the enthalpy change associated with a transition, the type of transition involved (exothermic or endothermic), and the quantity of material that undergoes a transition. As an example, we will consider the melting and crystallization of food components. When a material changes its physical state from solid-to-liquid (melting) or from liquid-to-solid (crystallization) it absorbs or gives out heat, respectively.
- A process that absorbs heat is an endothermic process, whereas a process that evolves heat is an exothermic process. Pure substances usually have very sharp melting or crystallization points and, therefore, all the heat is absorbed or evolved over a narrow range of temperature. Most foods are

complex materials and, therefore, do not exhibit sharp transitions from one phase to another. The amount of heat required for the phase change is called the latent heat and has the units of kJ/kg.

Latent heat = 335p/100 kJ kg^{-1}.

This equation and the ones given earlier for thermal conductivity and specific heat represent a considerable over-simplification so they should be used with caution, particularly in the region between "18°C to 0°C. Freezing of foodstuffs occur over a range of temperatures and not at any fixed point.

DEVICES FOR DETERMINATION OF HEAT PENETRATION

There are different types of thermometers available for measuring the temperature in a thermal process and, thereby, permitting the determination of heat penetration.

Mercury-in-glass (MIG) thermometer

Each retort system used for the thermal processing is equipped with a MIG thermometer. Aseptic processing systems may have a temperature indicating device other than MIG thermometer as the sole temperature indicator. The MIG thermometer is the reference instrument for all temperature readings, including vent temperature, come-up temperature and process temperature during processing.

It is important that the MIG thermometer be tested/calibrated at the operating temperature of the retort system (i.e., 115°C, 120°C, 125°C etc.) and if possible in the heating medium used in the retort. If the retort is operated at more than one processing temperature or over a wide range of temperatures the MIG thermometer should be checked at all of the temperatures normally used for processing. The MIG thermometers should be tested against a thermometer that can be traced back to a BIS Standard thermometer. The accuracy of the standard thermometer should be checked at least once every 3 years depending upon how it is handled and stored.

Temperature recording device

Each retort system is equipped with an accurate temperature-recording device. The recording device provides a continuous record of the temperature in the retort system during thermal processing. Common systems in use are circular or strip charts, which are marked with ink pens, electrical sparks, pressure pins, or which are created by graph plotters at the time temperature readings are received. Electronic temperature monitors and recorders are now available for the purpose and should be utilized for greater accuracy and precision avoiding human errors. A band or ribbon type surface pyrometer is used by processors to monitor container surface temperatures.

Temperature sensors

Temperature measurement can be accomplished by essentially five basic methods: (1) liquid-in-glass, (2) resistance thermometry, (3) thermoelectric thermometry, (4) optical/radiation pyrometry, and (5) bimetal. Investigators are most familiar with the liquid (mercury or alcohol usually) -in-glass and the bi-metal (dial gauge) types. It is possible now that investigators will encounter the use of the optical/radiation pyrometers as well.

- **Resistance thermometry**: A resistance thermometer is a temperature-measuring instrument consisting of a sensor (an electrical circuit element whose resistance varies with temperature), a framework on which to support the sensor, a sheath by which the sensor is protected, and wires by which the sensor is connected to a measuring instrument, which is used to indicate the effect of variations in the sensor resistance. Resistance thermometers provide absolute calibration of temperatures in that no reference junctions are involved, and no special extension wires are needed between the sensor and the measuring instrument (as with thermocouples).
- The sensors can be of two types: resistance temperature detectors (RTD's) and thermistors. The RTD sensing element is formed of solid conductors (usually in wire form) wound upon an insulating core. The insulating core is usually made of mica or ceramic. The conductors, which are wound in a helical coil to prevent mechanical restraints during thermal expansion, are generally made of platinum; however nickel and copper have been used. Platinum best meets the requirements because being a noble metal, it can be highly refined, it resists contamination, it is mechanically and electrically stable, and the relationship between temperature and resistance is quite linear.
- Thermistors (a contraction for "thermally sensitive resistors") are electrical circuit elements formed of solid semi conducting materials such as oxides of nickel, manganese, iron, cobalt, copper, magnesium, titanium, and other metals. The powdered metal is formed under pressure into the desired shape, usually a flat disc. The disc is sintered, leads are attached, and encapsulated in epoxy. The finished thermistor can also be encased in a sheath of plastic, stainless steel, copper or aluminum. Both the RTD and thermistor can be obtained in various configurations.
- **Thermoelectric thermometry**: The thermoelectric thermometer is a temperature measuring instrument consisting of two continuous, dissimilar thermocouple wires extending from a measuring junction to a reference junction with copper connecting wires to a potentiometer. Unlike the resistance types, where power must be supplied to the circuit, the thermocouple circuit generates a measurable low voltage output that is

almost directly proportional to the temperature difference between the "hot" junction and the "cold" junction. A unit change in this temperature difference will produce some net change in electromotive force (emf or voltage). Thermoelectric thermometry makes use of the known relationship between a difference in junction temperatures and the resulting emf developed by a thermocouple circuit. The temperature of one junction (reference junction, T1) is held at a constant known value. This is usually accomplished with an ice water (0°C) bath. The temperature of the other junction (measuring junction, T2) is determined by measuring the thermocouple circuit emf and referring to calibration tables for the particular thermocouple materials. The thermocouple junction usually is formed by twisting and fusing the two wires together or they may be butt-welded. The finished element may be used bare or enclosed in a sheath.

CHAPTER 7

PRESERVATION PROCESSES

INTRODUCTION

Foods are organic substances which are consumed for nutritional purposes. Foods are plant or animal origin and contain moisture, protein, lipid, carbohydrate, minerals, and other organic substances. Foods undergo spoilage due to microbial, chemical, or physical actions. Nutritional values, color, texture, and edibility of foods are susceptible to spoilage. Therefore, foods are required to be preserved to retain their quality for longer period of time. Food preservation is defined as the processes or techniques undertaken in order to maintain internal and external factors which may cause food spoilage. The principal objective of food preservation is to increase its shelf life retaining original nutritional values, color, texture, and flavor.

The history of 'Food Preservation' dates back to ancient civilization when the primitive troupe first felt the necessity for preserving food after hunting a big animal, which could not be able to eat at a time. Knowing the techniques of preserving foods was the first and most important step toward establishing civilization. Different cultures at different times and locations used almost the similar basic techniques to preserve food items.

Conventional food preservation techniques like drying, freezing, chilling, pasteurization, and chemical preservation are being used comprehensively throughout the world.

Scientific advancements and progresses are contributing to the evolution of existing technologies and innovation of the new ones, such as irradiation, high-pressure technology, and hurdle technology.

The processing of food preservation has become highly interdisciplinary since it includes stages related to growing, harvesting, processing, packaging, and distribution of foods. Therefore, an integrated approach would be useful to preserve food items during food production and processing stages.

Food preservation involves different food processing steps to maintain food quality at a desired level so that maximum benefits and nutrition values can be achieved. Food preservation methods include growing, harvesting, processing, packaging, and distribution of foods. The key objectives of food preservation are to overcome inappropriate planning in agriculture, to produce value-added products, and to provide variation in diet.

Food spoilage could be caused by a wide range of chemical and biochemical reactions. To impede chemical and microbial deterioration of foods, conventional and primitive techniques of preserving foods like drying, chilling, freezing, and pasteurization have been fostered. In recent years, the techniques to combat these spoilages are becoming sophisticated and have gradually altered to a highly interdisciplinary science. Highly advanced technologies like irradiation, high-pressure technology, and hurdle technology are used to preserve food items.

IMPORTANCE OF FOOD PRESERVATION

The evolution of, and continuous innovation to food preservation techniques over the years has been necessary for subsistence across the globe. Preventing the growth of bacteria, fungi, and other micro-organisms for the sake of storing food can be achieved through many diverse methods. Processes as simple as drying and salting have evolved into more modern techniques such as irradiation and high pressure food preservation. Innovations to preservation practices are instrumental to the agricultural and manufacturing industries, the military, and to an extent, the survival of many developing nations.

Increasing the shelf-life of food has many obvious benefits – a larger food supply leads to lower prices for consumers and perishable foods become more transportable, to name a few. New preservation technologies are not just working to increase longevity, they are also functioning to sustain the same qualities of the food that make it desirable in the first place. The physical characteristics and chemical composition of the food will no longer be compromised during the preservation process. Studies have indicated that meat products, for example pork and tuna, have the potential to remain edible for seven to eight years. With new techniques such as high-pressure processing (HPP) and innovations to packaging materials, meals maintain their same fresh qualities over the course of their stowage.

COMMON PRESERVATION TECHNIQUES

Common food preservation processes used comprehensively in the past, or presently

Drying, Salting, Pickling, Refrigeration, Freezing, Oxidation, Vacuum packaging, Canning/bottling, Smoking, Sugaring, Preservatives.

The earliest documented methods for preserving food are still very much in use today. Whether by natural occurrence, or human-induced, the processes of drying, freezing, and fermenting food have always served as viable solutions. However, these most basic practices have their drawbacks.

Drying and Fermentation

Drying is for the purpose of moderating the amount of water activity to prevent the growth of bacteria by aerobically respiring. Clearly, drying food does not fully prevent contamination, but in addition, the process changes the physical properties of the food. The acidic products that result from fermentation help to kill off contaminating microbes, but similarly to drying, fermentation changes the chemical properties of the food.

Refrigeration and Freezing

Refrigeration and freezing have both played a substantial role in the effort to preserve food. The ability to stunt microbial growth by storing food in inaptly cold environments has been very important to food transportation. However, more persistent microbes will inevitably contaminate frozen or refrigerated food over the course of time.

Canning and Bottling

Canning and bottling remains a very common technique for preservation. Food is cooked, sealed inside of a sterile container, and is boiled to eradicate any lingering bacteria. The process is very efficient, and it can leave the food edible for a few years. Canned food remains susceptible to anaerobic microbial growth, most notably that of *Clostridium botulinum*.

Preservatives

The use of preservatives in food has expanded a great deal in recent years. There are two forms of food preservatives: antimicrobial preservatives and antioxidants. Antimicrobial preservatives work to break down, and prevent the growth of bacteria and fungi, whereas antioxidants are used to avert the oxidation of the food. Some common antimicrobial preservatives include sulfites, sodium nitrite, and benzoic acid. Antioxidants generally include Butylated hydroxyanisole (BHA), Butylated hydroxytoluene (BHT), and *tert*-Butyl hydroquinone (TBHQ).

RECENT DEVELOPMENTS

Irradiation and Chemical Additives:

Some of the future methods of food preservation are irradiation and chemical additives. Although these methods are currently in use, they are expected to expand and develop further. Irradiation of food is the process of exposing food to ionizing radiation. This process can alter the bacteria, microorganism or virus' DNA, without harming the food. Irradiation is attractive because of its selective targeting. It is already used on non-food items. "The molecular bonds in the microbial DNA are the main targets of the irradiation, but DNA and RNA synthesis, denaturation of enzymes and cell membrane alterations may also be affected". The process of irradiation opens the possibility to process a large number of foods in great quantities, however it can be expensive.

The buildings for such a process require specific infrastructure and construction that is both expensive and time consuming. "The Food and Agriculture Organization (FAO), the International Atomic Energy Agency (IAEA), and World Health Organization (WHO) concluded in their report, that any food irradiated up to a maximum dose of 10 kGy is considered safe and wholesome". Essentially three things were concluded in their report:

- It won't lead to toxicological changes in the food that will negatively affect our health,
- The technology won't increase the microbial risk of the consumer, and
- Irradiation won't lead to nutritional losses.

However, most consumers are still radiation-phobic despite these positive results. They aren't willing to buy something that has, in their mind, been radiated and burned to safety. Another process is the addition of chemicals to preserve the freshness of the food. Consumers feel that this is a risk because of the unnatural state the food and their bodies are subjected to. New research has shown that even the most susceptible foods such as a ham and pepperoni sandwiches can be made to last years.

In the TIME article, researchers added chemicals to the sandwich to ensure the oxygen, moisture and other moist areas did not spread mold or microorganisms that would harm the freshness or integrity. Lauren Oleksyk , leader of the food-processing, engineering and technology team at the U.S. Department of Defense Combat Feeding Directorate mixed water-absorbing ingredients including glycerol and sorbitol into the filling. They also increased the use of fine, edible polymer films, which are undetectable in the mouth.

These were added to combat moisture that would be detrimental to the sandwich as it provides a good breeding ground for microorganisms and bacteria. In addition to the water absorbing ingredients, Oleksky's team also added oxygen eliminators. It sounds more intimidating than it is; essentially it is the process of adding oxygen-scavenging chemicals. Oxygen increases the possibility of spoilage to the food. By adding these chemicals, it limits the microorganism's chance to grow and thrive, and also inhibits the sandwich's ability to change chemically.

CLASSIFICATION OF FOODS

Foods can be broadly classified according to the shelf life, functions and nutrient value, and processing mechanisms. Different categories of foods are summarized in Table 1 and briefly discussed below.

Table 1 Classification of foods based on functions and nutrients

Functions	Nutrients
Body building and repairing foods	Milk, meat, fish, pulses, vegetables, and nuts
Energy-giving foods	Oil, butter, sugar, cereals, dry fruits, and starch foods
Regulatory foods	Water, raw vegetables, citrus fruits, and beverages
Protective foods	Milk, whole grain cereals, meat, vegetables, and fruits
Carbohydrate-rich foods	Rice, wheat, and starchy vegetables
Protein-rich foods	Milk, meat, fish, egg, and nuts
Fat-rich foods	Oils, butter, and egg yolk
Vitamin- and mineral-rich foods	Fruits and vegetables

Food categories based on shelf life

Food spoilage is a natural process; through this process, food gradually loses its color, texture, flavor, nutritional qualities, and edibility. Consumption of spoiled food can lead to illness and in the extreme situation to death. Considering the self life, food items can be classified as perishable, semi-perishable, and non-perishable.

Perishable

Foods that have shelf life ranging from several days to about three weeks are known as perishable. Milk and dairy products, meats, poultry, eggs, and seafood are the examples of perishable food items. If special preservation techniques are not apprehended, food items could be spoiled straight away.

Semi-perishable

Different food items can be preserved for long time (about six months) under proper storage conditions. These foods are known as semi-perishable. Vegetables, fruits, cheeses, and potatoes are few examples of semi-perishable food items.

Non-perishable

Natural and processed foods that have indefinite shelf life are called non-perishable food items. These foods can be stored for several years or longer. Dry beans, nuts, flour, sugar, canned fruits, mayonnaise, and peanut butter are few examples of non-perishable foods.

Food categories based on functions and nutrients

According to the functions to human body, food items can be categorized as: (a)

body building and repairing foods, (b) energy-giving foods, (c) regulatory foods, and (d) protective foods. Depending on the nutrition value, food items can be classified as: (a) carbohydrate-rich foods, (b) protein-rich foods, (c) fat-rich foods, and (d) vitamin- and mineral-rich foods. Table 1 presents different food items according to their functions and nutrients.

Food categories based on extent and purpose of processing

Different food processing techniques are used by the food industries to turn fresh foods into food products. Foods can be classified into three major groups based on the extent and purpose of food processing:

(a) unprocessed or minimally processed foods,

(b) processed culinary or food industry ingredients, and

(c) ultra-processed food products.

Classification of foods based on extent and purpose of processing is presented in Table 2.

FOOD SPOILAGE: MECHANISM

Food spoilage is the process in which food edibility reduces. Food spoilage is related to food safety. The primitive stage of food spoilage can be detected by color, smell, flavor, texture, or food. Different physical, microbial, or chemical actions can cause food spoilage. These mechanisms are not necessarily mutually exclusive since spoilage caused by one mechanism can stimulate another. Temperature, pH, air, nutrients, and presence of different chemicals are the major factors for food spoilage

PHYSICAL SPOILAGE

Food spoilage due to physical changes or instability is defined as physical spoilage. Moisture loss or gain, moisture migration between different components, and physical separation of components or ingredients are the examples of physical spoilage. The key factors affecting physical spoilage are moisture content, temperature, glass transient temperature, crystal growth, and crystallization.

Moisture content

A frequent cause of degradation of food products is the change in their water content. It may occur in the form of water loss, water gain, or migration of water. Moisture transfer in food is directly related to the water activity (a_w) of food item. Water activity (a_w) is a thermodynamic property which is expressed as the ratio of the vapor pressure of water in a system to the vapor pressure of pure water at the same temperature.

Table 2 Food classification based on the extent and purpose of processing

Food group	Extent and purpose of processing	Examples
Unprocessed or minimally processed foods	No processing or mostly physical processes used to make single whole foods more available, accessible, palatable, or safe	Fresh, chilled, frozen, vacuum-packed fruits, vege-tables, cereals; fresh frozen and dried beans and other pulses; dried fruits, unsalted nuts, and seeds; fresh,
		dried, chilled, frozen meats, poultry, and fish; fresh and pasteurized milk, yoghurt, eggs, tea, and coffee
Processed culinary or food industry ingredients	Extraction and purification of components of foods, resulting in producing ingredients used in the prepara- tion and cooking of dishes and meals, or in the formula- tion of ultra-processed foods	Vegetables, butter, milk cream, sweeteners, raw pastas, and noodles; food industry ingredients, such as high- fructose corn syrup, preservatives, and cosmetic additives
Ultra-processed food products	Processing of a mix of process culinary, or food industry ingredients and processed or minimally processed food stuffs in order to produce accessible, convenient, palatable, ready-to-eat or to-heat food products with longer shelf life	Breads, biscuits, cakes, and pastries; ice-cream, chocolates, cereal bars, chips; sugared fruits, milk drinks, and other soft drinks; pre-prepared meat, poultry, fish, and vegetable; processed meat including chicken nuggets, hot dogs, sausages, burgers; salted, pickled, smoked, or cured meat and fish; vegetables bottled or canned in brine; fish canned in oil

Equilibrium relative humidity at the same temperature may also be used in lieu of pure water vapor pressure. Water activity in food products reduces with temperature. In general, water activity of foods at normal temperature is 1.0, whereas at -20 and -40 °C temperatures the water activities are 0.82 and 0.68, respectively.

Temperature

The effect of temperature is the most significant factor in the case of fruit and vegetable spoilage. There is an optimum temperature range for slow ripening and to maximize post-harvest life. Slow ripening also requires an optimum relative humidity along with optimum air movement around fruit and vegetable. Apparently, these optimum conditions are called modified atmospheres (MA). Temperature usually besets the metabolism of the commodities and contemporarily alters the rate of attaining desired MA.

Low temperature can also have a negative effect on foods that are susceptible to freeze damage. At a lower temperature, when food products become partially frozen, breakage in cells occurs which damages the product. Most tropical fruits and vegetables are sensitive to chilling injury. This generally occurs before the food product starts to freeze at a temperature in between 5 °C and 15 °C.

Glass transition temperature

Glass transition temperature (T_g) effects the shelf life of food products. Solids in food items may exist in a crystalline state or in an amorphous metastable state. This phenomenon depends on the composition of solids, temperature, and relative humidity. The amorphous matrix may exist either as a very viscous glass or as a more liquid-like rubber. At glass transition temperature, changes occur from the glassy state to rubbery state. This is a second-order phase transition process, which is temperature specific for each food. The physical stability of foods is related to the glass transition temperature. Glass transition temperature (T_g) depends strongly on concentration of water and other plasticizers. When dry food products are kept in highly humid conditions, the state of food products changes due to glass transition phenomena.

Crystal growth and crystallization

Freezing can also contribute to food degradation. Foods, which undergo slow freezing or multiple freeze, suffer severely due to crystal growth. They are subject to large extracellular ice growth. Rapid freezing forms ice within food cells, and these foods are more stable than slow freezing processed foods. To minimize large ice crystal growth, emulsifiers and other water binding agents can be added during freezing cycles.

Foods with high sugar content can undergo sugar crystallization either by moisture accumulation or by increasing temperature. As a consequence, sugar comes to the surface from inside, and a gray or white appearance is noticed. Staling of sugar cookies, graininess in candies, and ice creams are the results of sugar crystallization.

Sugar crystallization can be delayed by the addition of fructose or starch. Moreover, above the respective glass transition temperature, time plays a crucial role in sugar crystallization process of food items.

MICROBIAL SPOILAGE

Microbial spoilage is a common source of food spoilage, which occurs due to the action of microorganisms. It is also the most common cause of foodborne diseases. Perishable foods are often attacked by different microorganisms. The growth of most microorganisms can be prevented or lingered by adjusting storage temperature, reducing water activity, lowering pH, using preservatives, and using proper packaging.

Microorganisms involved in food spoilage

Microorganisms involved in food spoilage can be divided into three major categories, which are molds, yeasts, and bacteria. Table 3 presents the active conditions of different microorganisms that affect foods.

Factors affecting microbial spoilage

There are intrinsic and extrinsic factors that can affect microbial spoilage in foods. The intrinsic properties of foods determine the expected shelf life or perishability of foods and also affect the type and rate of microbial spoilage. Endogenous enzymes, substrates, sensitivity of light, and oxygen are the primary intrinsic properties associated with food spoilage.

To control food quality and safety, these properties can be controlled during food product formulation. Intrinsic factors of food spoilage include pH, water activity, nutrient content, and oxidation-reduction potential. Extrinsic factors of food spoilage include relative humidity, temperature, presence, and activities of other microbes.

CHEMICAL SPOILAGE

Chemical and biochemical reactions occur naturally in foods and lead to unpleasant sensory results in food products. Fresh foods may undergo elementary quality changes caused by:

- microbial growth and metabolism which results in pH changes,
- toxic compounds, and/or
- the oxidation of lipids and pigments in fat which results in undesirable flavors and discoloration.

Chemical spoilage is interrelated with microbial actions. However, oxidation phenomena are purely chemical in nature and also dependent on temperature variations.

Oxidation

In presence of oxygen, amino acids convert into organic acid and ammonia. This is the elementary spoilage reaction in refrigerated fresh meat and fish.

$$\text{Amino acid} + O_2 \xrightarrow{\text{deaminase}} NH_3 + \text{Organic acid}$$

The term 'rancidification' is used to denominate lipids oxidation through which unsaturated fats (lipids) undergo reaction with oxygen. The consequences in food items are color alteration, off-flavor, and toxic substances formation. Rancidification can be catalyzed by the presence of metal oxides and exposure to light increases the reaction rate. After this reaction, carbonyl compounds, responsible for rancid taste of foods, are produced.

Proteolysis

Proteolysis, a ubiquitous and irreversible posttranslational modification, involves limited and highly specific hydrolysis of peptide and iso-peptide bonds of a protein. The entire phenomena require the presence of miscellaneous protease enzymes. Different specialized proteases play a key role in various regulatory processes. Moreover, highly specific proteolytic events are associated with normal and pathological conditions. Foods containing nitrogen compounds frequently incur this reaction. Proteins, after being incurred through proteolysis, eventually get converted into small-sized amino acids. The following reaction presents proteolysis mechanism:

$$\text{Proteins} \xrightarrow{\text{proteinase}} \text{Peptides} \xrightarrow{\text{peptidase}} \text{AminoAcids.}$$

Many of these peptides have stiff taste which can be bitter or sweet. Table 3 presents the taste of various amino acids.

Table 3 Taste of different amino acids

Taste	Amino acids
No taste/barely perceptible taste	D-Alanine, D-arginine, L-arginine, D-aspartate, L-aspartate, D-glutamate, L-histidine, D-isoleucine, L-isoleucine, D-lysine, L-lysine, D-proline, L-proline, D-serine, L-serine, L-threonine, D-valine, L-valine
Sweet taste	D-Tryptophan, D-histidine, D-phenylalanine, D-tyrosine, D-leucine, L-alanine, glycine
Bitter taste	L-Tryptophan, L-phenylalanine, L-tyrosine, L-leucine
Sulfurous taste	D-Cysteine, L-cysteine, D-methionine, L-methionine
Unique taste	L-Glutamic acid

Putrefaction

Putrefaction refers to the series of anaerobic reactions through which amino acids detour to a mixture of amines, organic acids, and stiff-smelling sulfur compounds, such as mercaptans and hydrogen sulfide. This is a biochemical phenomenon as the presence of bacteria is exigent all through the process. Along with amino acids, indole, phenols, and ammonia are also formed due to protein putrefaction. Most of these chemicals have displeasing odor. Putrefaction is quite common in meats and other protein-rich foods at temperatures greater than 15 °C. This elevated temperature facilitates microbial activities.

Maillard reaction

Non-enzymatic browning, which is also known also as Maillard reaction, is another primary cause of food spoilage. This reaction occurs in the amino group of proteins, or the amino acids present in foods. Color darkening, reducing proteins solubility, developing bitter flavors, and reducing nutritional availability of certain amino acids are the common outcomes of Maillard reaction. This reaction occurs during the storing of dry milk, dry whole eggs, and breakfast cereals.

Pectin hydrolysis

Pectins are complex mixtures of polysaccharides that make up almost one-third of the cell wall of dicotyledonous and some monocotyledonous plants. Indigenous pectinases are synthesized or activated during ripening of fruits and cause pectin hydrolysis which softens the structure of food. Damages of fruits and vegetables by mechanical means may also activate pectinases and initiate microbial attack. Pectin substances may also be de-esterified by the action of pectin methyl esterase.

This esterification process is initiated in situ on damaged tissues, firm fruits, and vegetables by strengthening the cell walls and enhancing intercellular cohesion via a mechanism involving calcium. Metal ions catalyze the decomposition of heat-labile fruit pigments, which consist of pectin ingredients. This process causes the color change in fruit jams or jellies. Therefore, jams and jellies are preserved in glass containers rather than metallic jars.

Hydrolytic rancidity

Hydrolytic rancidity causes lipid degradation by the action of lipolytic enzymes. In this reaction, free fatty acids are cleaved off triglyceride molecules in the presence of water. These free fatty acids have rancid flavors or odor. The released volatile fatty acids have a stiff malodor and taste; therefore, hydrolytic rancidity is extremely noticeable in fats, such as butter.

FOOD PRESERVING AND PROCESSING METHODS PHYSICAL METHODS OF FOOD PRESERVATION

Drying/Dehydration:

Drying or dehydration involves the removal of water from the food by controlled processes, which may be done by

- Evaporation due to heating of the product.
- Osmotic dehydration.
- Sublimation, or freeze drying.

There are two distinct stages in this process.

- In the first stage, surface water removal depends solely on the state of the air surrounding the food, such as its temperature, relative humidity and speed.
- In the second phase of drying, the moisture within the food moves to the surface. As the air is heated, its relative humidity decreases, resulting in more absorption of water.

The rate of drying depends on the time the moisture takes to get to the surface. The heating of the air around a food product causes faster drying.

Although in case of starchy foods, the shortest drying time may be referred, to as the common change of "case hardening". In such an event, water removal from the surface is much faster than the rate at which water migrates from the interior. The surface, therefore, dries into a hard layer, which actually prevents the migrating water from reaching the surface.

Factors on which the selection of a particular dryer/drying method depends include:

- form of raw material and its properties;
- desired physical form and characteristics of dried product;
- necessary operating conditions;
- operation costs.

Several types of dryers and drying methods, are commercially used to remove moisture from a wide variety of food products including fruit and vegetables. With this method of preservation, retains most of the flavours, giving less bulkiness to the product (reduced shipping costs) with extended shelf life.

There are five basic methods of drying which are as follows:

- Tunnel Dryer
- Solar Dryer
- Spray Dryer
- Vacuum Dryer
- Freeze Drying

Tunnel Drying

Many years ago a solar tunnel dryerwas developed at the university of Hohenheim, with the purpose of preserving agricultural products by the means of drying.

With a joint development of the Institute for Agricultural Technology in the Tropics and Subtropics at the University of Hohenheim and Innotechcompany, the solar dryers of this type are now in use in over 75 countries.

This device is employed for the hygienic preservation of foods. It forms a good alternative for seasonal agricultural surplus, deficient transport capabilities for fresh products and insufficient preservability of undried products. At the same ti me, the quality of goods dried in this way is substantially superior to traditional drying.

Fishermen in Bangladesh uses the solar dryer for the hygienic drying of their catch. Similarly spices are also dried in China etc.

The solar tunnel dryer is 20m long and 2m wide. It can be loaded with between 300 and 500kg of goods to be dried.

Sun Drying/Solar Drying

Sun drying and solar drying are obvious alternatives for drying due to the abundance of natural sunlight. Although the two terms are sometimes used interchangeably, sun drying refers to the removal of moisture by merely placing the commodity in the sun, e.g., on a barbecue, rack, etc. Sun drying of fruit crops is still practiced for certain fruit such as prunes, figs, apricots, grapes and dates.

The main problems for sun drying are dust, rain and cloudy weather. Therefore, drying areas should be dust-free and whenever there is a threat of a dust storm or rain, the drying trays should be stacked together and placed under cover. In order to produce dust-free and hygienically clean products, fruit and vegetable material should be dried well above ground level so that they are not contaminated by dust, insects, livestock or people.

All materials should be dried on trays designed for the said purpose. The most common drying trays have wooden frames with a fitted base of nylon mosquito netting. Mesh made of woven grass can also be used. The trays should be placed on a framework at table height from the ground to allow air to circulate freely around the drying material and also to keep the food product well away from dirt. The material should be stirred/turned over at least once an hour, during the initial period which will help the material dry faster and more evenly, prevent it sticking together and improve the quality of the finished product.

At night the trays should be stacked in a ventilated room or covered with canvas. Plastic sheets should NEVER be used for covering individual trays during sun drying.

Dry or nearly dry products can be blown out of the tray by the wind. However, this can be protected by covering the loaded tray with an empty one; this also gives protection against insects and birds.

Limitations of traditional sun drying include the following:

- Intermittent moisture loss as it is largely dependent on the weather.
- Drying rates are usually slow and do not result in high quality products.

- Moisture levels are too high for prolonged storage.
- Insect infestation.

Solar drying, involves capturing and concentrating solar energy for the purpose of removing water. This method has increased its popularity, although commercial solar dryers with high rates of efficiency are often quite expensive.

The advantages of solar drying over sun drying include:

- Faster drying rates as higher air temperatures are generated.
- Lower final moisture content of the finished product.
- Greater protection of the product from rain, dust, pests.
- Low insect and mould infestation due to higher temperatures.

Essentially, there are two types of solar dryers – direct and indirect. Regardless of which type is used, it is important to have information on the seasonal and daily variation of sunshine, humidity, temperature, wind speed and direction during drying.

Advantages of Drying

- **Long Shelf Life** – Most microorganisms responsible for food spoilage are unable to grow and multiply in the absence of moisture, also enzymes catalyzing the undesirable changes in foods are limited due to low moisture content, resulting into a longer shelf life of foods.
- **Reduced Weight** – Results in reduced transportation, storage and shipping costs.
- **Convenience** – The production of convenience items with novelty appeal for niche markets makes drying an attractive option.
- **Concentration of nutrients** – The removal of most of the water from a food results in a highly concentrated source of nutrients.
- **No refrigeration is required for dried products** – Savings in energy and storage costs together with the long shelf life provide a lucrative processing alternative for tropical countries.
- **Inexpensive** - The method is simple to carry out and inexpensive, and can be easily employed by grower, farmer, cooperative, etc., as it uses the natural resource/source of heat: sunlight.

Disadvantages of Drying

Disadvantages of drying are few, and mainly relate to oxidation, which usually accompanies drying. This results in losses of micronutrients such as carotene and ascorbic acid, with minimal loss in protein as a result of browning reactions, that reduces consumer appeal. There might also be changes in flavour and texture if drying is not properly controlled, particularly with regard to maximum temperatures.

Shade drying

Shade drying is carried out for products which can lose their colour and/or turn brown if kept in direct sunlight. Products which have naturally vivid colours like herbs, green and red sweet peppers, chilies, green beans and okra give a more attractive end-product when they are dried in the shade. Under dry conditions when there is a good circulation of air, shade drying takes little more time than is normally required for drying in full sunlight.

Spray drying

The development of spray drying equipment and technique evolved over a period of several decades from the 1870s through the early 1900s. Spray drying comes from an age of World War II, with the sudden need to reduce the transport weight of foods and other materials.

This technique enables the transformation of feed from a fluid state into dried particulate form by spraying the feed into a hot drying medium. It is a continuous particle processing drying operation. The feed can be a solution, suspension, dispersion or emulsion. The dried product can be in the form of powders, granules or agglomerates depending upon the physical and chemical properties of the feed, the dryer design and final powder properties desired.

Spray drying process mainly involves five steps:

- **Concentration:** Feedstock is normally concentrated prior to introduction into the spray dryer.
- **Atomization:** The atomization stage creates the optimum condition for evaporation to a dried product having the desired characteristics.
- **Droplet-air contact:** In the chamber, atomized liquid is brought into contact with hot gas, resulting in the evaporation of 95%+ of the water contained in the droplets in a matter of a few seconds.
- **Droplet drying:** Moisture evaporation takes place in two stages: 1) during the first stage, there is sufficient moisture in the drop to replace the liquid evaporated at the surface and evaporation takes place at a relatively constant rate and 2) the second stage begins when there is no longer enough moisture to maintain saturated conditions at the droplet surface, causing a dried shell to form at the surface. Evaporation then depends on the diffusion of moisture through the shell, which is increasing in thickness.
- **Separation:** Cyclones, bag filters, and electrostatic precipitators may be used to purify and cool the air so that it can be released to atmosphere. Spray

drying process have advantages that can be designed to virtually any capacity required. Feed rates range from a few pounds per hour to over 100 tons per hour. It can be used with both heat-resistant and heat sensitive products. Nearly spherical particles can be produced.

Vacuum drying

Since evaporation of water takes place more readily at lower pressures, drying under vacuum is faster. This method is more expensive than air-drying and is reserved for specialized products for sublimation from frozen foods.

Although the food structure is better conserved, the equipment and its maintenance are costly. Vacuum dehydration processes are useful for low moisture/ high sugar fruits like peaches, pears and apricots.

Freeze drying

Freeze-drying is a method of preservation that makes use of the physical principle known as sublimation, the process by which a solid passes directly to the gaseous phase without first melting. Freeze-drying is a desirable way of preserving food because at low temperatures (commonly around –10°C to –25°C) chemical reactions take place very slowly and pathogens have difficulty surviving. The food to be preserved by this method is first frozen and then placed into a vacuum chamber. Frozen water in the food then sublimates, leaving the moisture content in the final product of as low as 0.5%.

The steps of Freeze Drying include:

- **Freezing** of the product.
- **Vacuuming**– It ensures that the water does not pass through the liquid phase and goes straight to a gas.
- **Heating the product** – It increases the rate of the reaction.
- **Condensation** – Condenser plates move the vaporized solvent to a solid state.

Freeze drying of food products gives following advantages:

- Fresh-like flavors of the products
- Label friendly....all natural, no additives, preservatives or artificial flavoring
- Concentrated nutritional values
- Shelf stable
- Enhanced texture
- Vibrant color
- Intense fragrance

- High anti-oxidant values
- Organic compatible
- Superior alternative to drum and spray drying

Separation / Filtration

Separation operations are based on the principles that depend on the physical properties of the components of the liquid product and their differences. A separation process such as centrifugation relies on differences in solubility and specific gravity to separate blends of immiscible materials. In case of membranes, separation takes place with the use of difference in size of osmotic pressure of the components of liquid food.

In last two decades, separation processes have emerged in which membranes have played a prominent role in the isolation and enrichment of functional ingredients by concentrating, fractionating and purification of fluids or gases.

Membrane processing can be applied to vegetable protein concentrates and isolates derived from soybean, peas and canola. It permits concentration and separation without the use of heat. They make it possible to standardize the concentrations of specific components, and to decrease the concentration of undesirable components.

Table 4: Applications of Membranes in Food Industry.

Process	Application in Food Industry
Microfiltration	Removal of particles from beverages and dairy products, clarification of alcoholic beverages, waste water management.
Ultrafiltration	Concentration of egg white, blood serum proteins, whey protein concentrate, clarification of dextrose, waste water recovery and use, recovery of by-products from waste water, pre-concentration of dilute sugar solutions and syrup concentration.
Nanofiltration	Desalting of lactose, whey protein concentrates and antibiotics, recycling of caustic cleaning solutions, softening water to reduce scaling on equipment and heat exchange surfaces.
Reverse Osmosis	Production of pure water, concentration of maple sap, recovery of sugar from candy, recovery of oil-seed protein.

Concentration

Foods are concentrated for many of the same reasons that they are dehydrated. Concentration can be a form of preservation but this is true only for some foods. Concentration reduces weight and volume and results in immediate economic advantages.

Nearly all liquid foods which are dehydrated are concentrated before drying. This is because in the early stages of water removal, moisture can be more economically removed in highly efficient evaporators than in dehydration equipment. Further, increased viscosity from concentration often is needed to prevent liquids from running off drying surfaces or to facilitate foaming or puffing.

The more common concentrated fruit and vegetable products include items as fruit and vegetable juices and nectars, jams and jellies, tomato paste, many types of fruit purées used by bakers, candy makers and other food manufacturers.

The methods used to concentrate liquid foods depend on the composition of the product, the sensitivity of the components to decompose with heat, and the economics of the process.

Methods of concentration

- **Solar concentration** - One of the simplest methods of evaporating water. A typical example of this method is production at farm level in developing countries of fruit pastes/leathers (such as apricot or plum pastes).
- **Open Kettles** - Some foods can be satisfactorily concentrated in open kettles that are heated by steam, as in case of jellies and jams, tomato juices and purées and for certain types of soups. High temperatures and long concentration time should be avoided in order to avoid thickening and burn-on of product to the kettle wall.
- This type of evaporation is highly recommended for small scale operations in developing countries, widely used for jellies, jams and marmalades.
- **Flash evaporators-** Flashing is a process which gives rise to vaporization flow rate more significant thanthat obtained during simple evaporation. The industrial applications of flashing are varied. One of the most important concern is the desalination of seawater in multi-stage flash evaporators.
- **Vacuum evaporators-** Several vacuumised vessels in series are constructed, so that the product moves from one vacuum chamber to the next and thereby becomes progressively more concentrated in stages. In this way maximum use of heat energy is made. Such system is called a multiple effect vacuum evaporator and is a widely used system for concentrated tomato paste.
- **Freeze Concentration** - This process has been known for many years and has been applied commercially to orange juice. However, high processing costs largely due to losses of juice occlude [unclear] to the ice crystals.
- **Ultrafiltration and reverse Osmosis** - UF is used in a variety of industries mainly due to its low energy requirements, non-thermal nature and simplicity. It is ideal for removal of anti-nutritional factors (i.e. oligosaccharides, phytic

acid, and some trypsin inhibitors from vegetable protein processing) to produce purified protein isolates or concentrates with superior functional properties

Food products concentrated by reverse osmosis are generally more heat labile; they include high priced fruit juices, dairy products and pharmaceuticals. In most reverse osmosis applications, the liquid products are pre-filtered using membranes with larger pore openings, such as ultrafiltration and microfiltration membranes

Limitations

- Concentration exposes food to 100° C or above for prolonged periods, which can cause major changes in organoleptic and nutritional properties.
- Cooked foods and darkening of colour are two of the more common heat induced results.
- Concentration at a temperature of 100 °C or slightly above will kill many micro-organisms but cannot be depended on to destroy bacterial spores.
- When the food contains acid, such as fruit juices, the micro-organisms kill will be greater but sterility is doubtful.
- When concentration is done under vacuum, many bacterial types not only survive the low temperatures but multiply in the concentrating equipment.

Irradiation

Food irradiation is one of the food processing technologies available to the food industry to control organisms that cause food-borne diseases and to reduce food losses due to spoilage and deterioration. Food irradiation technology offers some advantages over conventional processes. Each application should be evaluated on its own merit as to whether irradiation provides a technical and economical solution that is better than traditional processing methods.

Applications

For products where irradiation is permitted, commercial applications depend on a number of factors including the demand for the benefits provided, competitiveness with alternative processes and the willingness of consumers to buy irradiated food products. There are a number of applications of food irradiation. For each application it is important to determine the optimum dosage range required to achieve the desired effect. Too high a dosage can produce undesirable changes in texture, colour and taste of foods.

The way in which the radiation dose absorbed is measured differs according to the source of radiation, and various dosimetry techniques exist.

Dose-range needed for effective treatment	kGy*
Inhibition of sprouting in potatoes and onions	0 03—0 1
Sterilization of insects and parasites	0 03—0.2
Killing of insects and parasites	0 05—5
Reducing by 10* the number of vegetative bacteria, moulds, and fungi	1—10
Reducing by 10' the number of dried orfrozen vegetative bacteria, fungi, and spores	2—20
Reduction by 10' the number of viruses	10—40
Sterilization of food	20-45

- 1 kGy (Gray) = 100 000 rad (= 1 Joule/kg).

Irradiation can extend the shelf-life of foods in a number of ways. By reducing the number of spoilage organisms (bacteria, mould, fungi), irradiation can lengthen the shelf life of fruits and vegetables.

Since ionizing radiation interferes with cell division, it can be used as an alternative to chemicals to inhibit sprouting and thereby extend the shelf life of potatoes, onions and garlic. Exposure of fruits and vegetables to ionizing radiation slows their rate of ripening. Strawberries, for example, have been found to be suitable for irradiation. Their shelf-life can be extended three-fold, from 5 to 15 days. Ionising radiation can also be used as an alternative to chemical fumigants for disinfestations of grains, spices, fruits and vegetables.

The Expert Committee concluded that irradiation to an overall dose of 10 kGy (kilograys) presents no toxicological hazard and introduces no special nutritional or microbiological problems, thus establishing the wholesomeness of irradiated foods up to an overall average absorbed dose of 10 kGy. Data were insufficient to formulate conclusions on applications of food irradiation above 10 kGy.

More than 30 countries have given clearances for the use of food irradiation to process some 40 food items and approximately 30 facilities world-wide treat food by irradiation processing. Approvals for additional items are being considered in many countries and many food irradiation facilities are being planned.

MAP / CAP

MAP and CAP can supplement proper temperature and relative humidity management in maintaining quality and reducing losses of tropical fruits. Its beneficial effects include reduction of respiration rate, inhibition of ethylene production and action, retardation of ripening and maintenance of nutritional quality.

Short term exposure of tropical fruits to O_2 levels below 1% and /or CO_2 levels below %12 can reduce incidence and severity of physiological disorders (such as chilling injury), pathogens and insects.

Controlled Atmosphere Packaging

A defined mix of gases is maintained or controlled over time by some external apparatus or internal chemical reactions. An example of a controlled atmosphere is the Transfresh container (used for ocean transport of fruit) which have a mechanical means of measuring the container atmosphere and adjusting the gas levels to maintain a predetermined mixture of CO_2, O_2 and N_2 during shipment. Placing an oxygen absorbing sachet inside a barrier package is an example of a controlled atmosphere package using a chemical reaction. The sachet absorbs any oxygen that transmits through the package barrier. Sulfur dioxide producing pads used in long term shipment or storage of table grapes to prevent growth of gray mold are another example of controlling package atmosphere.

Modified Atmosphere Packaging

Gas atmosphere is modified by (1) direct injection of gases (often CO_2 or nitrogen) into a package, (2) evacuating air from the package or (3) interaction between package contents and the air in the package causing the package atmosphere to modify over time. With proper packaging, the natural respiration of produce causes O_2 levels to drop and CO_2 levels to rise. Modified atmosphere packages have an atmosphere different from ambient air but, that atmosphere can change over time.

In the case of produce, package atmosphere is affected by the transmission rates of the packaging material and changes in storage temperatures. Higher temperatures lead to higher respiration rates, creating lower O_2 levels in the package atmosphere and higher concentrations of CO_2. Hence, the atmosphere inside the package is modified but not controlled.

Carbon Dioxide (CO_2)

Carbon Dioxide is a natural gas, which is found in small concentrations in the air. It inhibits the increase of most aerobic bacteria and mildew. CO_2 is the most important gas in the packaging of food under modified atmospheres. Higher the CO_2 concentration the longer the durability of the perishable food.

However fat and water absorb CO_2 gases very easily and excessive CO_2 concentrations causes quality failures regarding taste, loss of humidity and the concentration of the packaging (so called vacuum effect). If CO_2 is intended to regulate the growth of bacteria and mildew a concentration of at least 20% is recommended.

Nitrogen (N_2)

N_2 is an inert gas that is used to expel air especially Oxygen out of the packaging. It is also used as a filling gas that equalizes the effect of C_{O2} absorption by the perishable food. It reduces the vacuum effect and is also a natural component of the air.

Oxygen (O_2)

Oxygen is an essential gas for the respiration of all living beings and supports the decay of perishable food. It is the condition for the growth of aerobic micro organisms. In general Oxygen should be excluded for the MAP but in some cases a determined amount of Oxygen brings quite positive results.

- It keeps the natural colour of the perishable food (effect of freshness).
- It makes possible respiration, especially for fruits or vegetables.
- It inhibits the growth of anaerobic micro organisms in several kinds of fish and vegetable.

Drying

Drying or dehydration is the process of removing water from a solid or liquid food by means of evaporation. The purpose of drying is to obtain a solid product with sufficiently low water content. It is one of the oldest methods of food preservation. Water is the prerequisite for the microorganisms and enzymes to activate food spoilage mechanisms. In this method, the moisture content is lowered to the point where the activities of these microorganisms are inhibited. Most microorganisms can grow at water activity above 0.95. Bacteria are inactive at water activity below 0.9. Most of the microorganisms cannot grow at water activity below 0.88.

Drying has numerous advantages. It reduces weight and volume of foods, facilitates foods storage, packaging, and transportation, and also provides different flavors and smells. With all these benefits, drying is apparently the cheapest method of food preservation. However, this process also has limitations. In some cases, significant loss of flavor and aroma has been observed after drying. Some functional compounds like vitamin C, thiamin, protein, and lipid are also lost because of drying.

Classification of drying: Drying can be classified into three major groups: convective, conductive, and radiative. Convective drying is the most popular method to obtain over 90% dehydrated foods. Depending on the mode of operation, dryers can be classified as batch or continuous. For smaller-scale operations and short residence times, batch dryers are preferred. Continuous method of drying is preferential when long periodic operations are required and drying cost is needed to curtail.

Drying of different foods: Food items, such as fruits, vegetables, meats, and fishes, are processed by drying. Instant coffee and tea are also produced by spray drying or freeze drying.

Pasteurization: Pasteurization is a physical preservation technique in which food is heated up to a specific temperature to destroy spoilage-causing microorganisms and enzymes. Almost all the pathogenic bacteria, yeasts, and molds are destroyed by this process. As a result, the shelf life of food increases. This process was named after the French scientist Louis Pasteur (1895–1822), who experimented with this process in 1862. He used this process to treat wine and beer.

Pasteurization techniques: The efficiency of pasteurization depends on the temperature–time combination. This combination is mostly based on the thermal death-time studies of heat-resisting microorganisms. On the basis of temperature and heat exposure, pasteurization can be categorized as vat (batch), high temperature short time (HTST), and ultra-high temperature (UHT); HTST and UHT are continuous processes. Vat pasteurizer is suitable for small plants having the capacity of 500–100 gallons.

Vat pasteurization requires constant supervision to prevent overheating, over holding, or burning. High-temperature short-time (HTST) pasteurization is a continuous process pasteurizer equipped with sophisticated control system, pump, flow diversion devices or valves, and heat exchanger equipment. HTST pasteurization is also known as 'flash pasteurization'. Vat and HTST pasteurization perishes pathogenic microorganisms effectively. However, to inactivate thermo-resisting spores ultra-high temperature (UHT) pasteurization is more effective than VAT and HTST. During heat treatment of food items, minimal physical, chemical, or biological changes take place. After heating is done, the products are aseptically packaged in sterile containers. UHT pasteurized products have a longer shelf life than other pasteurized products.

High heat of pasteurization process may damage some vitamins, minerals, and beneficial bacteria during pasteurization. At pasteurization temperature, Vitamin C is reduced by 20 per cent, soluble calcium and phosphorus are reduced by 5 per cent, and thiamin and vitamin B12 are reduced by 10 per cent. In fruit juices, pasteurization causes reduction in vitamin C, ascorbic acid, and carotene. However, these losses can be considered minor from nutritional point of view.

THERMAL STERILIZATION

Thermal sterilization is a heat treatment process that completely destroys all the viable microorganisms (yeasts, molds, vegetative bacteria, and spore formers) resulting in a longer period of shelf life. Retorting and aseptic processing are two categories of thermal sterilization. Thermal sterilization is different from pasteurization.

RETORTING

Retorting is defined as the packaging of food in a container followed by sterilization. Foods with pH above 4.5 require more than 100 °C as sterilization temperature. The attainment of such temperature can be possible in batch or continuous retorts. Batch retorts are gradually being superseded by continuous systems. Hydrostatic retorts and rotary cookers are the most common continuous systems used in food industries.

Aseptic packaging

Aseptic packaging involves placing commercially sterilized food in a sterilized package which is then subsequently sealed in an aseptic environment. Conventional aseptic packaging utilizes paper and plastic materials. Sterilization can be achieved either by heat treatment, by chemical treatment, or by attributing both of them. Aseptic packaging is highly used to preserve juices, dairy products, tomato paste, and fruit slices.

It can increase the shelf life of food items to a large extent; as an example, UHT pasteurization process can extend the shelf life of liquid milk from 19 to 90 days, whereas combined UHT processing and aseptic packaging extend shelf life to six months or more. Packages used for aseptic processing are produced from plastics having relative softening temperature. Moreover, aseptic filling can accept a wide range of packaging materials including:

- metal cans sterilized by superheated steam,
- paper, foil, and plastic laminates sterilized by hot hydrogen peroxide, and
- a variety of plastic and metal containers sterilized by high-pressure steam.

Wide variation of packages thus enhances proficiency of aseptic packaging and diminishes cost.

The direct approach of aseptic packaging comprises of steam injunction and steam infusion. On the other hand, indirect approach of aseptic packaging includes exchanging heat through plate heat exchanger, scrapped surface heat exchanger, and tubular heat exchanger. Steam injection is one of the fastest methods of heating and often removes volatile substances from some food products. On the contrary, steam infusion offers higher control over processing conditions than steam injection and minimizes the risk of overheating products. Steam infusion is suitable to treat viscous foods.

Tubular heat exchangers are adopted for operations at higher pressures and flow rates. These exchangers are not very flexible to withstand production capacity alteration, and their use is only limited to low viscous foods. Plate exchangers, on the other hand, overcome these problems. However, frequent cleaning and sterilizing requirements have made this exchanger less popular in food industries.

FREEZING

Freezing slows down the physiochemical and biochemical reactions by forming ice from water below freezing temperature and thus inhibits the growth of deteriorative and pathogenic microorganisms in foods. It reduces the amount of liquid water in the food items and diminishes water activity. Heat transfer during freezing of a food item involves a complex situation of simultaneous phase transition and alteration of thermal properties. Nucleation and growth are two basic sequential processes of freezing. Nucleation means the formation of ice crystal, which is followed by 'growth' process that indicates the subsequent increase in crystal size.

Freezing time

Freezing time is defined as the time required to lower the initial temperature of a product to a given temperature at its thermal center. In general, slow freezing of food tissues results in the formation of larger ice crystals in the extracellular spaces, while rapid freezing produces small ice crystals distributed throughout the tissue. The International Institute of Refrigeration defines various factors of freezing time in relation to the food products and freezing equipment. Dimensions and shapes of the product, initial and final temperature, temperature of refrigerating medium, surface heat transfer coefficient of the product, and change in enthalpy and thermal conductivity of the product are the most important factors among them.

Individual quick freezing

Individual quick freezing (IQF) generally relates to quick freezing of solid foods like green peas, cut beans, cauliflower pieces, shrimps, meat chunks, and fish. On the other hand, freezing related to liquid, pulpy or semiliquid products, like fruit juices, mango pulps, and papaya pulps is known as quick freezing. The ice crystals formed by quick freezing are much smaller and therefore cause less damage to cell structure or texture of the food. Shorter freezing period impedes the diffusion of salts and prevents decomposition of foods during freezing. IQF also allows higher capacity for commercial freezing plants with the resultant cost reduction. However, higher investment is required to set up a quick freezing plant. Different quick freezing techniques, such as contact plate freezing, air-blast freezing, and cryogenic freezing, are used to process food items.

The comparison between different quick freezing techniques for fishery products is presented in Table 5.

Table 5 Different quick freezing techniques (fishery products)

Criteria	Contact plate freezing	Air-blast freezing	Cryogenic freezing
Capital cost	Low capital investment	Economic to construct and operate	High capital costs
Operating cost	Low operating cost	Higher operating cost	Higher operating cost
Heat transfer	Controlled heat transfer	Efficient heat transfer	Efficient heat transfer
Product line	Generally bulk freezing	Flexible	Flexible

Criteria	Contact plate freezing	Air-blast freezing	Cryogenic freezing
Required floor space	Large	Large	Small
Refrigeration plant	Required	Required	Not required
Maintenance cost	Low	Low	Minimum
Dehydration loss	High	High	Minimum
Product quality	Reasonably good product quality	Good product quality	Superior product quality

CHILLING

In chilling process, the temperature of foods is maintained between "1 and 8 °C. Chilling process reduces the initial temperature of the products and maintains the final temperature of products for a prolonged period of time. It is used to reduce the rate of biochemical and microbiological changes and also to extend shelf life of fresh and processed foods. In practice, freezing process is often referred to chilling, when cooling is conducted at <15 °C. Partial freezing is applied to extend the shelf life of fresh food items in modern food industries. This process reduces ice formation in foods, known as super chilling.

Chilling can be done by using various equipments, such as continuous air cooler, ice bank cooler, plate heat exchanger, jacketed heat exchanger, ice implementation system, vacuum attribution system, and cryogenic chamber. Chilling rate is mainly dependent on thermal conductivity, initial temperature of foods, density, moisture content, presence or absence of a lid on the food storage vessel, presence of plastic bags as food packaging equipment, and the size as well as weight of food units.

Advantages and disadvantages of chilling

Chilling storage is extensively used for its effective short-term preservation competency. Chilling retards the growth of microorganisms and prevents post-harvest metabolic activities of intact plant tissues and post-slaughter metabolic activities of animal tissues. It also impedes deteriorative chemical reactions, which include enzyme-catalyzed oxidative browning, oxidation of lipids, and chemical changes associated with color degradation.

It also slows down autolysis of fish, causes loss of nutritive value of foods, and finally bares moisture loss. Chilling is high capital intensive since this process requires specialized equipment and structural modifications. Chilling may reduce crispiness of selected food items. Chilling process also dehydrates unwrapped food surfaces, which is a major limitation of chilling process.

IRRADIATION

Irradiation is a physical process in which substance undergoes a definite dose of ionizing radiation (IR). IR can be natural and artificial. Natural IR generally includes X-rays, gamma rays, and high-energy ultraviolet (UV) radiation; artificially generated IR is accelerated electrons and induced secondary radiation. IR is used in 40 different countries on more than 60 different foods. The effects of IR include:

- disinfestation of grains, fruits, and vegetables,
- improvement in the shelf life of fruits and vegetables by inhibiting sprouting or by altering their rate of maturation and senescence, and
- improvement in shelf life of foods by the inactivation of spoilage organisms and improvement in the safety of foods by inactivating foodborne pathogens.

Regulatory limits of irradiation

The IR dose delivered to foods is measured in kilo grays (kGy). 1 gray is equivalent to ionizing energy dose absorbed by 1 kg of irradiated material. IR regulatory limits are set by the legislative bodies. Depending on the regulatory authority, these limits may be expressed as minimum dose, maximum dose, or approved dose range.

Effects of Irradiation

The nutritional parameters, such as lipids, carbohydrates, proteins, minerals, and most vitamins, remain unaffected by IR even at high doses. At a high dose, IR may cause the loss of some micronutrients, most notably vitamins A, B1, C, and E. According to FDA, IR has effects on food nutritive value that is similar to those of conventional food processing techniques.

HIGH-PRESSURE FOOD PRESERVATION

High hydrostatic pressure or ultra-high pressure processing (HPP) technology involves pressure attribution up to 900 MPa to kill microorganisms in foods. This process also inactivates spoilage of foods, delays the onset of chemical and enzymatic deteriorative processes, and retains the important physical and physiochemical characteristics of foods. HHP has the potential to serve as an important preservation method without degrading vitamins, flavors, and color molecules during the process.

Freshness and improved taste with high nutritional value are the peerless characteristics of HPP technology. This process is also environmental friendly, since energy consumption is very low and minimal effluents are required to discharge. The major drawback of this technology is the high capital cost. In addition, limited information and skepticism about this technology also limit the wide application of HPP processes.

Mechanism and working principle

HP process follows Le Chatelier's principle and isostatic principle. According to Le Chatelier's principle, biochemical and physicochemical phenomena in equilibrium are accompanied by the change in volume and hence influenced by pressure. Regardless of the shape, size, or geometry of the products, the isostatic principle relies on the instant and uniform pressure transmittance throughout food systems. HP processes affect all reactions and structural changes where a change in volume is involved. The combined effect of breaking down and permeabilization of cell membrane kills or inhibits the growth of microorganisms. Vegetative cells are inactivated at 3000 bar pressure (approximate) at ambient temperature, while spore inactivation requires much higher pressure in combination with the temperature rise to 60 °C to 70 °C. Moisture level is extremely important in this context since little effect is noticeable below 40% moisture content. Container processing and bulk processing are two methods of preserving foods under high pressure.

PULSED ELECTRIC FIELD

Pulsed electric field (PEF) food processing is defined as a technique in which food is placed between two electrodes and exposed to a pulsed high voltage field (20–40 kV/cm). Generally, the PEF treatment time is less than one second. Low processing temperature and short residence time of this process allow a highly effective inactivation of microorganisms. PEF processing is much effective to destroy gram-negative bacteria than gram-positive bacteria. Vegetative cells are much sensitive than spores to this process. All cell deaths occur due to the disruption of cell membrane function and electroporation. PEF technology retains taste, flavor, and color of the foods. Furthermore, this technique is not toxic. However, this process has no impact on enzymes and spores. It is also not suitable for conductive materials and only effective to treat liquid foods. This process is energy extensive and may possess environmental risks.

Preservation of liquid foods

Nonthermal food preservation processes, such as HPP and PEF, are reported to be more effective than thermal processing. Microbial inactivation achieved by PEF mainly depends on electric field strength (20–40 kV/cm) and number of pulses produced during processing. It has been found that most of the spoilage and pathogenic microorganisms are sensitive to PEF. However, it is noted that treatment of plant or animal cells require a high field strength and higher energy input, which increases the processing cost. In addition, this kind of field strength may destroy the structure of solid food. Therefore, PEF is more favorable to preserve liquid foods. Microbial inactivation by PEF has been found effective for fruit or vegetable juices, milk, liquid egg, and nutrient broth.

CHEMICAL AND BIOLOGICAL METHODS OF FOOD PRESERVATION

Preservation Using Chemicals

Salt and sugar

Salt and sugar have long been used as effective means of extending shelf life of various products as these solutes bind water, leaving less water available for the growth of microorganisms. Essentially the water activity (a_w) of the product is reduced, and since most microorganisms require a high water activity, they areunable to survive.

Salt and sugar in concentrated solutions have high osmotic pressure. When these are sufficient to draw water from microbial cells or prevent normal diffusion of water into these cells, a preservative condition exists. The critical concentration of sugar in water to prevent microbial growth will vary depending upon the type of micro-organisms and the presence of other food constituents, but usually 70% sucrose in solution will stop growth of all micro-organisms in foods. Less than this concentration may be effective but for short period of time unless the foods contain acid or they are refrigerated.

Salt becomes a preservative when its concentration is increased and levels of about 18% to 25% in solution generally will prevent all growth of micro-organisms in foods. Except in the case of certain briny condiments, however, this level is rarely tolerated in foods.

Acids

Acids such as citric acid, acetic acid (vinegar) and ascorbic acid are also known to confer protection against product deterioration. In these cases, the pH of the product is shifted to being low, that is, more acidic, where very few molds, yeast and bacteria are able to grow and multiply.

Acetic acid is a general preservative inhibiting many species of bacteria, yeasts and to a lesser extent moulds. It is also a product of the lactic-acid fermentation, and its preservative action even at identical pH levels is greater than that of lactic acid. The main application of vinegar (acetic acid) includes products such as pickles, sauces and ketchup.

Other acidulants

- Malic and tartaric (tartric) acids is used in some countries mainly to acidify and preserve fruit sugar preserves, jams, jellies, etc.
- Citric acid is the main acid found naturally in citrus fruits; it is widely used (in carbonated beverages) and as an acidifying agent of foods because of its unique flavour properties. It has an unlimited acceptable daily intake and is highly soluble in water. It is a less effective antimicrobial agent than other acids.

- Ascorbic acid or vitamin C, its isomer isoascorbic or erythorbic acid and their salts are highly soluble in water and safe to use in foods.

Preservatives

Chemical food preservatives are those substances which are added in very low quantities (up to 0.2%) and which do not alter the organoleptic and physico-chemical properties of the foods at or only very little.They are used to improve the colour and keeping qualities of the final product for some fruits and vegetables.

Food additives such as benzoate and sorbate are quite commonly used in the fruit drink industry to protect against microbial spoilage, while nitrites are used in meat processing. These chemicals work best at acidic pH ranges and when the products are pasteurized. A combination of heat and chemicals where applicable is usually more effective than either one on its own.

Treatment with preservatives takes place after blanching or, when blanching is not needed, after slicing. The composition and strength of the preservative solution vary for different fruit and vegetables. As a general rule, preservatives are not used for treating onions, garlic, leeks, chilies and herbs.

Preservation of food products containing chemical food preservatives is usually based on the combined or synergistic activity of several additives, intrinsic product parameters (e.g. composition, acidity, water activity) and extrinsic factors (e.g. processing temperature, storage atmosphere and temperature). This approach minimises undesirable changes in product properties and reduces concentration of additives and extent of processing treatments.

Lypophilic acid food preservatives

Benzoic acid

Benzoic acid in the form of its sodium salt, constitutes one of the most common chemical food preservative. Sodium benzoate is a common preservative in acid or acidified foods such as fruit juices, syrups, jams and jellies, sauerkraut, pickles, preserves, fruit cocktails, etc. Yeasts are inhibited by benzoate to a greater extent than are moulds and bacteria.

Sorbic acid

Sorbic acid is generally considered non toxic and is metabolized; among other common food preservatives the WHO has set the highest acceptable daily intake (25 mg/kg body weight) for sorbic acid. Sorbic acid and its salts are practically tasteless and odourless in foods, when used at reasonable levels (< 0.3 %) and their antimicrobial activity is generally adequate.

Sorbates

Sorbates are used for mould and yeast inhibition in a variety of foods including fruits and vegetables, fruit juices, pickles, sauerkraut, syrups, jellies, jams, preserves, high moisture dehydrated fruits, etc.

Potassium sorbate

Potassium sorbate, a white, fluffy powder, is very soluble in water (over 50%) and when added to acid foods it is hydrolysed to the acid form. Sodium and calcium sorbates also have preservative activities but their application is limited compared to that for the potassium salt, which is employed because of its stability, general ease of preparation and water solubility.

Gaseous chemical food preservatives

Sulphur dioxide and sulphites

Sulphur dioxide (SO_2) has been used for many centuries as a fumigant and especially as a wine preservative. It is a colourless, suffocating, pungent-smelling, non-flammable gas and is very soluble in cold water (85 g in 100 ml at 25°C).

The various sulphite salts contain 50-68% active sulphur dioxide. A pH dependent equilibrium is formed in water and the proportion of SO_2 ions increases with decreasing pH values. At pH values less than 4.0 the antimicrobial activity reaches its maximum. The antimicrobial action of sulphur dioxide against yeasts, molds and bacteria is selective, with some species being more resistant than others.

Sulphur dioxide and sulphites are used in the preservation of a variety of food products. In addition to wines these include dehydrated/dried fruits and vegetables, fruit juices, acid pickles, syrups, semi-processed fruit products, etc. In addition to its antimicrobial effects, sulphur dioxide is added to foods for its antioxidant and reducing properties, and to prevent enzymatic and non-enzymatic browning reactions.

Carbon dioxide

Carbon dioxide (CO_2) is a colourless, odourless, non-combustible gas, acidic in odour and flavour. In commercial practice it is sold as a liquid under pressure (58 kg per cm^3) or solidified as dry ice.

Carbon dioxide is used as a solid (dry ice) in many countries as a means of low-temperature storage and transportation of food products. Besides keeping the temperature low, as it sublimes, the gaseous CO_2 inhibits growth of psychrotrophic micro-organisms and prevents spoilage of the food (fruits and vegetables, etc.).

Carbon dioxide is used as a direct additive in the storage of fruits and vegetables. In the controlled/ modified environment storage of fruit and vegetables, the correct combination of O_2 and CO_2 delays respiration and ripening as well as retarding mould and yeast growth. The amount of CO_2 (%10-5) is determined by factors such as nature of product, variety, climate and extent of storage.

Chlorine

The various forms of chlorine constitute the most widely used chemical sanitizer in the food industry. These chlorine forms include chlorine (Cl_2), sodium hypochlorite (NaOCl), calcium hypochlorite ($Ca(ClO)_2$) and chlorine dioxide gas (ClO_2).

These compounds are used as water adjuncts in processes such as product washing, transport, and cooling of heat-sterilised cans; in sanitising solutions for equipment surfaces, etc.

General rules for chemical preservation

- Chemical food preservatives have to be used only at a dosage level which is needed for a normal preservation and not more.
- "Reconditioning" of chemical preserved food, e.g. a new addition of preservative in order to stop a microbiological deterioration already occurred is not recommended.
- The use of chemical preservatives MUST be strictly limited to those substances which are recognized as being without harmful effects on human beings' health and are accepted by national and international standards and legislation.

Preservation by Fermentation

- Food fermentation is usually carried out with the use of micro- organisms, producing alcoholic, acetic or lactic acid as end products as a result of their action on simple sugars present. These reactions take place in the absence of oxygen, that is anaerobic, and are responsible for the production of wine, vinegar, yoghurt and pickled vegetables among others. Having pure cultures of the desirable microorganism in the food products being fermented is critical to obtaining clean", characteristic flavours of end products.
- It is a process in which food spoils, but results in the formation of an edible product. Perhaps the best example of such a food is cheese. Fresh milk does not remain in edible condition for a very long period of time. Its pH is such that harmful pathogens begin to grow in it very rapidly. Early humans discovered, however, that the spoilage of milk can be controlled in such a way as to produce a new product, cheese.
- Bread is another food product made by the process of fermentation. The addition of yeasts brings about the fermentation of sugars present in the

mixture, resulting in the formation of a product that will remain edible much longer than will the original raw materials used in the bread-making process.

CONCLUSION

In conclusion, food preservation has been pivotal to our society since the beginning. Preservation has come from simple processes such as salting, to more complex preserving methods such as irradiation and chemical additives. Looking into the future, High Pressure Preservation seems to be the next logical step. In an increasingly health-conscious world, an uncontaminated, well preserved food source is the ultimate goal, and hopefully the costs of production will decrease for all of us to enjoy HPP foods.

CHAPTER

REFRIGERATION 8

INTRODUCTION

Refrigeration is a science that deals with absorption of heat at a temperature from surroundings at lower temperature and rejecting it to a relatively higher temperature at the cost of some external work following Clausius statement of second law of thermodynamics. Refrigeration sector is one of the largest economic sectors, covering precooling, cooling, air conditioning, and freezing applications for various commodities and people and appears to be an energy intensive sector by consuming about 20 per cent of the total electricity generated worldwide. The refrigeration sector includes air conditioners, household fridges, coolers, pre-coolers, refrigerators, heat pumps, and freezers for various applications, ranging from food cooling to space cooling. Thus, refrigeration has found its importance in many engineering and industrial sectors. One of them is processing, preservation and nourishment of food by storing them in places whose temperature is lower than that of the surroundings.

STORAGE LIFE OF FROZEN FOODS

According to the International Institute of Refrigeration, storage life is defined as the length of duration of a food product stored under specified conditions so that it is liable for consumption. Different food products have different storage lives. Also, it has to be considered that storage life of a particular food product may vary according to its location and other ecological aspects.

DIFFERENT REFRIGERATION PROCESSES INVOLVED IN FOOD PROCESSING

The principle behind preservation of foods by refrigeration is to reduce and maintain the temperature of the food in order to mitigate any detrimental or undesirable changes occurring in the food. These changes can be microbiological, physiological, biochemical or physical. This in turn can help in improving the nutritional quality of the food. There are several refrigeration processes involved in the processing and nourishment of food. The refrigeration techniques that are implemented one after the other have been illustrated below:

- **Pre-cooling:** Rapid cooling is necessary to retard the metabolism of food products and to increase their life span. The lesser the temperature is, the better the quality and longevity of the cuisines. Based on the type of the food product, different refrigeration techniques have been proposed. i.e. hydrocooling for small fruits, vacuum cooling for green leafy vegetables, blast-air system for cooling the surfaces of meat products, cooling of milk and other dairy products in specialised tanks, pulsed air in some other products and so on.
- **Chilling:** After precooling, the food product has to be chilled and maintained at the favourable temperature. Chilling is usually done in classic cold rooms equipped with sufficient ventilation. The chilling and conservation temperatures depend on the sensitivity of the product i.e. for high-sensitivity products such as mangoes, ginger, sweet potatoes temperatures below 8-12°C are not recommended, as they can endure metabolic disturbances that shorten the life of the goods.
- For medium-sensitivity products such as tangerines, green beans, potatoes, it is possible to lower the temperature not less than 4-6°C. For low-sensitivity products, a temperature of 2-3°C, down to just above the freezing point is recommended. In slaughterhouses, low temperatures that can be achieved by using blast chillers or cold storage rooms are recommended in order to inhibit the growth of microorganisms.
- For eggs, conservation at higher temperature could be enough as long as the shell remains unbroken and for dairy products an adequate temperature of nearly 8-10°C would be enough to inhibit the growth of pathogenic germs.
- **Freezing:** Freezing is a process of lowering the temperature of a product below its solidification point. Freezing hinders the metabolism of the fruit and vegetable products. Longer the product remaining frozen, longer is its durability for storing.
- **Super-cooling and Super-chilling:** These are new processes implemented for the storage of food products. Supercooling is lowering down the temperature of a food product just below its freezing point without any formation of ice and superchilling is partial freezing of a food at a temperature just below its freezing point.

Apart from the above-mentioned techniques, some other techniques are also adopted such as crystallisation of fat, cryoseparation of undesirable components; cryoconcentration and so on.

REFRIGERANTS USED IN FOOD PROCESSING INDUSTRIES

Refrigerants are the working fluids used in refrigeration systems. Selection of refrigerant plays a vital role in any refrigeration industry, so it has been a real challenge to develop and implement refrigerants having desirable thermodynamic properties and high energetic and exergetic performances.

Earlier, the refrigerants were selected based on their thermodynamic properties only, but eventually focus has also been made over the ozone depletion potential (ODP) and global

warming potential (GWP) of refrigerants. Some of the desirable properties of refrigerants needed in food processing industry have been listed below:

- It should have low flammability and toxicity.
- It should zero ODP value and GWP value as minimum as possible.
- It should be easy leak detectable.
- Selection of refrigerants has significant impact on safety, reliability, energy consumption, system performance and also on the environment.

PRINCIPLE OF REFRIGERATION

The principle of refrigeration is based on second law of thermodynamics. In refrigeration process, since the heat has to be transferred from a low temperature body to a high temperature body some external work has to be done according to the second law of thermodynamics as shown in Fig. 1. This external work is done by means of compressor.

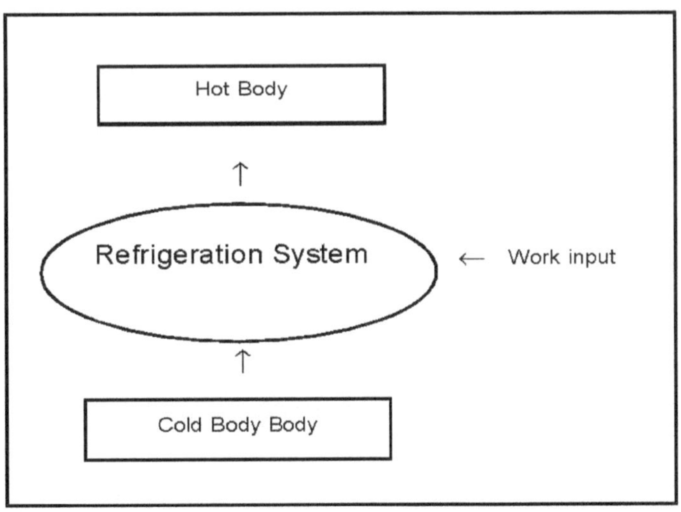

Fig. 1. The principle of refrigeration

IMPORTANCE OF REFRIGERATION IN DAIRY INDUSTRY

Refrigeration is a basic requirement for the processing and storage of milk and milk products as majority of dairy products are perishable in nature. The need of refrigeration is indicated below.

- Chilling of milk at producers' level by employing bulk milk coolers and at milk chilling centres is the first requirement in dairy industry. Immediate cooling of milk to about 2-3 ºC is very important to reduce the multiplications of micro-organisms and to get low bacterial count in the milk and milk products.
- Processing of milk using either batch pasteurizer or HTST plant requires chilled water or any other cooling medium for cooling of milk.
- Manufacture of many products requires refrigeration. e.g. butter, ice-cream etc.
- Storage of milk and milk products requires maintaining low temperature in the cold storages depending on the type of product to be stored. e.g. milk is stored at around 3-4 ºC while ice-cream is stored at -30ºC temperature.
- Transportation of many products requires refrigerated vehicles to maintain the quality of products.
- Low temperature storage is required at distribution of products as well as at the consumers' level.

A refrigeration system is an important and integral part of Dairy Industry. There are numerous types of refrigeration systems depending on different size, design, application, capacity of cooling or low temperature to be maintained, etc. But the principle of refrigeration in all these equipment is mostly common and that is 'vapour compression refrigeration'. The basic major components of refrigeration systems are compressor, condenser, expansion valve and evaporator. Some of the major refrigeration systems used in industry are mentioned as below:

- Ice Refrigeration
- Evaporative Refrigeration
- Refrigeration by Expansion of Air
- Refrigeration By throttling of the Gas
- Vapour Compression Refrigeration
- Vapour Absorption Refrigeration System
- Steam - Jet Refrigeration System
- Refrigeration by Using Liquid Gasses

- Dry ICE Refrigeration

Unit of Refrigeration

Capacity of refrigeration system is expressed as ton of refrigeration (TR). A ton of refrigeration is defined as the quantity of heat to be removed in order to form one ton (2000 lbs.) of ice at 0 ºC in 24 hrs, from liquid water at 0 ºC. This is equivalent to 12600 kJ/h or 210 kJ/min or 3.5 kJ/s (3.5 kW).

1 TR = 12600 kJ/h or 210 kJ/min or 3.5 kW

The capacity of refrigeration plant required in any dairy/food plant can be estimated based on the cooling load requirement of the plant.

PRINCIPLE OF VAPOUR COMPRESSION REFRIGERATION (VCR) SYSTEM

Any type of vapour compression refrigeration system has the same principle of working. A fixed quantity of refrigerant, which is any suitable gas, is filled in the closed system of four components, i.e., compressor, condenser, expansion valve and evaporator connected to each other through tubes. When the compressor is run, the refrigerant starts flowing through the system i.e., the system starts it's working. The compressor continuously sucks low pressure, low temperature refrigerant vapors from the evaporator and pump these to condenser at high pressure and high temperature condition.

While flowing through the condenser, the high temperature vapors release their heat to atmosphere and condense to high pressure liquid state. After condenser this high-pressure liquid enters the expansion valve where it is throttled to low pressure.

It is so constructed that a control quality of refrigerant flows (due to expansion valve) from one necessary steps to another at definite and predetermined pressure. On throttling the pressure and temperature of refrigerant (like ammonia, R-22 etc.) decreases and when this low pressure, low temperature throttled liquid flows through evaporator, it sucks heat and produce cooling.

On absorbing heat in evaporator all the low-pressure liquid evaporates to low-pressure, low-temperature vapors, which are again sucked by compressor. In this way all these processes go on continuously and as long as the compressor runs, the system produces cooling around the evaporator.

A block diagram of a vapour compression refrigeration system is shown in Fig. 2. The refrigeration system is an enclosed gas tight system of tubes and equipments as shown below:

- Evaporation
- Compression

- Condensation
- Pressure reduction or Expansion

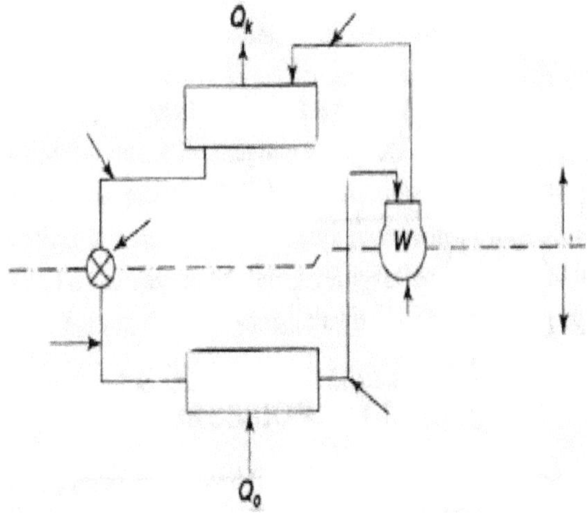

Fig. 2 : Block diagram of vapour compression refrigeration system

Components of the VCR system

Compressor

It is the main component of a Refrigeration System. It may be of many types but a reciprocating compressor is commonly used for different cooling applications in a dairy industry. In this type when an electric motor rotates the main shaft of compressor, a piston reciprocates i.e. moves up and down in the cylinder of compressor.

While moving down, the piston sucks the refrigerant vapours coming from the evaporator through suction valve and while moving upward, it compress (increase the pressure of) these vapours and discharge them to condenser through discharge valve. In this way, the basic working of a reciprocating compressor is exactly same as that of a cycle pump.

The purpose of the compressor is to draw low-pressure refrigerant vapour from the evaporator, and compress it so the vapour can be condensed back into a liquid by cooling with air or water.

The compressor is the workhorse of a refrigeration system and usually accounts for between 80% and 100% of the system's total energy consumption. It is important, therefore, that the system operates under optimum conditions. The amount of energy used by a compressor is affected by the:

- Type of compressor
- Compressor load

- Temperature difference of the system (i.e. the number of degrees by which the system is required to cool).

Condenser

The role of a condenser is to reject heat of compressed refrigerant to atmosphere by providing large area for heat exchange. It is named as condenser because by rejecting latent heat of refrigerant it condenses the refrigerant i.e. converts it from vapour to liquid state. In a Dairy Industry, either shell and tube type or evaporative type, also called as atmospheric condenser is used. Shell and tube type condenser is most satisfactory condenser in case there is no scarcity of cooling water. But its initial cost is high with separate cooling tower. On the other hand, when there is water shortage and cooling tower is being used, then generally the evaporative condenser is more suitable especially in case of large plants.

Evaporator

It is the most required important part of the refrigeration system. The refrigerant from the expansion valve comes in to the evaporator below the temperature required to be maintained in the evaporator and carries the heat from the evaporator.

The evaporators are mostly divided in to different categories as:

- Flooded Evaporators
- Dry-expansion Evaporators
  - Natural Convection Evaporators
  - Forced Convection Evaporators
- Shell and Tube Evaporator
- Shell and Coil Evaporator
- Double Pipe Evaporator
- Plate Evaporator
- Baudelot-Cooler

Pressure reduction or Expansion:

Expansion valve is one of the basic components of the refrigeration system. The refrigerant is in liquid stage under high pressure in the receiver and it is ready for reuse in the evaporator where much lower pressure is maintained. Some means of retaining of pressure difference is required. This is done by arranging a throttling device in the line between receiver and evaporator. This throttling maintains a higher pressure on one side and allows a lower pressure condition to exist on the other side when flow of refrigerant occurs.

It reduces the pressure of the refrigerant coming from the condenser and temperature also as per the requirement of the system. It also regulates the flow of the refrigerant as per the load on the evaporator. Refrigerant is a the heat transfer medium in vapour compression refrigeration cycle which absorbs the heat through evaporation at the evaporator and rejects the heat absorbed at evaporator plus the heat of work of compression at the condenser. Different devices which are used to perform above functions, are as under:

- Capillary Tube
- Pressure Control or Automatic Expansion Valve
- Thermostatic Expansion Valve
- High-side float Valve
- Low-side Float Valve
- Solenoid Valve

Regarding expansion valve, thermostatic expansion valve is most commonly used in a mid size refrigeration system. In small unit generally capillary tube is preferred because of its low cost, simple working and no maintenance required, etc.

Primary refrigerant

The working fluid of vapour compression refrigeration system is known as primary refrigerant which absorbs the heat through evaporation at the evaporator and again becomes liquid on cooling in the condenser. Primary refrigerants include only those working fluids which pass through the cycle of evaporation, compression, condensation and expansion. These refrigerants have very low boiling point. e.g. boiling point of ammonia is -33.3 º C at atmospheric pressure. There are many refrigerants which as discussed in another lesson.

Co-efficient of Performance (C.O.P.)

The performance of the refrigeration system is expressed as C.O.P. which is the ratio of refrigerating effect produced to the work of compression. The cooling effect (out put of the system) is produced at the evaporator and the refrigerant is compressed by the compressor using the electrical power (input to the system).

Absorption refrigeration

Absorption chillers allow cooling to be produced from heat sources such as clean fossil fuels, incinerated garbage, bio fuels, low-grade steam, hot water, exhaust gas or even solar energy, usually using a lithium bromide and water refrigerant. The COP of absorption refrigeration, however, is relatively low compared with vapour compression refrigeration systems with the best absorption chillers generating just over 1 kW of refrigeration for 1 kW of energy. The higher the temperature of the waste heat, therefore, the more effective the refrigeration will be. The advantages of absorption chillers are that they can utilize a waste heat source with lower greenhouse gas emissions compared to conventional vapour compression refrigeration systems.

Defrosting

Hot gas from the outlet of the refrigeration compressor can be used to defrost freezers, but the control must be accurate. The defrost water may then be used elsewhere in the plant. Once installed and optimized, a hot gas bypass defrost system can ensure frost-free evaporator operation. Once the evaporator is no longer covered in ice its cooling capacity will be increased.

Reducing load on refrigeration systems

Up to 10% of the power consumption in refrigeration plants can be from heat ingress through doorways in cool rooms. Many plants rely on good operator practice to keep doors closed, but this is not always effective. Automatically closing doors or an alarm system could be considered; and plastic strip curtains or swinging doors are useful at frequently opened entrances. Lights and fans also add to the heat load.

Sensors and timers can be used to ensure that lights are used only when necessary. Variable speed drives, coupled with a programmable controller, can cycle off fans and refrigerant feed during low load times. Cooling water loops using water at ambient temperature have also been used by some dairy processors to pre-cool high-temperature fluids (around 90°C) before chilling, thereby reducing the load on the refrigeration system.

REFRIGERATION CYCLE

The basis of mechanical refrigeration is the fact that at different pressures the saturation (or the condensing) temperatures of vapours are different as clearly shown on the appropriate vapour-pressure/temperature curve. As the pressure increases condensing temperatures also increase. This fact is applied in a cyclic process, which is illustrated in Fig. 3.

This is a thermodynamic diagram that looks very complicated at first sight but which in fact can make calculations straightforward and simple. For the present purposes the most convenient such chart is the one shown with pressure as the vertical axis (for convenience on a logarithmic scale) and enthalpy on the horizontal axis. On such a diagram, the properties of the particular refrigerant can be plotted, including the interphase equilibrium lines such as the saturated vapour line, which are important as refrigeration depends on evaporation and condensation.

To start with the **evaporator**; in this the pressure above the refrigerant is low enough so that evaporation of the refrigerant liquid to a gas occurs at some suitable low temperature determined by the requirements of the product. This temperature might be, for example -18°C in which case the corresponding pressures would be for ammonia 229 kPa absolute and for tetrafluoroethane (also known as refrigerant 134a) 144 kPa. Evaporation then occurs and this extracts the latent heat of vaporization for the refrigerant from the surroundings of the evaporator and it does this at the appropriate low temperature.

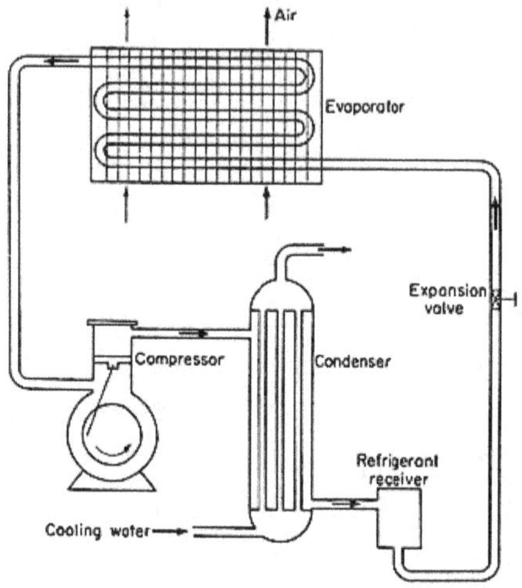

Fig. 3. Mechanical refrigeration circuit

This process of heat extraction at low temperature represents the useful part of the refrigerator. On the pressure/enthalpy chart this is represented by ab at constant pressure (the evaporation pressure) in which 1 kg of refrigerant takes in $(H_b - H_a)$ kJ . The low pressure necessary for the evaporation at the required temperature is maintained by the suction of the compressor.

The remainder of the process cycle is included merely so that the refrigerant may be returned to the evaporator to continue the cycle. First, the vapour is sucked into a compressor which is essentially a gas pump and which increases its pressure to exhaust it at the higher pressure to the **condensers.**

This is represented by the line *bc* which follows an **adiabatic compression line,** a line of constant entropy (the reasons for this must be sought in a section on refrigeration) and work equivalent to $(H_c - H_b)$ kJ kg^{-1} has to be performed on the refrigerant to effect the compression. The higher pressure might be, for example, 1150 kPa (pressures are absolute pressures) for ammonia, or 772 kPa for refrigerant 134a, and it is determined by the temperature at which cooling water or air is available to cool the condensers.

The condensing temperature, corresponding to the above high pressures, is about 30°C and so in this example cooling water at about 20°C, could be used, leaving sufficient temperature difference to accomplish the heat exchange in equipment of economic size.

This process of evaporation at a low pressure and corresponding low temperature, followed by compression, followed by condensation at around atmospheric temperature and corresponding high pressure, is the refrigeration cycle. The high-pressure liquid then passes through a nozzle from the condenser or high-pressure receiver vessel to the evaporator at low pressure, and so the cycle continues.

Expansion through the expansion valve nozzle is at constant enthalpy and so it follows the vertical line *ea* with no enthalpy added to or subtracted from the refrigerant. This line at constant enthalpy from point *e* explains why the point *a* is where it is on the pressure line, corresponding to the evaporation (suction) pressure.

By adjusting the high and low pressures, the condensing and evaporating temperatures can be selected as required. The high pressure is determined: by the available cooling-water temperature, by the cost of this cooling water and by the cost of condensing equipment. The evaporating pressure is determined by either the low temperature that is required for the product or by the rate of cooling or freezing that has to be provided.

Low evaporating temperatures mean higher power requirements for compression and greater volumes of low-pressure vapours to be handled therefore larger compressors, so that the compression is more expensive. It must also be remembered that, in actual operation, temperature differences must be provided to operate both the evaporator and the condenser.

There must be lower pressures than those that correspond to the evaporating coil temperature in the compressor suction line, and higher pressures in the compressor discharge than those that correspond to the condenser temperature.

Overall, the energy side of the refrigeration cycle can therefore be summed up:

- heat taken in from surroundings at the (low) evaporator temperature and pressure $(H_b - H_a)$,
- heat equivalent to the work done by the compressor $(H_c - H_b)$ and
- heat rejected at the (high) compressor pressure and temperature $(H_c - H_e)$.

A useful measure is the ratio of the heat taken in at the evaporator (the useful refrigeration), $(H_b - H_a)$, to the energy put in by the compressor which must be paid for $(H_c - H_b)$. This ratio is called the **coefficient of performance** (COP).

The unit commonly used to measure refrigerating effect is the **ton of refrigeration** = 3.52 kW. It arises from the quantity of energy to freeze 2000 lb of water in one day (2000 lb is called 1 short ton).

REFRIGERANTS

Although in theory a considerable number of fluids might be used in mechanical refrigeration, and historically quite a number including cold air and carbon dioxide have been, those actually in use today are only a very small number. Substantially they include only ammonia, which is used in many large industrial systems, and approved members of a family of halogenated hydrocarbons containing differing proportions of fluorine and chosen according to the particular refrigeration duty required. The reasons for this very small group of practical refrigerants are many.

Important reasons are:

- actual vapour pressure/temperature curve for the refrigerant which determines the pressures between which the system must operate for any particular pair of evaporator and condenser temperatures,
- refrigerating effect per cubic metre of refrigerant pumped around the system which in turn governs the size of compressors and piping, and the stability and cost of the refrigerant itself.

Ammonia is in most ways the best refrigerant from the mechanical point of view, but its great problem is its toxicity.

In working spaces, such as encountered in air conditioning, the halogenated hydrocarbons (often known, after the commercial name, as Freons) are very often used because of safety considerations.

They are also used in domestic refrigerators. Environmental problems, in particular due to the effects of any chlorine derivatives and their long lives in the upper atmosphere, have militated heavily, in recent years, against some formerly common halogenated hydrocarbons. Only a very restricted selection has been judged safe.

MECHANICAL EQUIPMENT

Compressors are just basically vapour pumps and much the same types as shown in Fig. 4.3 for liquid pumps are encountered. Their design is highly specialized, particular problems arising from the lower density and viscosity of vapours when compared with liquids. The earliest designs were reciprocating machines with pistons moving horizontally or vertically, at first in large cylinders and at modest speeds, and then increasingly at higher speeds in smaller cylinders. An important aspect in compressor choice is the compression ratio, being the ratio of the absolute pressure of discharge from the compressor, to the inlet suction pressure. Reciprocating compressors can work effectively at quite high compression ratios (up to 6 or 7 to 1). Higher overall compression ratios are best handled by putting two or more compressors in series and so sharing the overall compression ratio between them.

For smaller compression ratios and for handling the large volumes of vapours encountered at low temperatures and pressures, rotary vane compressors are often used, and for even larger volumes, centrifugal compressors, often with many stages, can be used. A recent popular development is the screw compressor, analogous to a gear pump, which has considerable flexibility. Small systems are often "hermetic", implying that the motor and compressor are sealed into one casing with the refrigerant circulating through both. This avoids rotating seals through which refrigerant can leak. The familiar and very dependable units in household refrigerators are almost universally of this type.

Refrigeration Evaporator

The evaporator is the only part of the refrigeration equipment that enters directly into food processing operations. Heat passes from the food to the heat-transfer medium, which may be air or liquid, thence to the evaporator and so to the refrigerant. Thermal coupling is in some cases direct, as in a plate freezer. In this, the food to be frozen is placed directly on or between plates, within which the refrigerant circulates.

Generally, however, the heat transfer medium is air, which moves either by forced or natural circulation between the heat source, the food and the walls warmed by outside air, and the heat sink which is the evaporator. Sometimes the medium is liquid such as in the case of immersion freezing in propylene glycol or in alcohol-water mixtures. Then there are some cases in which the refrigerant is, in effect, the medium such as immersion or spray with liquid nitrogen. Sometimes there is also a further intermediate heat-transfer medium, so as to provide better control, or convenience, or safety. An example is in some milk chillers, where the basic refrigerant is ammonia; this cools glycol, which is pumped through a heat exchanger where it cools the milk. A sketch of some types of evaporator system is given in Fig. 4.

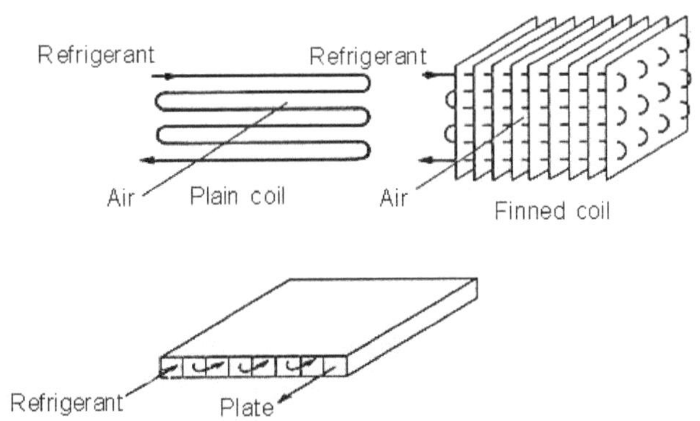

Fig. 4. Refrigeration evaporators

In the freezer or chiller, the heat transfer rates both from the food and to the evaporator, depend upon fluid or gas velocities and upon temperature differences.

The values for the respective **heat transfer coefficients** can be estimated by use of the standard heat transfer relationships. Typical values that occur in freezing equipment are given in Table 1.

Table 1 Expected Surface Heat Transfer Coefficients (h_s)

	J m^{-2} s^{-1} °C^{-1}
Still-air freezing (including radiation to coils)	9
Still-air freezing (no radiation)	6
Air-blast freezing 3 m s^{-1}	18
Air-blast freezing 5 m s^{-1}	30
Liquid-immersion freezing	600
Plate freezing	120

Temperature differences across evaporators are generally of the order 3 - 10°C. The calculations for heat transfer can be carried out using the methods which have been discussed in other sections, including their relationship to freezing and freezing times, as the evaporators are just refrigerant-to-air heat exchangers.

The evaporator surfaces are often extended by the use of metal **fins** that are bonded to the evaporator pipe surface. The reason for this construction is that the relatively high metal conductance, compared with the much lower surface conductance from the metal surface to the air, maintains the fin surface substantially at the coil temperature. A slight rise in temperature along the fin can be accounted for by including in calculations a fin efficiency factor. The effective evaporator area is then calculated by the relationship

$$A = A_p + fA_s \text{ R} \quad (1)$$

where A is the equivalent total evaporator surface area, A_p is the coil surface area, called the primary surface, A_s is the fin surface area, called the secondary surface and f (phi) is the fin efficiency.

Cold Storage

For cold storage, the requirement for refrigeration comes from the need to remove the heat:

- coming into the store from the external surroundings through insulation
- from sources within the store such as motors, lights and people (each worker contributing something of the order of 0.5 kW)
- from the foodstuffs.

Heat penetrating the walls can be estimated, knowing the overall heat-transfer coefficients including the surface terms and the conductances of the insulation, which may include several different materials.

CHAPTER

EVAPORATION 9

INTRODUCTION

Evaporation is a common unit operation in food processing industries in which partial removal of water takes place from an aqueous solution. The removal of water in the form of vapour is achieved by the process of vaporization or boiling of aqueous solution. Evaporation is confined to the preservation of liquid foods by concentrating them as a result, microbiological damage is checked and shelf life increased. One more advantage is the reduction in mass and volume of material, resulting in reduced cost of packaging, storage and transportation. The application of evaporation in food industries includes concentration of aqueous solution of sugar, milk, sodium chloride, fruit juices etc. In all the cases, the concentrated solutions are the required products and removed water is discarded.

But, in processes like desalination, water is the required product. Sometimes, evaporation is used as an intermediate operation mainly before drying, particularly in spray drying. This helps increased economy of dryer, as the removal of huge amount of water in processes like spray drying is a costly affair.

Frequently in the food industry a raw material or a processed food contains more water than is required in the final product. When the foodstuff is a liquid, the easiest method of removing the water, in general, is to apply heat to evaporate it. Evaporation is thus a process that is often used by the food technologist.

The basic factors that affect the rate of evaporation are the:

- rate at which heat can be transferred to the liquid,
- quantity of heat required to evaporate each kg of water,
- maximum allowable temperature of the liquid,
- pressure at which the evaporation takes place,
- changes that may occur in the foodstuff during the course of the evaporation process.

Considered as a piece of process plant, the evaporator has two principal functions, to exchange heat and to separate the vapour that is formed from the liquid.

Important practical considerations in evaporators are the:

- maximum allowable temperature, which may be substantially below 100°C.
- promotion of circulation of the liquid across the heat transfer surfaces, to attain reasonably high heat transfer coefficients and to prevent any local overheating,
- viscosity of the fluid which will often increase substantially as the concentration of the dissolved materials increases,
- tendency to foam which makes separation of liquid and vapour difficult.

WORKING PRINCIPLE

Evaporation is a process in which simultaneous heat and mass transfer process occur resulting in the separation of vapour from a solution. Evaporation occurs where molecules obtain enough energy to escape as vapour from a solution. The rate of escape of the surface molecules depends primarily upon the temperature of the liquid, the temperature of the surroundings, the pressure above the liquid, surface area and rate of heat propagation to the product. It is thus a thermal separation or thermal concentration process that begins with a liquid product and ends up with a more concentrated, but still liquid and still pumpable concentrate as the main product from the process.

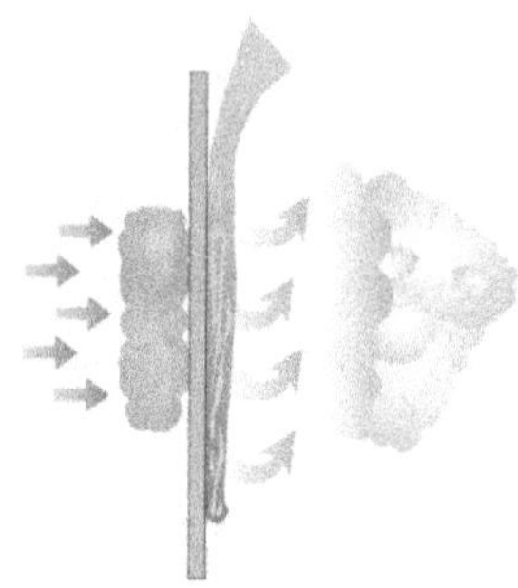

Fig. 1. Working Principle

REASON FOR INSTALLATION OF EVAPORATORS

There are several reasons for concentrating food liquids:

- To reduces costs for storage and transportation.
- To induce crystallization.
- It may be used as an economical preparatory step for subsequent dehydration as in the cases of spray drying, vacuum drying, freeze-dried, crystallization and mixing.

To reduce water activity to increase microbiological and chemical stability.

Recover valuable substances and by-products from waste streams.

Components of Evaporator System

- **A heat source:** It is generally comprised of a heat exchanger for the evaporation of water.
- **An evaporation vessel:** It drives out water from the product as vapor.
- **A vapor separation vessel:** It separates vapor and product.
- **A vacuum system:** It draws water vapor out of the separation vessel. This vacuum also reduces pressure in the evaporation vessel, which reduces the boiling point.

Evaporators may operate singly, or several may operate in series. Each one is referred to as an effect and in multiple-effect systems, the product output from one effect is the feed for the following effect. Similarly, higher-temperature vapor driven out of the product in one effect is used to heat a lower-temperature product in another. Efficiency is gained by using multiple-effect systems.

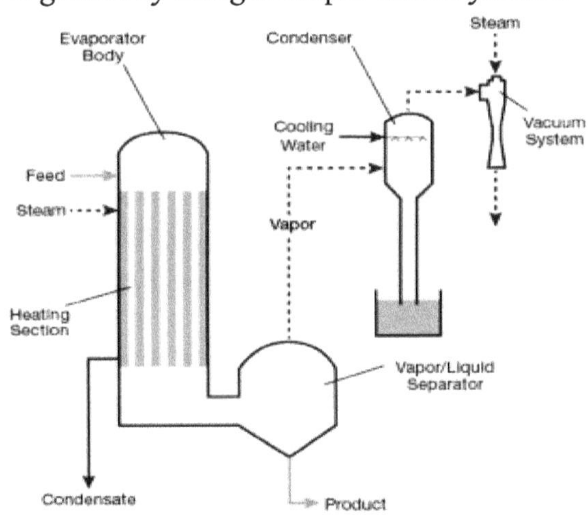

Fig. 2. Components of Evaporator System

TYPES OF EVAPORATORS USED

Circulation or Vertical Tube Evaporators

Tubes carrying the steam internally are placed vertically in the bottom of the cylindrical evaporator, chamber. It is popularly known as Calandria evaporator. As a single pass through the tube heat exchanger is usually not sufficient to achieve the required degree of concentration, the product is returned through the circulation tube. The tubes are arranged vertically inside which feed solution passes. The vapour condensed outside the tube bundles (Fig. 3b). The lighter liquid because of boiling and decreased density moves up and heavier liquid moves down through a central larger pipe. Heat transfer increases due to natural circulation of liquid. Viscous liquids cannot climb on the tubes, so viscous products are not evaporated in vertical short tube evaporators. The length of tubes varies from 0.5 to 2.0 m with diameter of tube ranging between 25 to 75 mm. The central tube occupies 25 to 40% of the total tube area. This evaporator is generally use for evaporation of sugar and salt solutions.

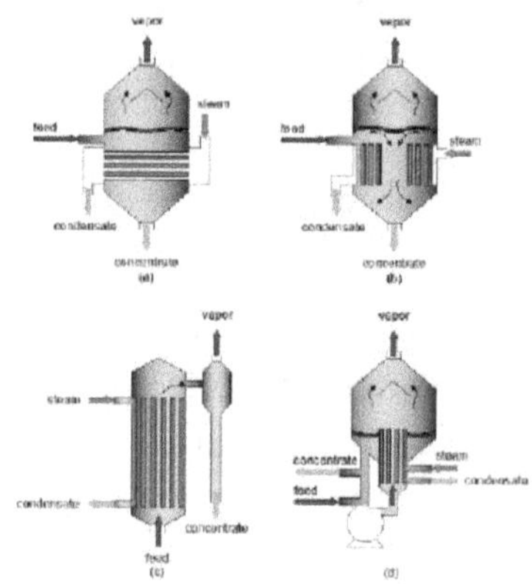

Fig. 3. Evaporator types. (a) Horizontal-tube, (b) Vertical tube, (c) Long-tube vertical and (d) Forced circulation

Horizontal Tube Evaporators

In certain conditions, if, for, example, there is limited vertical space for installation, the tubular heat exchanger can be used as a horizontal evaporator. To ensure that both the top and the bottom tubes are covered with liquid, the evaporator must be operated in a flooded condition, i.e., the feed inlet must be positioned above the uppermost tube. Steam enters a chest on one end of the tubes, moves through the tubes and the condensate is removed from the chest at the opposite end while the vapour is removed from the top of the cylindrical chamber.

Plate Evaporators

The plate evaporator is characterized by a large heat exchange surface occupying a relatively small space. Like the plate heat exchanger, it is constructed from profiled plates with the condensing steam used as a heating medium and the evaporating product passing between alternate pairs. The capacity of a plate evaporator can be altered by changing the number of plates.

Forced Circulation Evaporators

When highly viscous products are to be evaporated, the forces which normally move the liquid along are not sufficient to transport the product satisfactorily. Centrifugal or positive pumps are used for the forced circulation of viscous liquids. The operation is only economical if the product has been pre-concentrated on a simpler evaporator.

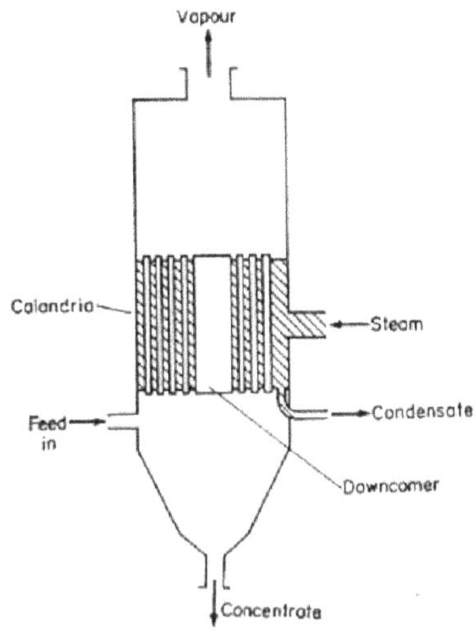

Fig. 4. Evaporator

Falling Film Evaporators

The heat exchange surface in a falling film evaporator consists of a bundle of tubes down which the product flows. Depending on the design, the tubes are 4-10 m long and have a diameter of 25-80 mm. The falling film evaporator has become of great importance to the dairy industry. The relatively short contact times and the possibility of accurate temperature control ensure mild evaporating conditions.

HEAT TRANSFER CALCULATION IN SINGLE EFFECT EVAPORATOR

The typical evaporator is made up of three functional sections: the heat exchanger, the evaporating section, where the liquid boils and evaporates, and the separator in which the vapour leaves the liquid and passes off to the condenser or to other equipment. In many evaporators, all three sections are contained in a single vertical cylinder. In the centre of the cylinder there is a steam heating section, with pipes passing through it in which the evaporating liquors rise. At the top of the cylinder, there are baffles, which allow the vapours to escape but check liquid droplets that may accompany the vapours from the liquid surface.

A diagram of this type of evaporator, which may be called the conventional evaporator, is given in Fig. 4. In the heat exchanger section, called a calandria in this type of evaporator, steam condenses in the space surrounding the tubes and the liquid being evaporated boils on the inside of the tubes and in the space above the upper tube plate.

The resistance to heat flow is imposed by the steam and liquid film coefficients and by the material of the tube walls. The circulation of the liquid greatly affects evaporation rates, but circulation rates and patterns are very difficult to predict in any detail. Values of overall heat transfer coefficients reported for evaporators are of the order of 1800-5000 Jm-2 s -1°C -1 for the evaporation of distilled water in a vertical tube evaporator with heat supplied by condensing steam.

However, with dissolved solids in increasing quantities as evaporation proceeds leading to increased viscosity and poorer circulation, heat transfer coefficients in practice may be much lower than this. As evaporation proceeds, the remaining liquors become more concentrated and because of this the boiling temperature rises. The rise in the temperature of boiling reduces the available temperature drop, assuming no change in the heat source. And so the total rate of heat transfer will drop accordingly. Also, with increasing solute concentration the viscosity of the liquid will increase, often quite substantially, and this affects circulation and the heat transfer coefficients leading again to lower rates of boiling.

Yet another complication is that overall heat transfer coefficients have been found to vary with the actual temperature drop, so that the design of an evaporator on theoretical grounds is inevitably subject to wide margins of uncertainty.

Heat transfer in evaporators is governed by the equations for heat transfer to boiling liquids and by the convection and conduction equations. The heat must be provided from a source at a suitable temperature and this is condensing steam in most cases. The steam comes either directly from a boiler or from a previous stage of evaporation in another evaporator. Major objections to other forms of heating, such as direct firing or electric resistance heaters, arise because of the need to avoid local high temperatures and because of the high costs in the case of electricity. In some cases, the temperatures of condensing steam may be too high for the product and hot water may be used. Low-pressure steam can also be used but the large volumes create design problems.

Calculations on evaporators can be carried out combining mass and energy balances with the principles of heat transfer.

EXAMPLE 8.1. Single effect evaporator: steam usage and heat transfer surface

A single effect evaporator is required to concentrate a solution from 10% solids to 30% solids at the rate of 250kg of feed per hour. If the pressure in the evaporator is 77kPa absolute, and if steam is available at 200kPa gauge, calculate the quantity of steam required per hour and the area of heat transfer surface if the overall heat transfer coefficient is 1700Jm^{-2} s^{-1}°C^{-1}. Assume that the temperature of the feed is 18°C and that the boiling point of the solution under the pressure of 77kPa absolute is 91°C. Assume, also, that the specific heat of the solution is the same as for water, that is 4.186×10^3 Jkg^{-1}°C^{-1}, and the latent heat of vaporization of the solution is the same as that for water under the same conditions.

From steam tables, condensing temperature of steam at 200kPa gauge (300kPa abs.) is 134°C and latent heat 2164kJkg^{-1}; the condensing temperature at 77kPa (abs.) is 91°C and latent heat is 2281 kJ kg^{-1}.

Mass balance (kgh^{-1})

	Solids	Liquids	Total
Feed	25	225	250
Product	25	58	83
Evaporation			167

Heat balance

Heat available per kg of steam

= latent heat + sensible heat in cooling to 91°C

$= 2.164 \times 10^6 + 4.186 \times 10^3 (134 - 91)$

$= 2.164 \times 10^6 + 1.8 \times 10^5$

$= 2.34 \times 10^6$ Jkg^{-1}

Heat required by the solution

= latent heat + sensible heat in heating from 18°C to 91°C

$= 2281 \times 10^3 \times 167 + 250 \times 4.186 \times 10^3 \times (91 - 18)$

$= 3.81 \times 10^8 + 7.6 \times 10^7$

$= 4.57 \times 10^8$ J

Now, heat from steam = heat required by the solution,

Therefore quantity of steam required per hour

$= (4.57 \times 10^8)/(2.34 \times 10^6)$

= 195 kgh^{-1} Quantity of steam/kg of water evaporated

= 195/167

= 1.17 kg steam/kg water.

Heat transfer area

Temperature of condensing steam = 134°C.

Temperature difference across the evaporator = (134 - 91) = 43°C.

Writing the heat transfer equation for q in Js^{-1},

$q = UA\ DT$

$(4.57 \times 10^8)/3600 = 1700 \times A \times 43$

$A = 1.74 m^2$

Area of heat transfer surface = **1.74m²**

It has been assumed that the sensible heat in the condensed steam (cooling from 134°C to 91°C) is recovered, and this might in practice be done in a feed heater. If it is not recovered usefully, then the sensible heat component, about 8%, should be omitted from the heat available, and the remainder of the working adjusted accordingly.

Mass balance on a single-effect evaporator

$M_F = M_P + M_V$ (1)

$M_S = M_C$ (2)

Drawing a control volume around the evaporator allows us to perform an energy balance of the system, in which h is the enthalpy of the respective streams:

$M_F h_F + M_S h_S = M_P h_P + M_V h_V + M_C h_C$ (3)

The energy balance assumes that no work is performed on the system ($W = 0$) and heat loss is negligible ($Q = 0$). In real applications, heat loss is typically estimated to be %2 of the vapor energy input to each effect (2). Latent heat from saturated steam drives the evaporation of water, and depending on the initial feed temperature, may also be used to raise the temperature to boiling.

The product concentration, where x_F and x_P are the mass fractions of the respective streams, can be calculated from:

$M_F x_F + M_P x_P$ (4)

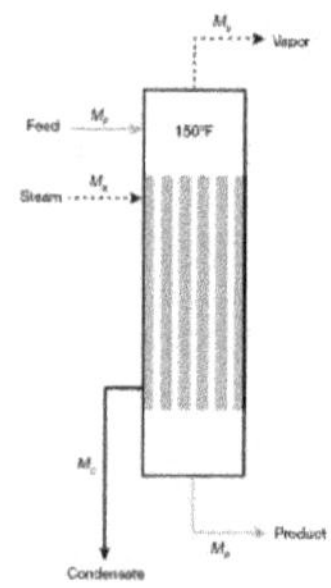

Fig. 5. An energy balance for this evaporator assumes no work is performed on the system and heat loss is zero. To get a closer approximation of the evaporator rating, however, the heat loss is estimated to be 2% of the vapor energy input to each effect.

FEATURES OF THE EVAPORATION SYSTEM IN FOOD AND BEVERAGE INDUSTRY

- The feed liquid of the falling film evaporator is added from the top of the evaporator, and falls in a film shape along the tube wall under the action of gravity, and during this process, it evaporates and thickens, and a concentrated liquid is obtained at the bottom. Falling film evaporator can evaporate materials with higher concentration and higher viscosity.

- Since the flows as a film in the single-pass evaporator, the heat transfer coefficient is high.
- Short residence time, not easy to cause material deterioration, suitable for processing heat-sensitive materials.
- The liquid hold-up is small, and the falling film evaporator can operate quickly according to changes in energy supply, vacuum, feed volume, concentration, etc.
- Since the process fluid only flows under the action of gravity, instead of being driven by high temperature difference, low temperature difference evaporation can be used.
- Falling film evaporator is suitable for evaporating and concentrating foaming materials, and the material and liquid integration process does not cause much impact and avoids the formation of foam.

STEAM ECONOMY

Another way to measure and compare the performance of evaporators is to consider the amount of water evaporated per amount of steam consumed. This metric is called the steam economy (SE) and is commonly expressed as the ratio of water evaporated to steam consumed.

In theory, the evaporation of 1 kg of water requires slightly more than 1 kg of steam, since the latent heat of vaporization decreases as pressure increases. As the steam's latent heat is transferred to the feed, some of the heat raises the initial feed temperature to the boiling point and the remaining latent heat evaporates water. Thus, the steam economy will always be less than one. Depending on the initial feed temperature and system losses, the steam economy of a single-effect evaporator typically is 0.75–0.95 kg water/kg steam.

Improving steam economy

Consider the example of the single-effect evaporator (Fig. 5). For heat transfer to occur, the temperature of the saturated steam must be higher than the boiling temperature of the feed. If the latent heat of the water vapor leaving the evaporator can be reused, the steam economy can be increased. But, because heat transfer requires a temperature gradient, the latent heat of the water vapor can only be transferred to a lower-temperature fluid.

The two primary means of improving the steam economy of an evaporator both involve manipulating temperature via pressure change.

Multiple-effect evaporators

If the water vapor from the first effect of a multiple-effect evaporator can be introduced into the steam chest of a second effect operating at a lower boiling point, the latent heat in the water vapor can be reused. Lowering the vapor pressure of the second effect relative to the first effect lowers the boiling point of the second effect. This arrangement of reusing vapor latent heat is called multiple-effect evaporation.

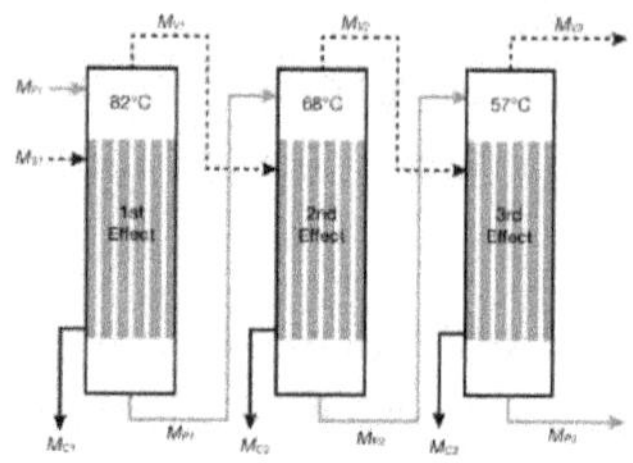

Fig. 6. The steam economy of an evaporator can be improved by adding more effects. The first effect has the highest boiling temperature, while the last effect has the lowest. Each successive effect uses the vapor from the previous effect as a heat source.

Figure 6 shows the mass flows in a three-effect evaporator. The vapor from the first effect (M_{V1}) becomes the heating source for the second effect (M_{S2}). This is possible because the second effect's boiling temperature has been lowered to 68°C (at 29 kPa), which is less than the saturated temperature of the first effect, 82°C (at 51.8 kPa). The latent heat is reused again as M_{V2} becomes the heat source of the third effect, where the boiling point temperature has been lowered to °57C (at 17.5 kPa). The steam economy of multiple-effect evaporators can be estimated by:

$$SE = eN \tag{5}$$

Evaporators are generally rated based on their evaporation rate — the amount of water they evaporate per hour (*e.g.*, kg/hr, ton/hr). The mass balances for the single-effect evaporator in Figure 5, which include the mass flowrates for feed (M_F), product (M_P), vapor (M_V), steam (M_S), and condensate (M_C), are:

Recompression

The second method for reusing the latent heat of vapor is to increase its condensing temperature by compressing it. This increases the vapor's condensing pressure and temperature and creates the temperature gradient necessary for latent heat reuse within the evaporator system.

Thermal vapor recompression (TVR)

Thermal vapor recompression (TVR) uses a steam ejector to compress the vapor leaving the evaporator vessel. In this application, the steam ejector is often called a thermocompressor; its primary purpose is to compress vapor, but its secondary benefit is creating vacuum within the evaporator vessel. As a thermocompressor, the ejector uses high-pressure steam (0.8–2.1 MPa) as the motive fluid to draw in the low-pressure vapor from the evaporator and produce a combined vapor of suitable pressure and temperature for evaporator heating.

Figure 7 illustrates one possible configuration of a three-effect evaporator with TVR. In this example, the thermocompressor raises the saturation temperature of the vapor from 71°C to 88°C so it can be reused to heat the second effect operating at 80°C.

Steam ejectors are better suited to handling large volumes of low-density vapor than mechanical vacuum pumps and are easier and cheaper to maintain. As a rule of thumb, TVR improves an evaporator's steam economy by an amount equivalent to the addition of another effect, but at a much lower cost.

Mechanical vapor recompression (MVR)

Mechanical vapor recompression uses a mechanical compressor or fan to compress low-pressure vapor leaving an evaporator to a higher pressure and temperature for reuse as the evaporation heat source. The internal reuse of the vapor also eliminates the need for an external condenser, which most evaporator systems need to condense the vapor leaving the last effect. Because compression ratios of 1.2–1.4 are sufficient to produce a condensing temperature increase of 6–12°C, centrifugal fans are used for their large volumetric capacity, economy of operation, and relatively low cost. Steam turbines are typically used to drive mechanical fans at installations where the cost of steam is relatively low and low-pressure steam can be used for process heating. Otherwise, electric motors with variable-frequency drives (VFDs) are used to drive mechanical fans.

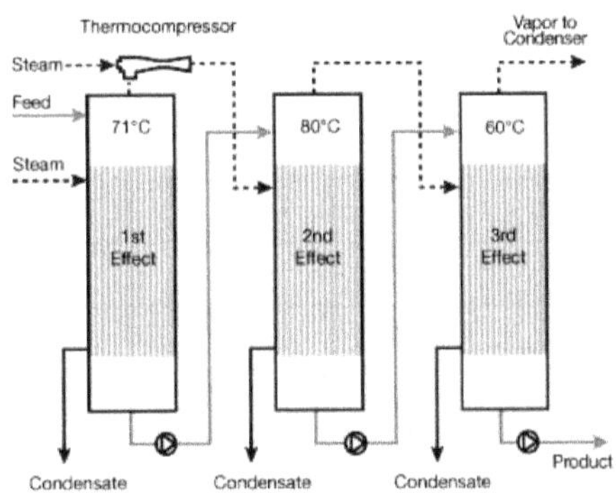

Fig. 7. Thermal vapor recompression (TVR) can be applied to an evaporator system to improve the steam economy. A thermocompressor raises the saturation temperature of the vapor exiting an effect to be reused to heat the next effect.

Figure 8 shows a single-effect, vertical long-tube evaporator with MVR. An MVR evaporator requires steam to bring the system up to boiling temperature during startup, but requires very little steam once operating at steady state, with only the work of the compressor adding energy to the system.

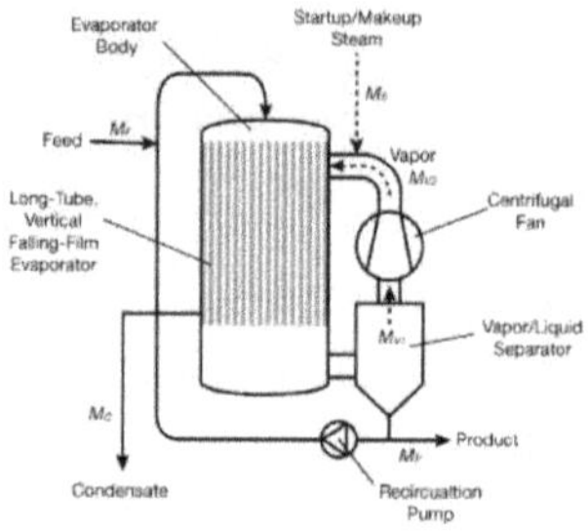

Figure 8. Mechanical vapor recompression (MVR) improves steam economy by compressing low-pressure steam exiting the evaporator and recycling it as the heat source.

COMPARING STEAM ECONOMIES

Table 2 compares the steam economies of multiple-effect, TVR, and MVR evaporators *(5)*. Steam economy increases steadily in multiple-effect evaporators as the number of effects increases. Adding a thermocompressor to a three-effect system can approximately double the economy, but that pales in comparison to the potential steam economy of an evaporator with MVR — which is 5–10 times that of the three-effect evaporator.

Feed that enters the first effect at a temperature above the boiling point first flashes some vapor before heat transfer within the evaporator even begins. Thus, the steam economy will be higher than if the initial feed temperature was below the boiling point, which would require heat transfer to the liquid before evaporation can occur.

Table 2. Comparison of evaporator steam economies.

Type of Evaporator	Steam Economy, kg water/kg steam
Single-effect	0.90–0.98
Double-effect	1.7–2.0
Triple-effect	2.4–2.8
Six-effect	4.6–4.9
Triple-effect with TVR	4–8
Single-effect MVR	10–30

Note that the steam economy of an evaporator is highly dependent on the initial temperature of the feed as it enters the evaporator system and the properties of the liquid being concentrated.

Boiling point rise, also called boiling point elevation (BPE), is the difference between a solution's boiling point and the boiling point of pure water at the same pressure. Some solutions, especially as they become more concentrated, may require a large temperature gradient (16–27°C) to sufficiently transfer enough heat to reach boiling conditions. High temperature gradients limit the number of effects that can be applied, thereby limiting the potential steam economy increase of a system. Also, single-stage thermal and mechanical compressors are economically limited to compression ratios of about two, which produces a temperature increase of 5–20°C. Thus, the potential to increase the steam economy by adding effects, TVR, and MVR may be constrained by the thermal and rheological properties of the liquid being concentrated.

EVAPORATORS IN THE FOOD INDUSTRY

The economics of a process dictate the evaporator type and configuration best suited to a particular application. The three most common types of evaporators used in food concentration are long-tube vertical evaporators, forced-circulation evaporators, and plate-type evaporators.

Falling-film evaporators are constructed around a vertical bundle of large-diameter long tubes, with a liquid distributor on top and a vapor/liquid separator section below. Liquid is fed into the top of the heat exchanger tubes and flows as a film down the inner walls. Heat transferred from the condensing steam through the tube walls creates boiling conditions, and the vapor generated escapes into the inner void. As the film flows downward and water evaporates, the film becomes thinner and the liquid becomes more viscous. Since the film flows by gravity down the long tube, residence times in the heat exchanger section are short (0.5–2 min), which is beneficial to heat-sensitive foods, such as fruit juices and milk.

Forced-circulation evaporators use a centrifugal pump to recirculate the liquid through the heat exchanger (*i.e.,* steam chest), where the liquid is heated by steam. As liquid is pumped upward through the heat exchanger, the static head pressure suppresses vapor formation until the heated liquid enters the vapor/liquid separator, where a portion flashes to vapor. The centrifugal pump allows the residence time in the heat exchanger to be controlled based on the heat-sensitivity of the liquid being concentrated. This feature is important when concentrating pulpy, viscous liquids, such as tomato juice and mango juice.

Plate-type evaporators operate similarly to long-tube vertical evaporators, but with the film flowing on plates rather than tubes. Their advantages over tubular-type evaporators include compact size that allows installation within buildings and their ability to be disassembled for easier cleaning.

Conclusion

The primary means of increasing the steam economy of an evaporator is to reuse the latent heat of the water vapor. A multiple-effect evaporator uses the water vapor from one effect as the heating medium for the next effect, which operates at a lower boiling point. The latent heat in water vapor can also be reused by thermally or mechanically compressing the vapor to a higher pressure and temperature. These three methods of improving efficiency have their limits, however, which are specific to the physical properties of the liquid feed being evaporated.

CHAPTER

PSYCHROMETRY 10

INTRODUCTION

Psychrometry is the science of studying thermodynamic properties of moist air and the use of these to analyze humid air conditions and processes. The term derives from the Greek *psuchron* meaning "cold" and metron meaning "means of measurement"

The capacity of air for moisture removal depends on its humidity and its temperature. The study of relationships between air and its associated water is called psychrometry.

Humidity (Y) is the measure of the water content of the air. The **absolute humidity**, sometimes called the humidity ratio, is the mass of water vapour per unit mass of dry air and the units are therefore kg kg^{-1}, and this will be subsequently termed just the humidity.

Air is said to be saturated with water vapour at a given temperature and pressure if its humidity is a maximum under these conditions. If further water is added to saturated air, it must appear as liquid water in the form of a mist or droplets. Under conditions of saturation, the partial pressure of the water vapour in the air is equal to the saturation vapour pressure of water at that temperature.

The total pressure of a gaseous mixture, such as air and water vapour, is made up from the sum of the pressures of its constituents, which are called the partial pressures.

Each partial pressure arises from the molecular concentration of the constituent and the pressure exerted is that which corresponds to the number of moles present and the total volume of the system. The partial pressures are added to obtain the total pressure.

PSYCHROMETRIC PROPERTIES

Dry-Bulb Temperature - Td_b

Dry-Bulb Temperature, usually referred to as **the** air temperature, is the air property that is most commonly used. People referring to air temperature normally referring to Dry Bulb Temperature.

Dry-Bulb Temperature can be measured by using a normal thermometer. With Dry-Bulb Temperature the sensible heat content in the air can be determined along the bottom axis of the psychrometric chart. The vertical lines extending upward from this axis are constant-temperature lines.

Wet-Bulb Temperature - T_{wb}

Wet-Bulb Temperature is associated with the moisture content of the air.

Wet Bulb Temperature can be measured with a thermometer that has the bulb covered with a water-moistened bandage with air flowing over the thermometer.

Wet-Bulb Temperatures are always lower than dry bulb temperatures with less than 100% relative humidity in the air. The Wet-Bulb Temperature and the Dry-Bulb Temperature will be identical with 100% relative humidity in the air (the air is at the saturation line).

On the chart, the Wet-Bulb Temperature lines slopes a little upward to the left, and the temperature is read at the saturation line.

Relative Humidity - *RH*

Relative Humidity is the ratio of the mass of water vapor contents in the humid air - m_w - to the maximum possible mass of vapor - m_{wmax} - at the actual pressure and temperature. Relative humidity can also be expressed as the ratio of water vapor pressure - p_w, to the water vapor pressure of saturated air at the same temperature - p_{ws}.

Relative humidity is expressed as a percentage.

The moisture-holding capacity of air increases with air temperature. In practice the relative humidity will indicate the moisture level of the air compared to the maximum moisture-holding capacity of air at saturation.

Note: The moisture holding capacity of air increases dramatically with temperature! - important for drying processes.

The relative humidity lines in the psychrometric chart are curved lines that move upwards to the right. The line representing saturated air where the relative humidity - *RH* is %100 - is the uppermost curved line in the chart.

Dew Point Temperature - T_{dp}

Dew Point is the temperature at which water vapor starts to condense in the air - the temperature at which air becomes completely saturated. Above this temperature the moisture stays in the air.

The Dew Point Temperature can be read in the psychrometric charts by following the horizontal line from the state-point to the saturation line. The Dew Point Temperature is represented along the 100% relative humidity line.

Specific Volume of Humid Air - v

Specific Volume represents the space occupied by a unit weight of dry air (ft^3/lb, m^3/kg). Specific volume is indicated along the bottom axis of the psychrometric chart with the constant-volume lines slanting upward to the left.

Moisture Content and Humidity Ratio - x

Moisture Content and Humidity Ratio is the amount of water vapor by weight in dry air.

The moisture content of air is expressed as the weight of water vapor per unit weight of dry air (lb_{H2O}/lb_{air}, kg_{H2O}/kg_{air}). Humidity ratio is indicated along the right-hand axis in psychrometric charts.

Enthalpy - h

Enthalpy is the measure of the total thermal energy in air.

Energy content is expressed as energy per unit weight of air (Btu/lb_{air}, J/kg_{air}).

Enthalpy in the psychrometric chart can read from where the appropriate wet-bulb line crosses the diagonal scale above the saturation curve. Air with the same amount of energy may either be drier hotter air (higher sensible heat) or cooler moister air (higher latent heat).

EXAMPLE 1. Partial pressure of water vapour

If the total pressure of moist air is 100 kPa (approximately atmospheric) and the humidity is measured as 0.03 kg kg^{-1}, calculate the partial pressure of the water vapour.

The molecular weight of air is 29, and of water 18

So the mole fraction of water = (0.03/18)/(1.00/29 + 0.03/18)

= 0.0017/(0.034 + 0.0017)

= 0.048

Therefore the water vapour pressure

= 0.048 x 100 kPa

= 4.8 kPa.

The **relative humidity (RH)** is defined as the ratio of the partial pressure of the water vapour in the air (p) to the partial pressure of saturated water vapour at the same temperature (p_s). Therefore:

$RH = p/p_s$

and is often expressed as a percentage $= 100\, p/p_s$

EXAMPLE 2. Relative humidity

If the air in Example 7.5 is at 60°C, calculate the relative humidity.

From steam tables, the saturation pressure of water vapour at 60°C is 19.9 kPa.

Therefore the relative humidity $= p/p_s$

$= 4.8/19.9$

$= 0.24$

or **24%.**

If such air were cooled, then when the percentage relative humidity reached 100% the air would be saturated and this would occur at that temperature at which $p = p_s = 4.8$ kPa.

Interpolating from the steam tables, or reading from the water vapour pressure/temperature graph, this occurs at a temperature of 32 C and this temperature is called the **dew point** of the air at this particular moisture content. If cooled below the dew point, the air can no longer retain this quantity of water as vapour and so water must condense out as droplets or a fog, and the water remaining as vapour in the air will be that corresponding to saturation at the temperature reached.

The humidity Y can therefore be related to the partial pressure p_w of the water vapour in air by the equation:

$Y = 18\, p_w / [29(P - p_w)]$ (1)

where P is the total pressure. In circumstances where p_w is small compared with P, and this is approximately the case in air/water systems at room temperatures, $Y \approx 18p_w/29P$.

Corresponding to the specific heat, c_p, of gases, is the **humid heat,** c_s of moist air. It is used in the same way as a specific heat, the enthalpy change being the mass of dry air multiplied by the temperature difference and by the humid heat. The units are J kg^{-1} C^{-1} and the numerical values can be read off a psychrometric chart. It differs from specific heat at constant pressure in that it is based only on the mass of the dry air. The specific heat of the water it contains is effectively incorporated into the humid heat which therefore is numerically a little larger than the specific heat to allow for this.

Wet-bulb Temperatures

A useful concept in psychrometry is the wet-bulb temperature, as compared with the ordinary temperature, which is called the **dry-bulb temperature**. The wet-bulb temperature is the temperature reached by a water surface, such as that registered by a thermometer bulb surrounded by a wet wick, when exposed to air passing over it.

The wick and therefore the thermometer bulb decreases in temperature below the dry-bulb temperature until the rate of heat transfer from the warmer air to the wick is just equal to the rate of heat transfer needed to provide for the evaporation of water from the wick into the air stream.

Equating these two rates of heat transfer gives

$$h_c A(T_a - T_s) = lk'_g A(Y_s - Y_a)$$

$$h_c(T_a - T_s) = lk'_g(Y_s - Y_a)$$

where a and s denote actual and saturation temperatures and humidities; h_c is the heat-transfer coefficient and k'_g the mass transfer coefficient from the air to the wick surface; λ is the latent heat of evaporation of water.

As the relative humidity of the air decreases, so the difference between the wet-bulb and dry-bulb temperatures, called the wet-bulb depression, increases and a line connecting wet-bulb temperature and relative humidity can be plotted on a suitable chart.

When the air is saturated, the wet-bulb temperature and the dry-bulb temperature are identical.

Therefore if $(T_a - T_s)$ is plotted against $(Y_s - Y_a)$ remembering that the point (T_s, Y_s) must correspond to a dew-point condition, we then have a wet-bulb straight line on a temperature/humidity chart sloping down from the point (T_s, Y_s) with a slope of:

$$-(\lambda k'_g / h_c)$$

A further important concept is that of the adiabatic saturation condition. This is the situation reached by a stream of water, in contact with the humid air, and where the temperature of the air and the humidity follow down a line called the adiabatic saturation line. Both ultimately reach a temperature at which the heat lost by the humid air on cooling is equal to the heat of evaporation of the water leaving the stream of water by evaporation.

Under this condition with no heat exchange to the surroundings, the total enthalpy change (kJ $kg^{-1 \text{ dry air}}$)

$$\Delta H = c_s(T_a - T_s) + \lambda(Y_s - Y_a) = 0$$

$c_s = - h_c/k''_g$

where c_s is the humid heat of the air.

Now it just so happens, for the water/air system at normal working temperatures and pressures that for practical purposes the numerical magnitude of the ratio:

$c_{s=-}h_c/k''_g$ (known as the **Lewis number**) ≈ 1 (2)

This has a useful practical consequence. The wet bulb line and the adiabatic saturation line coincide when the Lewis number = 1.

It is now time to examine the chart we have spoken about. It is called a psychrometric chart.

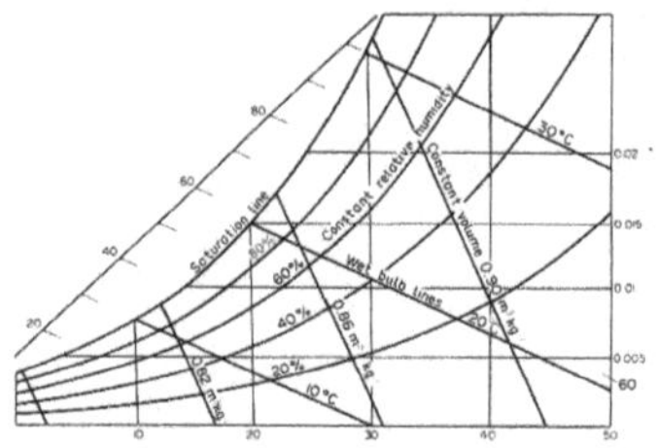

Fig. 1 Psychrometric chart

Psychrometric Charts

In the preceding discussion, we have been considering a chart of humidity against temperature, and such a chart is given in skeleton form on Fig. 1.

The two main axes are temperature (dry bulb) and humidity (humidity ratio). The saturation curve (T_s , Y_s). is plotted on this dividing the whole area into an unsaturated and a two-phase region. Taking a point on the saturation curve (T_s, Y_s) a line can be drawn from this with a slope:

$- (lk'_g/h_c) = (\lambda/c_s)$

running down into the unsaturated region of the chart (that "below" the saturation line). This is the wet bulb or adiabatic cooling line and a net of such lines is shown. Any constant temperature line running between the saturation curve and the zero humidity axis can be divided evenly into fractional humidities which will correspond to fractional relative humidities [for example, a 0.50 ratio of humidities will correspond to a 50% RH because of eqn. (1) if P » p_w].

This discussion is somewhat over-simplified and close inspection of the chart shows that the axes are not exactly rectangular and that the lines of constant dry-bulb temperature are not exactly parallel. The reasons are beyond the scope of the present discussion but can be found in appropriate texts such as Keey. The chart also contains other information whose use will emerge as familiarity grows.

APPLICATIONS OF PSYCHROMETRIC CHART DRYING

Drying is perhaps one of the oldest and simplest methods of food preservation which involves the removal of water from the food. It is a complex operation which involves transient transfer of heat and mass. It depends on the understanding of the relationship between the moisture contained in food and the moisture present in the drying medium.

The heating of air supplies heat for evaporation and carring away the evaporated moisture from the product. Drying is a humidification and air cooling process. The term humidification describes the interface transfer of mass and energy between gas and pure liquid when they are brought into contact. Thus, hygrothermal properties of humid air are required for the design calculations of dryers.

Psychrometric conditions are playing an important role in the drying and storage of food products. To control the drying rate and quality of dried products, proper conditions of drying are necessary. Besides that, specific psychrometric conditions have been identified for optimum storage of many fresh and dried products.

Psychrometric chart is extensively used to analyze product drying and air cooling because it is a simple and quick alternative to more costly methods. Drying with air is a cost effective method to reduce the moisture of a biological material. The addition of a small amount of heat greatly improves the air drying potential. It is useful to understand the changes in its drying potential as moisture is added or removed by the food materials, people, animals, etc. since air at a particular temperature and pressure has a limit to its drying capability 8. Psychrometrics has been applied in foods drying, spray drying and solar drying operation.

DRYING OF FOODS

Drying experiments and computer simulations are needed to determine the capacity of high temperature grain dryer installations such as batch-in-bin and cross-flow dryers. However, these experiments require high cost and consume a lot of time, while computer simulations involve complicated models and computation knowledge. Therefore, new and improved methods are necessary to provide a simple and rapid procedure that can predict the effect of the drying temperature and the airflow rate on dryer capacity.

Researchers have developed a simplified procedure for capacity calculation in batch-in-bin and cross-flow grain dryers by constructing assumptions in the heat and mass transfer equations. Psychometric chart is used in their calculation procedure as the grain drying process in batch-in-bin and cross-flow dryers can be illustrated in the chart. It allows the calculation of the exit moisture content for batch-in-bin dryer by superimposing the curves of the drying constant and equilibrium moisture content on the chart. They reported that for cross-flow grain dryers, the procedure for the capacity calculation is an extension method for the average moisture content calculation in a stationary bed of grain.

Others described the basic behaviors observed when air of high or low temperature and humidity is passed through grain. Properties for air and grain are derived from published data in order to apply this theory to the flow of air through a grain bed. The analysis has been very helpful in predicting the performance of aeration cooling systems. An important consequence is that in aeration the majority of the grain bed is not cooled to the inlet air dry-bulb temperature but comes to a temperature which is also dependent on the air and grain moisture contents. Since the moisture front closely follows a constant air enthalpy (and thus wet-bulb temperature) line, they argued that wet-bulb temperature is a more satisfactory criterion than dry-bulb temperature for the control of grain aeration systems.

The moisture sorption isotherm data of macaroni stored in a chamber the relative humidity of which is regulated by an atomizing humidifier were determined at 25°, 35° and 45°C and relative humidity ranging from 10 to 90%. The moisture sorption properties of foods have been shown to be influenced by food composition, processing treatment, temperature and relative humidity. Besides, moisture sorption isotherms were determined for spray dried tomato pulp and dehydrated tomato slices.

SPRAY DRYING

Psychrometry is an interesting and useful tool in the control and optimization of the spray drying process. In practice, psychrometric chart plays a vital role in the determination of psychrometric properties reflecting the quality of the drying air. Some investigators used the psychrometry as a methodology tool for optimizing the spray drying process. They studied the evolution of the psychrometric variables and developed mathematical models which describe the influence of the nature of the product on the drying mechanism. The existence of these models has successfully made the optimization possible.

Scientists pointed out the importance of controlling the outlet air relative humidity in order to control the outlet air temperature. This will eventually optimize the moisture content of skim milk powder produced by the spray drying process. The relative humidity of the outlet air depends on the temperature and absolute humidity. If the temperature and absolute humidity are kept constant, the outlet air relative humidity, moisture content and water activity of powder will not change.

SOLAR DRYING

Dehydration is a preserving method for the farm produce which involves the removal of moisture to prevent the growth of moulds, bacteria and insects that are the basis of product spoilage. However, dehydration of vegetables and other food crops by the traditional method of open air sun drying is not satisfactory because the products become infested with bacteria and insects and deteriorate rapidly at the high ambient temperatures and relative humidity.

Due to this reason, solar drying systems have been designed and developed to improve the quality of dried crops such as cassava, okro, pepper and groundnuts. To construct a solar dryer, psychometric analysis on the local weather conditions and thermal processes is necessary.

In his effort to construct a solar drying system based on convective heat flows, researcher used the psychrometric chart to determine the thermal exchanges within the dryer. During the dehydration process of cassava, he used the psychrometric chart to help predicting the temperature changes, moisture and humidity variation, quantity of thermal energy and the specific volume of dry air required for the process. These data was used to provide general information about the drying process in the solar drying system.

In the development of a solar-assisted dryer, researcher used the psychrometric equations to evaluate the energy requirement for the drying of onion. He used the psychrometric equations to compute the average air properties between the five trays in his dryer by considering the dry- and wet-bulb temperatures measured at a distance below the bottom tray and above the top tray.

Researchers used the psychrometric methods to determine the drying energy requirements and pressure drop across the bed of mango in their solar air heater integrated drying chamber for food dehydration.

Others used the psychrometric chart in their effort to analyse the single layer drying process of potato slices via a cyclone type dryer. In the energy and exergy analyses of the drying process of olive mill wastewater using indirect type natural convection solar dryer, researchers analyzed the wet air feature during the drying process by determining the values of the psychrometric variables.

In their research, they considered wet air as a one-phase homogeneous system with two components which is governed by the ideal gas laws for fluid mixtures. Therefore, they applied the equations for the determination of the psychrometric properties by considering wet air as an ideal gas.

CHILLING AND REFRIGERATION

Most of the fresh vegetables are produced for fresh market and it requires proper postharvest control system to maintain their quality and to prolong their postharvest life. A better understanding of pyschrometry, therefore, is needed to improve the postharvest cooling and storage conditions for perishable products. The use of pyschrometric properties, in this purpose, is to analyze process conditions which involve moist air, such as storage of fruits and vegetables and transportation of foods.

CHILLING AND REFRIGERATED STORAGE OF FRUITS AND VEGETABLES

The air temperature, where the freshly harvested vegetables are exposed to, has a very significant effect on the postharvest life by controlling the flesh temperature of vegetables. Proper control of storage temperature is important to maintain a maximum shelf-life and quality. Storage temperature above the optimum will result in unacceptably high respiration rate. Indeed, these commodities are more easily susceptible to ethylene and disease damage. Below the optimum, they are more easily damaged by freezing or chilling.

Besides, humidity ratio also needs to be controlled. During refrigerated storage, moisture removal rate of drying process is reduced by decreasing the difference in vapor pressure (humidity ratio) and surrounded air. Vapor pressure is proportionate to the absolute humidity (humidity ratio). By referring to a psychrometric chart, the total removed moisture can be reduced by cooling the product close to surrounding air temperature and maintaining the surrounding air condition to saturation.

Besides that, condensation problem may happen on the commodities, containers and storage room walls if there is a great temperature fluctuation in the air or in the returned air over the evaporator cooling coil. To solve the condensation problem, the temperature difference between the return air and the coil must be reduced by using sufficiently large coil surface area. This is the best way to maintain high humidity levels but will also increase the cooling system cost.

In addition, according to researchers, an interaction occurs between the refrigerated display cases and space conditioning systems and the store air conditions. The conditioned store air exchanges heat and moisture with the refrigerated cases. It is found that removal of additional moisture can lead to energy cost savings.

Drier store air will reduce the latent load on the refrigeration compressors by reducing the moisture entering the display cases. This will lead to less condensation and frost formation, reductions in defrost cycles, decreases in anti-sweat heater energy requirements and improvements in temperature stability of products.

Outdoor temperature and relative humidity has a significant effect on the energy usage of refrigerated cases. This was studied by expanding that nominal operating point (temperature and relative humidity) into an operating envelope on the psychometric chart. These design envelopes illustrate the tradeoffs that can be made in refrigerated case loads between sensible (temperature) and latent (moisture) load components still be within the 100% design band. It further demonstrates that reductions in loads compared to design conditions can be achieved at lower temperatures and humidity within the refrigerated case design limits.

Fruits and vegetables that have higher respiration rates will result in higher water loss. The rate of water loss is generally determined by the difference in the water pressure between the air in tissues and the air in the surrounding atmosphere. This difference is known as vapor pressure deficit (VPD), which is influenced by the psychometric properties of air (temperature and humidity) and the produce (temperature and equilibrium humidity). For a product, higher water loss will have a higher deficit. Psychrometric chart can be used to assess the effect of changing air conditions on VPD and hence the water loss during storage, pre-cooling and other handling operations related to water vapor transport.

Refrigerated chamber for transportation of foods: Some researchers 23 reported that fruit logistics requires an increasing number of measurements to be performed in refrigerated chambers and during transportation. Psychrometric equations were used for quick determination of changes in the absolute water content of air, allowing estimation of future water loss, and detection of condensation on the product.

They applied psychrometric data from ASABE (American Society of Agricultural and Biological Engineers) in fruit logistics monitoring. In fact, fruits and vegetables are more easily faced to a variety of risks during transportation or storage. The causes of quality deterioration of fresh fruits and vegetables normally are water loss and condensation that reduce their marketability. Therefore, fruit logistics requires various measurements to be performed by the sensors in refrigerated chambers and during transportation. The instrumented sensors are used to measure the temperature and humidity whereby this technology allows on-line monitoring of the perishable goods.

Pyschrometric equations can be used regarding the information provided by the sensors. Indeed, psychrometric equations can be applied if there is a change in the absolute humidity of air. Besides, it allows estimating future water loss and detecting condensation by modeling the evaporation and condensation of water related to product in the refrigerated chamber.

AIR BLAST COOLING

One of the critical processing procedure for the safety and quality of cooked meat is the cooling of cooked meat. A rapid cooling treatment after cooking can minimise or reduce the growth of the surviving organisms 24. Air blast cooling is a rapid conventional method due to forced convection at the surface. Air blast cooling utilises forced convection at the meat surface where the temperature difference between the cold air and meat surface is the driving force for heat transfer.

Air blast freezing involved the removal of heat. Transient convective heat transfer occurs between a fluid medium and the food item in many food processing applications, including air blast freezing. In many food processing applications, including air blast freezing, transient convective heat transfer occurs between a fluid medium and the food item. The cooling at the surface of a product occurs by convection; the cooling inside the product occurs by conduction. Water is immobilized by the structure of the material during the heat transfer process, but it is removed by conduction with variable thermal properties.

According to researchers, in the meat processing industry, air blast chilling has been usually used for chilling, which involves heat conduction within the meat and forced convection and radiation heat transfer between the cold air and the surface of meat joint. Due to the difficulty in calculation, moisture evaporation on the surface of the meat can be significant and is often ignored. The chilling rate depends on the combined heat losses caused by convection, radiation and evaporation. The moisture evaporates from the surface of meat, which can cause weight loss from the meat due to the removal of latent heat of evaporation. Therefore, in such circumstances the evaporation and radiation must be taken into consideration in the simulation methodology to achieve an accurate prediction of temperature and weight loss during the chilling process.

Researchers developed a CFD model to simulate the air flow condition in the air blast chiller with two fans for cooked meat joints and to further predict chilling rate and weight loss of the meat. It is in order to examine the effects of variable operating conditions such as local surface heat transfer coefficient and the fluctuation of inlet temperature on the prediction accuracy.

Researchers also reported that the finite element method has been used to model the three-dimensional transient heat transfer of roasted meat during air blast cooling process for evaluating the cooling performance.

Moisture loss rate was considered in the model to account for the weight loss during air blast cooling. The model includes the variations in physical properties of the roasted meat and in operating conditions during the air blast cooling process. Relative humidity has a significant influence on the weight loss. A higher weight loss is achieved when the relative humidity becomes lower. The air temperature has similar effect as the relative humidity.

The weight loss is greatly lower when the air temperature decreased and also accelerates the decrease of the surface temperature.

Storage of grains

Aeration is a great method of modifying the conditions within the grain store so that desirable properties, such as high viability of seeds, whiteness of rice, baking properties of wheat and low free fatty acid content of oilseeds, can be maintained throughout the storage period.

For grain storage technologists to access the effect of aeration on the temperature and humidity condition within a bulk of stored grain, it is essential to have a good understanding of the properties of moist air. Properties that are useful in grain storage application include the wet-bulb temperature of the air, which is important on determining the conditions in a bulk of aerated grain.

It has been reported that refrigerated aeration is an effective non-chemical method of controlling insect pests in bulk grain without damage to grain quality 34. This must be done under grain conditions that prevent the growth of moulds or other pests.

Grain condition, in terms of temperature and moisture content, comes into equilibrium with the surrounding air, and conditions where insects and moulds are viable can be found from the psychrometric chart. Frozen storage of foods: Temperature fluctuations are unavoidable during frozen storage, particularly in small cold stores. This problem is even more serious in display cabinet, during transport and even in home freezers.

It is widely believed that this phenomenon will lead to unfavorable effects on food products. For unwrapped food, the effect of temperature changes depends on the sources of these changes and it can be estimated by a psychrometric chart.

Water-rich foodstuff such as meat and fish are subjected to moisture gain or lost when they undergo environmental changes that are related with the variation in temperature and humidity. Pham36 showed that the psychrometric chart can graphically represent processes such as cooling and freezing of meat, temperature fluctuation of food products due to air infiltration or heat conduction during storage and transportation by unrefrigerated containers.

During these processes, the moisture gain or lost of the product is clearly characterized by the chart. Pham36 also developed equations to estimate the weight and moisture change of a product due to temperature changes. He reported that the psychrometric chart helps to provide data (such as temperature, enthalpy and humidity of the air) needed for the calculation and enables the direction and maximal possible moisture change to be calculated from a minimum set of data.

The psychrometric chart plays the main role in designing some of food processes where there is heat and mass transfer between foods and surrounding air. Some of these food processes are drying, spray drying, solar drying, chilling and refrigerated storage of fruits and vegetables, refrigerated chamber for transportation of foods, air blast cooling, storage of grain and frozen storage of foods. It could be inferred that the most applications of psychrometric chart were used in drying and also cold storage.

CHAPTER

MASS TRANSFER 11

INTRODUCTION

Drying is defined as the removal of moisture from a product, and in most practical situations the main stage during drying is the internal mass transfer. The mechanisms of water transfer in the product during the drying process can be summarized as follows:

- Water movement due to capillary forces
- Diffusion of liquid due to concentration gradients
- Surface diffusion
- Water vapor diffusion in pores filled with air
- Flow due to pressure gradients
- Flow due to water vaporization–condensation

In the pores of solids with rigid structure, capillary forces are responsible for the retention of water, whereas in solids formed by aggregates of fine powders, the osmotic pressure is responsible for water retention within the solids as well as in the surface. The type of material to be dried is an important factor to consider in all drying processes, since its physical and chemical properties play a significant role during drying due to possible changes that may occur and because of the effect that such changes may have in the removal of water from the product.

A hygroscopic material is one that contains bound water that exerts a vapor pressure lower than the vapor pressure of liquid water at the same temperature. It is expected that products made mainly of carbohydrates will behave in a hygroscopic way, since the hydroxyl groups around the sugar molecules allow formation of hydrogen bonds with water molecules. The interaction between the water molecules and the hydroxyl groups causes solvation or solubilization of sugars. In water soluble proteins, as in most of the globular proteins, the polar amino acids are uniformly distributed in the surface, while the hydrophobic groups are located towards the inside of the molecule. This arrangement allows formation of hydrogen bonds with water, which explains the solubility of this type of proteins.

Dehydration involves the simultaneous transfer of heat, mass and momentum in which heat penetrates into the product and moisture is removed by evaporation into an unsaturated gas phase. Owing to the complexity of the process, no generalized theory currently exists to explain the mechanism of internal moisture movement. Although it is now accepted that in most practical situations of air drying of foods the principal rate-determining step is internal mass transfer, there is no agreement on the mechanism of internal moisture movement. In the case of capillary-porous materials such as fruits and vegetables, interstitial spaces, capillaries and gas-filled cavities exist within the food matrix and water transport takes place via several possible mechanisms acting in various combinations. The possible mechanisms include:

- liquid diffusion caused by concentration gradients,
- liquid transport due to capillary forces,
- vapour diffusion due to shrinkage and partial vapour-pressure gradients (Stefan's law),
- liquid or vapour transport due to the difference in total pressure caused by external pressure and temperature (Poiseuille's law),
- evaporation and condensation effects caused by differences in temperature,
- surface diffusion in liquid layers at the solid interface due to surface concentration gradient,
- liquid transport due to gravity.

Most foods are classified as capillary porous rigid or capillary porous colloids. Therefore, it is often proposed that a combination of capillary flow and vapour diffusion mechanisms should be used to describe internal mass transfer. Water activity, rather than moisture content, influences biological reactions. In the regions of water adsorption on polar sites or when a mono-molecular layer exists, there is little enzyme activity. Enzyme activity begins only above the region of mono-molecular adsorption. When the moisture content of a substrate is reduced below 10 %, microorganisms are no longer active. It is necessary however to reduce the moisture content to below 5 % in order to preserve nutrition and flavour.

WATER ACTIVITY

Water activity and moisture content of the material are very important in food unit operations. During processing and storage, many chemical and physical factors are influenced by the water activity and moisture content level. Chemical changes that are enhanced by water activity include enzymatic reactions, nonenzymatic browning, and microbial activity. In many food products, enzymes are not inactivated during the heating process.

Consequently, enzymatic reactions can take place at even low moisture contents. Water activity also affects the nonenzymatic browning reactions in foods. When water is present, carboxyl and amino compounds are involved as reactants, products, or catalysts in the browning process. Bacterial growth is also affected at fairly high water activity levels. If water activity is maintained at a value below 0.90, most bacteria remain dormant. Most yeasts and molds, however, can grow and multiply at water activity levels as low as 0.80.

Physical changes such as texture and aroma can depend greatly on water activity. Textural changes are most often seen in freeze drying and subsequent storing of foods, particularly meats and fish. The water activity in dried foods can also affect the retention of aroma. Different foods that are stored together will be altered if their individual relative humidities (water activities) are different. Under the action of a driving force created by a difference in water activities changes in the moisture and water activities of these foods will follow their own isotherm curve until an equilibrium water activity is achieved.

Water activity (a_w) is measured as the equilibrium relative humidity (ERH), the percent relative humidity (RH) of an atmosphere in contact with a product at the equilibrium water content. a_w is also ratio of the partial pressure of water in the headspace of a product (P) to the vapour pressure of pure water (P_0) at the same temperature

$$a_w = ERH = P/P_0$$

The relationship between a_w and the rate of deteriorative reactions is very important to design dehydration systems. Reducing a_w below 0.7 prevents microbiological spoilage. However, although microbiological spoilage does not occur at $a_w = 0.7$, prevention of other deteriorative reactions needed to preserve a dried food product successfully reduction of a_w to below 0.3.

MASS TRANSFER

During dehydration, water is vapourized only from the surface. The transfer of water vapour from the wet surface to a stream of moving air is analogous to convection heat transfer, therefore, a mass transfer coefficient is used. Moisture flux is proportional to the driving force which is the difference in vapour pressure on the surface and the vapour pressure of water in air surrounding the surface. At the same time that water is removed from the surface, water diffuses from the interior of a solid towards the surface. The later is a general form of diffusion which is analogous to conduction heat transfer. The differential equations for conduction also apply to diffusion, but mass diffusivity is used in place of thermal diffusivity.

Example 1

A dry food product has been exposed to a 30% RH environment at 15°C for 5 h without a weight change. The moisture content has been measured and is 7.5% (wet basis). The product is moved to a 50% RH environment and a weight increase of 0.1 kg/kg product occurs before equilibrium is reached.

- Determine water activity of the product in the first and second environment.
- Compute moisture contents of the product on dry basis in both environments.

Solution

Equilibrium RH = 30% in first environment.

Product moisture content is 7.5% (wet basis) in first environment.

Weight gained in second environment = 0.1 kg / kg product.

The water activity is equilibrium RH divided by 100. so water activities are 0.3 and 0.5 in first & second environment respectively.

MASS TRANSFER IN DRYING

In heat transfer, heat energy is transferred under the driving force provided by a temperature difference, and the rate of heat transfer is proportional to the potential (temperature) difference and to the properties of the transfer system characterized by the heat-transfer coefficient. In the same way, mass is transferred under the driving force provided by a partial pressure or concentration difference. The rate of mass transfer is proportional to the potential (pressure or concentration) difference and to the properties of the transfer system characterized by a mass-transfer coefficient.Writing these symbolically, analogous to $q = UA\ \mathrm{D}T$,

we have $\mathrm{d}w/\mathrm{d}t = k'_g A\ \mathrm{D}Y$ (1)

where w is the mass being transferred kg s^{-1}, A is the area through which the transfer is taking place, k'_g is the mass-transfer coefficient in this case in units kg m^{-2} s^{-1}, and Y is the humidity difference in kg kg^{-1}.

Unfortunately the application of mass-transfer equation is not as straightforward as heat transfer, one reason being because the movement pattern of moisture changes as drying proceeds. Initially, the mass (moisture) is transferred from the surface of the material and later, to an increasing extent, from deeper within the food to the surface and thence to the air. So the first stage is to determine the relationships between the moist surface and the ambient air and then to consider the diffusion through the food. In studying the surface/air relationships, it is necessary to consider mass and heat transfer simultaneously. Air for drying is usually heated and it is also a major heat-transfer medium. Therefore it is necessary to look carefully into the relationships between air and the moisture it contains.

CHAPTER

FOOD FREEZING 12

INTRODUCTION

Water and foods freeze differently. When water freezes and ice crystals form there is actually a small amount of heat produced. The temperature of the water remains at 32°F as the ice crystals grow, and then a rapid temperature drop occurs as the ice crystals cool. Because foods are a complex mixture of many substances, their freezing point is below the freezing point of water. A slow temperature drop occurs as ice crystals form within the food. The temperature then drops more quickly as the crystals cool inside the food. Each food will freeze differently based on things like the amount of water, sugar, muscle tissue or air in the food.

Rapid freezing improves the quality of the food. The faster food freezes, the smaller the crystals that form. Small crystals do less damage to cell walls. Slow freezing produces large ice crystals that punch through cell membranes. As a result, when foods with large ice crystals thaw, there is more dripping and loss of liquid. Small crystals are unstable and over time grow to form larger crystals. This is especially noticeable in the freeze-thaw cycle in frost-free freezers. During the defrosting period, the food thaws slightly allowing ice crystals to grow and rupture more cells when they refreeze. Opening the freezer door also allows heat to enter the freezer and repeat this freeze thaw cycle. A chest type freezer may be better in reducing heat loss from opening the door because warm air rises.

Freezing is a quick and convenient way to preserve fruits and vegetables at home. You can produce frozen fruits and vegetables of high quality and maximum nutritional value if done correctly. Our directions are based on:

- The chemical and physical reactions that take place during the freezing process.
- The effect of freezing on the tissues of fruits and vegetables.
- Food microbiology.

Chemical changes during freezing

Fresh fruits and vegetables, when harvested, continue to undergo chemical changes that can cause spoilage and deterioration of the product. This is why these products should be frozen as soon after harvest as possible and at their peak degree of ripeness.

Fresh produce contains chemical compounds called enzymes, which cause the loss of color, loss of nutrients, flavor changes, and color changes in frozen fruits and vegetables. These enzymes must be inactivated to prevent such reactions from taking place.

Blanch vegetables to deactivate enzymes

- Enzymes in vegetables are deactivated by the blanching process.
- Blanching is the exposure of the vegetables to boiling water or steam for a brief period of time.
- Then the vegetables must be rapidly cooled in ice water to prevent them from cooking.
- In most cases, blanching is absolutely essential for producing quality frozen vegetables.
- Blanching also helps destroy microorganisms on the surface of vegetables and makes some vegetables, such as broccoli and spinach, more compact.

Add ascorbic acid to fruit to control enzymes

- The major problem associated with enzymes in fruits is browning and loss of vitamin C.
- Because fruits are usually served raw, they're not blanched like vegetables.
- Enzymes in frozen fruit are controlled by using chemical compounds that interfere with deteriorative chemical reactions.
- The most common control chemical is ascorbic acid (vitamin C). Ascorbic acid may be used in its pure form or in commercial mixtures with sugars.
- Soaking the fruit in dilute vinegar solutions or coating the fruit with sugar and lemon juice doesn't prevent browning as effectively as treatment with ascorbic acid.

Limit air during freezing

Air exposure can cause frozen products to develop rancid oxidative flavors. Keep frozen food in airtight containers or use wrapping material that does not permit air to pass into the product. Also, remove as much air as possible from the freezer bag or container to reduce the amount of air in contact with the product.

Changes in texture during freezing

Water makes up over 90 percent of the weight of most fruits and vegetables. Water and other chemicals are held within the fairly rigid cell walls that give structure and texture to the fruit or vegetable. When you freeze fruits and vegetables you actually are freezing the water in the plant cells.

When the water freezes, it expands and the ice crystals cause the cell walls to rupture. So the texture of thawed produce is much softer than when it was raw. This is particularly true of food that is usually eaten raw. For example, when a frozen tomato is thawed, it becomes mushy and watery.

Celery and lettuce are not usually frozen because of this and we suggest that you serve frozen fruits before they have completely thawed. Partially thawed fruit is more appetizing when the effect of freezing on the fruit tissue is less noticeable.

Textural changes due to freezing are not as apparent in products that are cooked before eating because cooking also softens cell walls. These changes are also less noticeable in high starch vegetables, such as peas, corn and lima beans.

Freeze food as quickly as possible

- Avoid mushy fruits and vegetables by freezing produce as quickly as possible.
- In rapid freezing, a large number of small ice crystals are formed. Small ice crystals produce less cell wall rupture than slow freezing, which produces only a few large ice crystals.
- Some home freezer manuals recommend that you set the freezer temperature at the coldest setting several hours before placing foods in the freezer.
- Check your freezer manual to see if it shows the location of the coldest shelves in the freezer. You should place unfrozen products on these shelves.

DON'T OVERLOAD THE FREEZER

All freezer manuals give guidelines for the maximum number of cubic feet of unfrozen food that can be frozen at one time. This is usually 2 to 3 pounds of vegetables to each cubic foot of freezer space per 24 hours. Overloading the freezer with unfrozen products will result in a long, slow freeze and a poor-quality product.

Changes caused by fluctuating temperature

Fluctuating temperatures in the freezer can cause water vapor to form on the surface of the container. This is sometimes found in commercially frozen foods that have been improperly handled.

Store frozen fruits and vegetables at 0 degrees F or lower

To maintain top quality, frozen fruits and vegetables should be stored at zero F or lower. A freezer thermometer can help you determine the actual temperature of your freezer. If your freezer has a number of temperature settings, such as from 1 to 9, check the manual to see what settings are recommended for different uses.

Storing frozen foods at temperatures higher than zero F can shorten the shelf life of frozen foods. Don't attempt to save energy in your home by raising the temperature of frozen food storage above zero F.

FREEZER BURN

- Moisture loss, or ice crystals evaporating from the surface area of a product, produces freezer burn – grainy, brownish spots where the tissues become dry and tough.
- These areas often develop unpleasant flavors.
- Packaging designed specifically for freezing foods will prevent freezer burn.

Effects of microbes in the freezer

- Freezing doesn't destroy the microorganisms that may be present on fruits and vegetables.
- While blanching destroys some microorganisms and many die during freezer storage, sufficient numbers are still present to spoil the product when it thaws.
- Carefully inspect any frozen products that have accidentally thawed by the freezer going off or the freezer door being left open.

NUTRIENT VALUE OF FROZEN FOODS

Freezing, when properly done, can preserve more nutrients than other methods of food preservation. To maintain top nutritional quality in frozen fruits and vegetables it is essential to follow directions for pretreatment of the vegetables, to store the frozen product at zero degrees F and to use it within suggested storage times.

IMPORTANCE OF FREEZING AND REFRIGERATION

It is imperative to understand the importance of freezing and maintaining a cold chain in food production and distribution. It is also desired to know why it is needed and how it works. Let us first understand what spoilage is and how it happens:

Food Spoilage

Any commodity or food product spoils in the following three ways:

Microbial Spoilage

Microbes are part of nature and are present in every environment - water, soil and air. Foods coming from land or water contain microorganisms specific to their environment, however, most of the pathogenic organisms (disease-causing ones to human beings) are terrestrial in nature. Foods of terrestrial origin may bring in contaminations from the farm itself like root vegetables and tubers. Fruit and leafy vegetables may get contamination from the soils in contact, dust, bird droppings, human excreta, through insects found in that environment. Water used in irrigation is also a potent contamination source. Farmworkers and food handlers during post-harvest operations can contribute to contamination if it is carried out under unhygienic conditions.

Seafood immediately after catch is considered free from pathogenic contamination. This is a generalized concept and has several exceptions.

There are various species of salt-loving pathogenic microorganisms who live and multiply in saline environments (*Vibrio paraheamolyticus*). Seafood get contaminated from pathogenic organisms during their handling by fishermen either in the boats or in the landing centers or beaches or from dirty, contaminated contact surfaces, during transport, processing plants, storage facilities, processing workers etc.

Saprophytic microorganisms, responsible for biodegradation of dead animals and plants, are present in all the environments and on the body surface of all the animals and plants. They live as commensals on the body surface without damaging the host and start working immediately after its death.

Bacteria are categorized into three groups. The first group is thermophils. They are heat-loving ones, like *Clostridium botulinum*, *Clostridium tetani*, *Listeria monocytogenes*. They can even withstand boiling water temperatures.

The second group is called mesophils who survive and multiply in room temperatures with optimum to 30 to 40°Celsius. Most of the diseases causing bacteria fall in this category. Multiplication of the mesophilic bacteria is directly proportional to the temperature.

Higher is the temperature, higher will be their metabolism, and a higher rate of their multiplication therefore higher will be the spoilage of the commodity. In average each microbe multiplies in every four minutes. Lowering temperature reduces their metabolism and therefore multiplication and this gets stopped when the temperature of the product reaches to –12°Celsius.

The third group of bacteria called cryophils or psycrophils can withstand and multiply even in sub-zero temperatures. They are however not considered very significant, as they are not known for pathogenesis, although they can render a food inedible after prolonged abuse.

Vegetables carry their contamination from the soil and dust apart from handling, contact surfaces and storages as in the case of seafood. Soil is the mother of all microbes and vegetables are grown directly in the soil. Vegetables may therefore carry much more bacterial load than any other food. Plant products are rich in cellulose and are rich in proteins.

Autolytic Spoilage (enzymatic)

Animals, vegetables and fruits have their own enzymes, which are essential for digestion and other metabolic activities while they are alive. After death, these enzymes start digesting their own tissues and carry out break downs of proteins, carbohydrates and fats and spoil the product. Enzymatic reactions are also like microbial ones and are directly proportional to atmospheric temperatures, thus with a reduction in the temperature enzymatic spoilage slows down and almost stops (approx 90%) at –18°C and 100% at below –60°C.

Oxidative Spoilage

All products either from plant or animal origin contain some percentage of oil and fats. In chemical language, they are called triglycerides (three molecules of fatty acids bonded with one molecule of Glycerol) and partially are unstable in nature. Triglycerides keep breaking in higher temperatures into glycerol and fatty acids, reaching to highest levels at cooking or rather frying temperatures. As they are free from the bond they are called free fatty acids. Free fatty acids are highly reactive with oxygen which they derive from the surrounding air and get converted into peroxides and other compounds which are not in good taste and aroma and sometimes unhealthy too.

Once again these fatty acids are available for the reaction when they are in liquid form. Once they are frozen they take solid form and are less reactive. In fruits and vegetables, there are various other compounds necessary for human health and nutrition. Terpentenoids (most of the aromatic substances), flavonoids (various colors in the fruits and vegetables), chlorophyll & carotenes, who provide special aroma, flavour and colour to a fruit and vegetable also undergo oxidation and render a food unsuitable for human consumption.

Refrigeration

All these three types of putrefactions processes are directly or indirectly related with temperature abuse and can be controlled by keeping the food in a temperature-controlled atmosphere.

The raw material may undergo various handling, pre-processing and processing stages. During these processes product may be subjected to mechanical, chemical and physical stresses that may affect the rate of metabolism and rate of putrefactive processes hence the quality of product. Therefore it is imperative that raw material is handled properly to minimize physical damages, bruises and temperature abuse.

Physical damages and bruises may facilitate entry of microorganisms and temperature abuse will enhance rate of putrefaction. Once a raw material is degraded there is no method to reverse the process and "GARBAGE IN IS GARBAGE OUT". Therefore raw material in all stages i.e. storage, pre-processing and processing should be kept under controlled temperature. Most of the animal origin products keep better at or close to 0°C (0 to 4 for operations ease) whereas plant origin products keep better at around 10°C.

Animal origin raw materials are dead with the death of animal. but most of the plant origin products are however alive even after they are plucked from the plants. They need respiration, transpiration and utilize their stored energy for keeping themselves alive. Therefore their requirements for storage are different than the animal origin products. They may need more windy and more humid environment to facilitate proper respiration and to prevent loss of biological water through Transpiration. Lower temperature helps to reduce the rate of respiration and transpiration and keeps vegetables and fruits fresh and juicy.

FREEZING

We can control all three type of spoilage by freezing a commodity to a temperature of –18°C,. Even at this temperature some reactions are going on. They are however very slow and do not have any impact on the edible qualities till at least up to 18 months, if temperature is maintained at -18°C continuously. At any stage if product temperature is allowed to go up all above three processes will resume their activities and render the product inedible or spoiled.

In case of fresh foods also same is true. Therefore till product is taken for consumption, it should be stored under refrigeration if time required for storage is less or it should be frozen and stored to a temp of –18°C.

QUALITY OF FROZEN FOOD VS FRESH

The quality of frozen foods had been always under doubt. In fact unfrozen products the so-called "FRESH" may not be always fresh. The unsold fruits, vegetables and meat products are kept for next day sales by vendors and this process continues till either it is sold or till gets spoiled. As cold chain infrastructure in India is still not available widely, fruits and vegetables are left at ambient temperatures for next day sales.

Vegetables and fruits keep metabolizing. most of their stored energy and nutrients and might have consumed most of it before a customer has purchased these after few days. If same vegetable has been frozen immediately after harvesting and stored at desired temperatures, it will still be close to fresh preserving fully its chemical, organoleptic and nutritional values. In similar way meat products too keep deteriorating slowly even if they are kept under refrigeration.

PRINCIPLE, METHODS AND APPLICATIONS

Introduction to Freezing

Freezing preservation is one of the most beneficial preservation methods. It involves, conversion of liquid content of food into ice crystals, which lowers down water activity and microbial growth is arrested due to cold shock. Pure water is frozen at 0°C but since fruits and vegetables contain number of dissolved solids like sugars, acids, they freeze at below 0°C.

Process of Freezing

During freezing the commodity cools down below their freezing point but donýÿt freeze this phenomenon is called as ***super cooling***. It is shown by AB phase of curve (Fig. 1). At super cooled storage nuclei formation (***nucleation***) which is the first and most important step in ice-crystal formation in in freezing process. Here the temperature of water will be lower than 0°C but it will remain in liquid form. At this stage, further lowering of temperature result in the formation of ice crystals. The second step is called ***crystal growth stage.***

The release of heat of crystallization further enhances temperature (BC). Since food molecules contain substantial amount of so Slute hence, a progressive freezing occurs as depicted in Fig. 10.1. Various water molecules gathers around nuclei and due to subsequent addition, crystal growth occurs. Nucleation may be either due to chance orientation of molecule or due to induction of nuclei from outside, but in fruits & vegetable mostly chance nucleation occur. In next step, crystal growth around these nuclei occurs and as a result of ice-crystal formation, heat of crystallization is generated, which cause increase in temperature of commodity. This T_{es} is shown by BC lines. So, time taken by freezing curve from initial cooling to E point of curve is known as ***thermal arrest time.*** It determines how quick or slow freezing process is. After this point more ice crystal formation takes place and temperature lowers down.

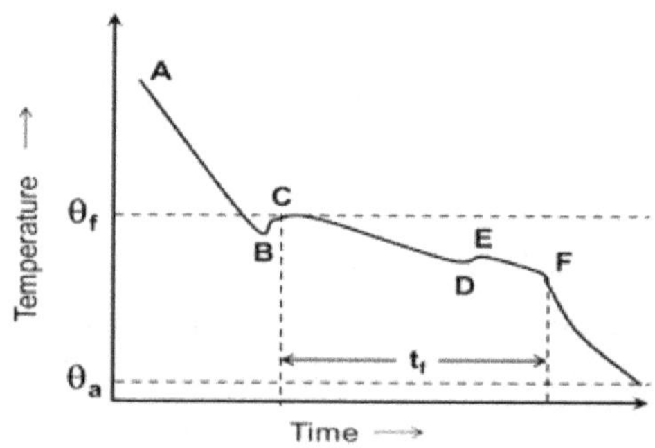

Fig. 1 Schematic diagram of freezing process

Advantages of Freezing

- No nutrient loss
- Retain freshness of commodity.
- Retain colour and flavor constituents.
- No microbial contamination.
- No respiration, hence longer shelf-life.

Effect of Freezing

Freezing process is divided into two broad categories *viz.* slow freezing and quick freezing.

- *Slow freezing*: when thermal arrest time is more than 30 min.
- *Quick freezing*: Thermal arrest time is less than 30 min.

In slow freezing, less number of nuclei is formed and as a result of slow freezing more concentrated solution is left in inter-cellular spaces which causes osmotic effect and liquid comes out from cells. This affects turgidity of cell and they collapse and on thawing cannot regain their original shape. Also, crystals forms are larger in size and pierce the cell membrane, puncture it and damage the cells. Whereas in quick freezing large numbers of nuclei are formed, hence having large numbers of crystals of smaller size evenly distributed within the cell and in the intercellular space. Since process is very quick, hence no concentration effect occur and commodities retain their original shape.

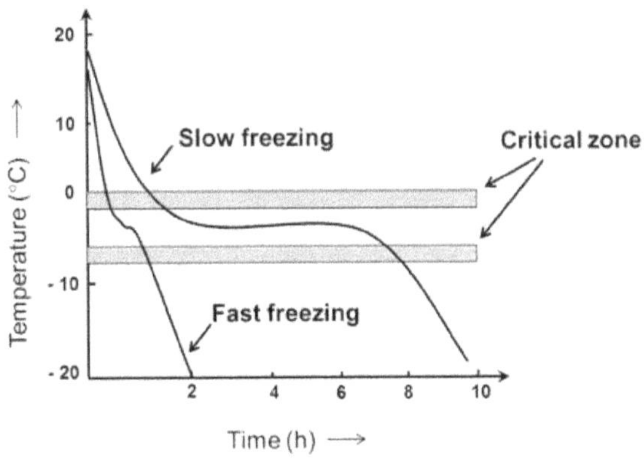

Fig. 2 Schematic diagram of temperature changes of food through the critical zone during freezing process

Freezing Methods

Mode of heat transfer in freezing food product is convection. Following points should be considered while selecting a freezing method:

a) Product dimension

b) Shape

c) Specific heat

d) Thickness of pieces

e) Freezing rate

f) Packaging

g) Food product components.

Air freezing

This is an oldest method of freezing and utilizes cool air having a temperature of -18 to -40°C as freezing method. Different types of air freezing are:

a) Tunnel freezing

b) Fluidized bed Freezing

c) Air blast freezing

Air-blast freezers recirculate air over foods at between °30-C and °50-C at a velocity of 6.0-1.5 ms^{-1}. The high air velocity reduces the thickness of boundary air films. Air flow is either parallel or perpendicular to the food and is ducted to pass evenly over all food pieces.

Air freezing may result in

- Excessive drying
- Costly
- More efficient & more rapid heat transfer
- Less product dehydration & less frequent requirement of frosting.
- Short freezing time so less moisture loss.

Plate freezing

Packaged or fresh commodities are placed over the surface of plate cooled by refrigerant in a cylindrical scraped ýÿ surface heat exchanges. Double plates are specially used in retail storage. Plate freezing is a slow freezing process and packages must be of uniform thickness.

Liquid immersion freezing

Certain liquids are used as refrigerant which are known as cryogens. Example: Liquid Nitrogen, Liquid NO_2, Liquid Ammonia, etc.

Cryogenic freezing

Freezers of this type use a change of state in the refrigerant (or cryogenic) to absorb heat from the freezing food. The heat provides the latent heat of vaporization or sublimation of the cryogen. The cryogen is in intimate contact with the food and rapidly removes heat from all surfaces to produce high heat transfer coefficients and rapid freezing. The two most common refrigerants are liquid nitrogen and solid carbon dioxide. The main advantages of cryogenic freezing are as follows:

- Short freezing time due to high heat transfer
- Reduction in flavor loss
- Reduction in drip loss
- Reduction in oxidative changes
- Improved texture of the product
- Suitable for freeze sensitive products

The main disadvantage of cryogenic freezing is relatively high cost of cryogens.

Changes Associated With Ice Formation

Volume changes

The volume of ice is 9% greater than pure water when water is transformed into ice at 0°C, and hence upon freezing there is expansion of foods. However, few exceptions also exist. For example, highly concentrated sucrose solution. The degree of expansion depends upon (a) composition, (b) fraction of water that fails to freeze, and (c) temperature range.

- ***Moisture*** High moisture contents in foods produce greater changes in volume.
- ***Cell arrangement in fruits and vegetables*** Fruits and vegetables have intercellular air spaces which absorb internal increases in volume without large changes in their overall size.
- ***Concentration of solutes*** High concentrations reduce the freezing point and foods do not freeze or expand at commercial freezing temperatures.
- ***Freezer temperature*** It determines the amount of ice and hence the degree of expansion.
- ***Temperature range*** Observed changes depend upon the temperature range to which food product is exposed.
- Cooling of the specimen: contraction
- Ice formation: Expansion
- Cooling of ice-crystals: contraction
- Solute crystallization: contraction
- ***Crystallized components***: These include ice, fats and solutes, contract when they are cooled and this reduces the volume of food.

Concentration of non-aqueous system

During freezing water freezes first. So nearly all the dissolved substances are therefore concentrated in the diminished quantity of water. So in some manner it is similar to dehydration. Unfrozen phases have changed physico-chemical properties like pH, titrable acidity, ionic strength, viscosity, freezing point, or potential. Dissolved gases will be removed/expelled from the space. Water structure & water-solute interaction may be drastically increased, macromolecules will come together and many detrimental reactions may occur. As freezing progresses, concentration of a particular solute increases and eventually reaches or exceeds their respective saturation concentration crystallization.

METHODS OF FREEZING

There are various methods of freezing and various kinds of refrigerants used by the food industry industry.

Contact Freezing

Contact freezers are used for freezing products in bulk and in blocks. The product is filled in trays or pans with water that acts as a freezing media. It gives better contact with freezers plates and gives fast freezing. After freezing product gets embedded inside ice which also acts as a barrier in between product inside and oxygen in the air to arrest putrefactive oxidation and dehydration from dry cold store air. Ammonia (NH_3), Freon 104, Carbon dioxide (CO_2) gases are generally used as refrigerants in this system.

The problem in this method of freezing is that customer has to thaw whole block even when he needs only a fraction of it. Customer is forced to refreeze remaining quantity, which is not a recommended practice as product quality suffers during refreezing and product deteriorates gradually on repetition of this process. However it is a good freezing method for keeping raw materials in bulk for later commercial uses.

Blast Freezing

In blast freezing it is air, which is cooled and blasted or circulated over the product. Cold air is blown over the product, which cools the product and takes off its heat. This is a kind of exchange. Cold air is blown from one end generally top of the chamber and hot air is sucked from below to pass through the cooling coils. Cooling coils have circulating refrigerant inside which takes heat from the air in order to get converted into gaseous form. The refrigerant gas is passed through the condensers and converted again into liquid and sent again to cooling coils to cool air.

This process is repeated again and again till core of the product attains temperature of –18°C. This is a conventional and one of the most economical methods of freezing but takes longer time for freezing. Moreover product is gradually cooled from ambient temperature to freezing temperature, passing through full range degree by degree, causing slow freezing, higher drip loss and up to some extent some quality loss also.

Therefore conventional blast freezers have gone through several technological upgradations and various quick freezing models (IQF Freezers) are available now which include continuous belt freezers, spiral freezers, flow freezers, fluidized belt freezers etc. where freezing time is cut down only to minutes from hours.

However, most of the upgraded freezers are meant for individual quick freezing (IQF) type of production where product is small in size and thickness and the individual piece is frozen separately. All retail consumer packs are preferred this way as it brings convenience to customer for end use. Customer can pick up whatever quantity, he /she needs.

Brine Freezing

Though brine-freezing technology is considered to be old and obsolete freezing method yet it is still in use for commercial ice production in several developing countries. Freezers have a tank fitted with cooling coils. Brine made of common salt (sodium chloride) or calcium chloride is filled into the tank and cooled with the help of passing refrigerant. The product to be frozen is submerged in the cold circulating brine and gets frozen. The product however comes in the contact of brine that alters its natural taste and makes it salty or even bitter if brine is made of calcium chloride.

Cryogenic Freezing

This is the most efficient methods of freezing, however, it is not commercially viable. Liquid Nitrogen is used as a refrigerant and sprayed directly over the product. The temperature of liquid nitrogen is –90°C, therefore it freezes the product instantly without giving any time and chance for putrefactive and biochemical processes to set in. The cost of nitrogen is much higher in comparison to other refrigerants and its single-use makes it unviable. Availability, handling and storage of nitrogen are other hurdles making its use more difficult. It is therefore not adopted for commercial food production. however, its use in medical surgeries, preservation of organs for transplant, etc is still continuing.

Slow Freezing vs Quick Freezing

As mentioned earlier product can be frozen by several means some of them very slow like keeping some commodity in cold store directly or in deep freezers, moderately slow like in blast freezer or frozen quickly in IQF and Cryogenic freezers. When the process is very slow like freezing in cold stores, it may take several days or even weeks to freeze a commodity depending on its size, thickness and girth.

A commodity is considered frozen when its temperature reaches to –18°C or below at its core. Biochemical changes and putrefactive activities keep working although at a slow pace and if time is longer especially when a commodity is thick or too big, they may render it unsuitable for human consumption by the time its core temperature reaches to –18°C. Another activity, which attributes to the above, is the loss of nutrients and drip loss during slow freezing due to the translocation of electrolytes.

Loss of Nutrients Due to Slow Freezing

Plant and animal tissues have 80 to 90% water. Biological water filled inside cells and intercellular spaces (plasma), blood, lymph everywhere is water and they are carriers of nutrients, chemicals, proteins, carbohydrates, salts and minerals into the cells and excretes back carbon dioxide and other break down products from metabolic and catabolic activities. When temperature of a commodity reaches below 0°C (–2 to –4°C), water wants to get frozen and convert into ice as freezing point of water is 0°C. Salts, minerals and other nutrients are dissolved in biological water and form a kind of mild brine, which freezes at a much lower temperature.

When freezing is slow it gives enough time for some of the water to get dissociated from brine and converted into ice. This process starts in intercellular plasma in the intercellular spaces. Separation of a portion of water from brine increases its ionic concentration, which sets up a process called ex-osmosis as a result electrolytes, and other nutrients flow out from the cells (intracellular plasma).

This process may become vigorous when freezing is very slow and may damage cell structures very profusely and as a result all vital nutrients and chemicals are lost into intercellular spaces that flow out when the product is thawed for end-use. This is also the reason for drip or weight loss during freeze-thaw process.

Therefore the freezing process has to be designed to cross –2 to –4 degree critical temperature range soonest possible without giving much time to set above nutrient migration process, this defines quick freezing method.

CHAPTER

MEMBRANE SEPARATION 13

INTRODUCTION

The use of membrane technology as a processing and separation method in food industry is gaining wide application. Membrane separations can be used either as alternatives to conventional techniques or as novel technology for processing new ingredients and foods. Membrane separations are considered green technologies. In many cases, membrane processes are more advantageous than traditional technologies. For example, using cold pasteurization and sterilization with suitable membranes instead of high temperature treatment for the removal of microorganisms is more economical in terms of energy consumption. Using membrane filtration to remove microorganisms for shelf-life extension of foods instead of using additives and preservatives also create a green image for the processed foods as well as for the processing procedure.

Concentration by membrane filtration instead of thermal evaporation does not employ severe heating and that it preserves the natural taste of food products and the nutritional value of heat-sensitive components. The recovery of valuable components in diluted effluents and wastewater treatment applications are among the most useful and currently active aspects of membrane technology. Pressure-driven membrane processes, namely MF, UF, NF and RO facilitate separation of components with a large range of particle sizes. It is for this reason that they find wide range of applications in food processing industry. The first part of this manuscript is to give introduction about very basic knowledge in membrane separation technology. More importantly, this review presents up-to-date commercial and potential applications of pressure-driven membrane separation processes in dairy processing industry.

THEORY OF MEMBRANE SEPARATIONS

Pressure-driven membrane processes: The pressure-driven membrane processes include microfiltration (MF), ultrafiltration (UF), nanofiltration (NF), and reverse osmosis (RO). When a feed is introduced to a membrane separation system it is separated into retentate (sometimes called concentrate), the fraction that is retained by the membrane, and permeate (also called filtrate), the fraction that passes through the membrane. The products of interest can either be in the retentate or in permeate or in both streams.

The word 'pressure-driven' means that the main driving force for separation of these processes is transmembrane pressure (TMP), which is the pressure discrepancy between retentate sides and permeate side. Generally speaking, the pore sizes (or MWCO-molecular weight cut-off in cases of NF and RO) of membranes decrease in the order from MF to RO).

However, the separation principle is not based on the pore sizes alone. Especially in UF and NF the charge of the molecules/solutes and their affinity for the filtering membrane are also important.[1] MF is normally used for separation of suspended particles and microorganisms from soluble components in feed. UF can be applied to separate soluble macromolecules such as proteins and peptides. NF is applied for partial demineralization and, at the same time, concentration. Basically, NF allows monovalent salts to pass through while it retains multivalent salts.

Operational TMP values increase from MF to RO. For example, in MF applications the applied TMP is rarely higher than 3 bar. While those for UF are normally in range of 3-7 bar and for NF 10-30 bar. RO membranes, theoretically, allow only water to permeate. This means that RO processes works against chemical potential difference, namely osmotic pressure. For this reason, the operational TMP applied in RO is normally much higher (e.g., 10-75 bar) than in other pressure-driven separation processes. RO membranes, generally, reject 95% NaCl.

DEFINITIONS AND TERMS

To be easier to follow the following parts, several common definitions and terms are given here. In *dead-end* filtration, which is normally applied in laboratory sample preparation, both pressure vector and feed flow are normal to the membrane while in *crossflow* (CF) or also called *tangential* filtration, the feed is pumped parallelly with the membrane and it is possible to recirculate the retentate back to the feed flow.

Compared to dead-end filtration, CF filtration is, due to the tangential movement of the feed, characterized with lower extent of *concentration polarization* and *membrane fouling*. These twophenomena are two major obstacles causing reduction in membrane separation performance.

Concentration polarization expresses the raise-up in concentration of a macromolecular solute (retained by the membrane) at the surface of the membrane compared to that in the bulk phase. Membrane fouling indicates the deposition of solutes/particles on the surface (*external fouling*) and/or into the membrane pores (*internal fouling*). The components causing membrane fouling are called *foulants*. *Diafiltration* (DF) is carried out by diluting theretentate with a pure solvent, usually water or a buffer, and re-concentrating the diluted retentate. The main purpose of DF is to increase the purity of the membrane-rejected components (retentate components) and DF can be performed either in batch or in continuous manner.

Membrane separation is a process of separating food components by using semi-permeable membranes, basing on the molecular size and molecular weight of the components. The driving force of the separation process is, for example, differences in concentration or pressure between the two sides of the membrane.

MEMBRANE PROCESSING METHODS

The different membrane processing methods are as follows:

Reverse osmosis (RO). Concentration of solution by removal of water

- Nanofiltration (NF). Concentration of organic components by removal of part of monovalent
- ions like sodium and chlorine (partial demineralization).
- Ultrafiltration (UF). Concentration of large and macro molecules
- Microfiltration (MF). Removal of bacteria, separation of macromolecules

The difference in application of these technologies is shown in Fig. 1.

There are also other methods for selective separation as follows.

- Pervaporation
- Ion exchange
- Electro-dialysis

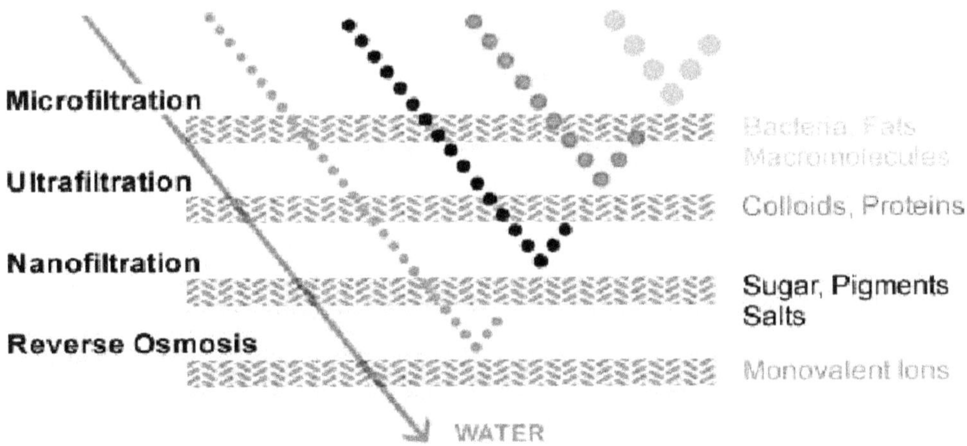

Fig. 1. Application of different membrane processing technologies

REVERSE OSMOSIS

Reverse osmosis has become a standard process in the food industry. It is used to purify water for plant operations, to concentrate cheese whey proteins or milk in the dairy industry, for sugar concentration in the cereal processing industry, for concentration of juices, and for wastewater treatment in meat and fish processing industries. Reverse osmosis deserves a special look because of its suitability for a wide variety of applications.

Natural osmosis is a phenomenon in which a liquid passes through a semi-permeable membrane from a dilute solution to a more concentrated one. In reverse osmosis, pressure is applied to the more concentrated solution forcing water to flow towards the dilute solution.

In this process, the reject water is typically 30% to 50% of the feed flow. Thus, at maximum efficiency, every 100 gallons of water entering the system would produce 50 gallons of purified water. RO is helping to save companies both energy and water by helping to limit evaporation steps or by treatment of wastewater streams.

Many foods require the removal of large amounts of water to concentrate the product for more efficient packaging or shipping. Although evaporation is common, it requires substantial amounts of energy requirements compared with RO. In the U.S. the corn wet-milling industry alone consumes about 93.7 trillion Btu/year, which is equal to 90% of the energy consumed in grain milling operations. The amount of water evaporated in this industry is around 35 billion lb/year for steep water, and around 12 billion lb/year for sweet water. The energy required for RO is approximately 110 kJ/kg water versus 700 kJ/kg for the most efficient evaporator, resulting in substantial savings.

Wastewater is a problem in any industry. Reducing the level of the dissolved solids and biological oxygen demand (BOD) is sometimes the only way processing plants can discharge water safely. Also, removing dissolved solids from water allows it to be reused, which not only cuts down on water usage but on the amount that is discharged. The collected solids also may be recovered if they are of value. For example, an RO system can recover proteins, sugars, and enzymes from wastewater that may be reused within plant operations.

Membrane technology has made a tremendous impact on the food industry over the last several years. The separation of materials for different applications has become an important industrial operation. Considerable progress continues to be made in membrane technology, and newer applications for existing systems are being discovered as the trend is to create integrated systems which utilize several different membrane types within a process.

It is a membrane separation process, driven by a pressure gradient, in which the membrane separates the solvent (generally water) from other components of a solution. The solvent flow is opposite to the normal osmotic flow.

The membrane pore size is very small allowing only small amounts of very low molecular weight solutes to pass through the membranes.

Applications

Broadly used to separate water from low molecular weight solutes (salts, aroma compounds etc.), which have high osmotic pressure.

- Concentrate and purify fruit juices, enzymes, fermentation liquors and vegetable oils; pre-concentrate juices and dairy products before evaporation.
- Concentrate wheat starch, citric acid, egg white, milk, coffee, syrup, natural extracts and flavors.
- To clarify wine and beer (De-alcoholization for low alcohol wines and beers)
- To determine and purify water from boreholes or rivers or desalination of sea water.
- Water and waste water purification.
- Concentration of whey during cheese manufacture.

RO materials and equipment

- Polymer having high permeability for water, high rejection for soluble salts and durability.
- Cellulose acetate is outstanding being inexpensive.
- Polyacrylonitrile, polyamides, polyurethanes are also used, but they have low temperature resistance and low pH range.
- Polysulphones and ceramic materials can withstand high temperature and wider pH range.
- Operating pressure ranging from 40 to 80 bar @ 450 to 2400 lit/h of flux.

NANOFILTRATION (NF)

- New class of pressure-driven membrane processes that lies between RO and UF
- Pressure range: 10-50 bar (lower pressure than RO)
- Rejects ions with more than one negative charge (such as SO_4^{2-}, CO_3^{2-})
- Also known as loose reverse osmosis

Applications

- Removes materials having molecular weight between 300 to 1000 Da.
- Employed for pre concentration for RO.
- Removal of inorganic salts Na, K, Cl, urea, lactic acid
- Partial demineralization
- Membranes that leak particle species with a radius in the nanometer range
- Reduction of salty taste as well as reduction of salt from cheese making
- Pretreatment for electro dialysis, ion exchange
- Acid removal

Separation of dispersed particles such as colloids, fat globules etc.

Ultrafiltration

Ultra-filtration is similar to reverse osmosis with low pressure of operation. In UF, the membrane pore size is larger allowing some components to pass through the pores with the water. It is a separation/ fractionation process.

Membrane materials

Membranes are most commonly made of organic polymers like cellulose and its derivative, polyamides, polyolefines, polycarbonate, polysulfones, chlorine and fluorine substituted hydrocarbons and Inorganic membranes based on oxides of silicium, zirconium, titanium, and aluminum.

The principal requirements of membrane are:

- High permeate flux
- Chemical stability and inertness
- Good mechanical strength
- Thermal stability
- Good retention capability
- Smooth, fouling resistant surface
- Resistance against microbial action
- Affordable cost
- Long service life

MEMBRANE OPERATING MODES AND CONFIGURATIONS

Membrane processes can be performed in dead end or cross flow mode. In dead end filtration mode, some particles are trapped in membrane form a filter cake. The main problem of this system is the rapid formation of cake layer over the membrane surface which is responsible for decline of membrane flux. In cross flow mode, solution is fed parallel to membrane surface at higher pressure that divide the feed in to permeate and retentate. The turbulent flow helps in reduction of cake layer formation that improves the membrane flux.

Membrane modules are commercially available in several configurations namely, plate-and-frame, spiral-wound, hollow-fiber and tubular. The choice of configuration is based on pressure drop, ease of cleaning and replacement, surface/volume ratio and resistance to fouling. In the plate-and-frame module, a series of membrane discs separated by spacers are held between two end plates. This arrangement resembles the plate heat exchanger. The feed solution enter one end of module, spread all over the membrane surface and permeate is collected in permeate channel.

MEMBRANE BASICS

Separation is based on principles that rely on the chemical and physical properties of particles and molecules. For example, centrifugation uses the physical property of weight to separate solids from liquids. Another example, ion exchange, relies on the principle of charge to separate different species from one another. Other principles such as vapor pressure, solubility and diffusion also can perform separations. Membranes use the principle of size to separate different materials.

Membrane filters are very thin microporous sheets of film attached to a thicker porous support structure. At its most basic, a membrane serves as a sieve, separating solids from liquids forced through it. Not only can membranes separate solids from liquids, they can separate soluble molecules and ionic particles of different sizes from each other.

Membranes in the process industry use tangential flow filtration, or crossflow filtration. In tangential flow filtration (TFF) the flow of the feed stream runs parallel to the surface of the membrane at high velocities.

As the fluid passes across the membrane surface it acts to sweep the retained components, or retentate, from the membrane surface and back into the bulk solution to avoid plugging the pores. This is opposed to traditional through-flow or dead-end filtration, in which the feed stream flows perpendicular to the surface of the membrane, the object of which is to form a dry cake.

In TFF, the fraction that contains the solutes and lower molecular weight components that can pass through the membrane is known as permeate. The retentate is continuously recirculated, passing over the membrane surface until the desired effect—such as the concentration or clarification of a desired product—is achieved. The rate of permeation is known as flux and gives an indication of membrane performance.

TFF operations can perform both concentration and clarification applications. In concentration, the membrane retains the desired product, and liquid is removed as permeate. The retentate becomes more and more concentrated as permeate is removed.

In clarification applications, the desired product passes through the membrane and is collected as permeate, perhaps leaving insoluble materials or other undesired compounds in the retentate. Both concentration and clarification operations are used extensively in the food industry, primarily to process juice and other beverages.

MEMBRANE CONSTRUCTION

Membranes are fabricated from many different types of materials. Initially, reverse osmosis and ultrafiltration membranes were cellulose-based, but are now made from polymers based on engineering plastics. Typical materials have included polyacrylonitrile (PAN) and blends of PAN with polyvinyl chloride (PVC), aromatic polyamid, and polyvinylidene fluoride (PVDF).

Today, most polymeric membranes used in the food industry are made of polysulfone (PS) or polyethersulfone (PES). As required by other processing equipment in the food industry, all membrane surfaces, backings, spacers, and support structures that make contact with food products must meet Title 21, Part 177, of the Code of Federal Regulations, generally recognized as safe (GRAS), or otherwise approved by the FDA for food contact.[5] Materials also are chosen for their cleanability and ability to withstand a variety of conditions under which it might perform.

Membranes can be divided into four basic categories: reverse osmosis (RO), nanofiltration (NO), ultrafiltration (UF), and microfiltration (MF). Each of these categories is distinguished by the size of the species they retain. Retention is based on the pore size of the membrane.

Reverse osmosis has the tightest pore construction, and can separate in the ionic range. Nanofiltration, a newer category of membranes, operates similar to reverse osmosis but has a somewhat looser construction, allowing monovalent ions and some divalent ions to pass. Ultrafiltration is used to separate different-size molecules such as proteins and other macromolecules. Microfiltration membranes have the largest pore sizes of all categories, and is primarily used for removing suspended solids and bacteria.

Another difference between these types of filtration is the pressures at which they operate. RO membranes operate up to 1500 psi, NO membranes operate up to 300 psi, UF membranes operate from 10 to 200 psi, while MF membranes operate in the range of 1 to 25 psi. Pore size differences between the membranes determine the operating pressure. Higher pressures are necessary to force liquid through membranes with smaller pore sizes.

FILTRATION FOUNDATION

The module design—or support structure—of a membrane is critical to its performance. Some factors to consider include flux (the rate of permeation), the solids content of the process fluid, cost, cleanability and scalability. Food industry applications make use of four basic module designs: spiral-wound, tubular, hollow-fiber, and plate-and-frame styled systems.

Spiral-wound membranes cover more than 60% of food-industry applications, mainly for dairy and other soluble protein processing, polysaccharide gum concentration, and in most RO and NO applications. Fluid is pumped into the spacer channel parallel to the membrane surface and permeate passes through the membrane and into a porous permeate channel until it reaches a perforated tube at the center which acts as a permeate carrier.

Although resistance to fouling is good, these membranes have difficulty handling viscous material, or material with a high solids content.

Tubular systems account for about 10% to 15% of food industry applications. Tubular designs have a porous outer structure with a semi-permeable membrane coating on the inside of the tube.

The shell of the module collects permeate while retentate discharges at the end. Tubular designs are easy to clean and can be visually inspected. They can handle liquids with high solids content and larger suspended particulates better than some of the other membrane designs. The membrane area, however, is typically small. Tubular membranes are suitable for beverage clarification or the reverse osmosis of pulp containing juices.

Plate-and-frame and hollow-fiber systems are among the miscellaneous configurations that make up the remaining percentage of designs used in food industry applications. In plate-and-frame styles, flat-sheet membranes are affixed to both sides of a porous plate and sandwiched in a holder. As indicated in figure 5, the feed stream enters the system and is directed via several channels to sweep over the surface of the membrane.

The fluid flows from these channels into the same outlet line and leaves as retentate. The permeate passes through the membrane into channels separate from the retentate and leaves as permeate. Several membranes and their supports can be stacked together in the holder to increase the overall membrane surface area. The advantage of this system is that if one membrane fails, it typically can be replaced at low cost.

Also, plate-and-frame systems offer increased diversity. Once the initial capital cost to acquire the hardware is accomplished, a variety of membranes may be used in them. For example, both ultrafiltration and microfiltration processes may be done on the same unit merely by swapping out the proper membranes. Plate-and-frame systems have been used for the dealcoholization of beer in Europe and Australia, and also have been used for some high-viscosity concentration applications in the dairy industry.

Hollow-fiber membranes are similar to tubular membranes except that the hollow fibers are much smaller. The inside diameter of the fiber may range from 0.5 to 1.1 mm as opposed to 12.5 to 25 mm for the tubular design. The feed stream flows through the inside of the fibers and the permeate is collected in the containment shell. Hollow-fiber elements may have of hundreds of fibers all oriented in parallel. This allows for a high packing density and resistance to blockage of the flow channels. Also, this membrane can be backwashed to aid in cleaning. The strength of the fibers, however, is a limitation and low transmembrane pressures must be used to avoid bursting the fibers. The entire element must be discarded if even one fiber breaks.

MEMBRANE PROCESSING IN FOOD PROCESSING

Availability of few food materials in naturally pure form underlines the importance of different separation and purification processes in transforming food or food substances into safe and direct consumable form.For selected applications, membrane processing is superior to other conventional methods owing to its inherent advances such as low energy consumption, fewer and mild processing steps, greater separation efficiency, improved final product quality, eco-friendly and 'cleaner' processing i.e. better quality products at lowest cost with highest environment protection through minimum/low waste generation. Now a days, these processes have been become an integral part of different industries like functional food and nutraceuticals industry, agricultural, dairy and food and bio-product industries.

MEMBRANE APPLICATION IN BEER, WINE AND VINEGAR PRODUCTION

Currently, membranes are being used in this area for clarification of beer, wine and vinegar as well as dealcoholization of beer. In the classical beer production process, as an alternate, reverse osmosis is used for beer dealcoholization prior to its clarification by microfiltration before pasteurization and bottling.

Cross-flow MF is used to recover beer from the yeast-beer mixture (settled on fermentation tank bottom) and also to concentrate yeasts. The membrane system investment is usually compensated by the beer recovery. Moreover, MF now successfully employed for beer clarification i.e. for removal of yeasts, microorganisms and hazewithout adversely affecting its sensorial attributes. RO is used to tailor the alcohol content in beer to produce alcohol free or reduced alcohol (8-10 times less alcohol content) beer without affecting its flavour.

It is carried out in four steps i.e. beer is pre-concentrated in permeate (water + alcohol) and retentate (concentrated beer + flavour) followed by diafiltration of the permeate and tailoring of alcohol content as per the desired tasteusing desalted and deoxygenized water; post treatment which includes the addition of hops and syrups in alcohol tailored beer to compensate the taste losses arises from the loss of taste carrier alcohols. These steps are performed <7°C that results in the production of improved quality beer rather than conventional process.

Reverse osmosis is used for the must correction, rejuvenation and dealcoholization of wine, while MF/UF is used for the wine clarification during wine manufacture. Must composition is balanced with the help of RO, which increase sugar content in wine through concentration and at the same time enhances tannins and organoleptic components by 5-20%.

It is an efficient method to maintain the quality of must obtained from the grapes harvested during rain (i.e. with moisture content) but the same is less effective in maintaining the quality of must made from the grapes of stalled maturity due to cold weather.

The use of thisparticular method is depends on theregulationof various countries. During wine clarification, MF and UF providesan alternative to the classical fining substances, filters and also reduces the number of processing steps. Researchers reported that 0.20-0.50 μm and 0.45- 0.65 μm pore size MF membranes are usually used during the filtration of white and red wines, respectively.

Rejuvenation of older wines is accomplished through RO process which removes the water, alcohol and other substances responsible for adverse aroma followed by the addition of mineral free water during diafiltration. Alcohol content of the wines is also tailored in the same manner as discussed above for beer.

Clarification of vinegar employing UF not only provides wider vinegar range with same sensorial attributes without turbidity,but also reduces number of operations and reduces storage time.

MEMBRANE APPLICATION IN FRUIT JUICE AND BEVERAGES PRODUCTION

MF and UF are the economic and efficient alternatives to the classical filtration methods available for clarification of different juices. The use of membrane processing in beverage industry have severalinherent advantages likeimproved product quality, reduced cost of production, better working environment, cleaner production with less waste generation and improved product safety.

Researchers reported several merits of membrane processes over classical beverages or fruit juice fining agents such as reduced energy consumption and cost, eradication of different filtration aids, mild thermal treatments that safeguard the product against thermal degradation, easier process and compact equipment design, fewer wastes, reduced chemicals requirement andenhanced productivity (96– 98%, juice recovery).

Moreover, fruit juice clarification with MF and UF have substantial potential of cost saving. Annual production costs of a UF plant (capacity- 250 m3 of juice/ day) was estimated 79% lesser than a conventional plant.

MF (pore sizes, 0.05–10 μm) is resembles classical coarse filtration, employed to filter juicesin fibrous enrichedpulp (in retentate)as well as microbes free clarified fraction (in permeate). Moreover, it is known as cold - sterilization process. UF process is the alternate of the fining step used in classical method. It is wider membrane pores than MF yet retails the bacteria, fats, proteins and colloids but allow to pass the smaller molecules like vitamin, minerals and sugars. Prior to UF process, both pre and enzyme treatment of the juice is essential for better yield, high capacity and excellent product quality. Juices are first concentrated in RO, which decreases the moisture content to its half value but retains major amount of sugar (98– 99%) and volatile flavours (80-90%) in the retentate (having 20-25°Brix) and same then concentrated > 75 °Brix in classical evaporation. This combination results in 60-75% saving of energy than traditional evaporation.

MEMBRANE APPLICATIONS IN FUNCTIONAL FOODS AND NUTRACEUTICALS

Global market of functional foods and nutraceuticals is rising rapidly. The future of recently evolved bio-activemolecules and nutraceutical ingredients is usually decided by theseparation ability and efficiency with the retention of the desired characteristics/biological activity of these ingredients by the employed separation process with minimum inputs (i.e. energy + capital + labor cost). Separation as well as purification of such ingredients is highly expensive. Although, a number of challenges have to overcome to raise the concentration of a target substance in the end product from its raw material yet, several methods are accomplishing such tasks successfully. On commercial scale, membrane processing is now considered as low cost and effective tool to concentrate and purify several bioactive substances from different feed streams.

Various membrane processes alone or in combination are used to produce numerous nutraceuticals and bioactive substances from lipids, carbohydrates and proteins protein based sources. Moreover, this technology offers many key advantages during processing of such ingredients which includes elimination of classical evaporation (that reduces their activity and highly energy consuming); reduces overall water needs by re-utilizing waste water; enhances profit by formation of new products and needs lesser floor space and investments.

MEMBRANE APPLICATION IN VEGETABLE OIL PROCESSING

Traditional oil refining methods have several demerits like nutrient loss, needs higher amount of energy, water and chemicals; greater losses of neutral oil as well as the generation of more effluent. Vegetable oil refining employing membranes has been not only reported as an easier/simpler process but also the same offers different advantages like better nutrient retention, reduced demand of energy and added chemicals, milder process parameters and cleaner production. Several researchers have been used membrane processes for the recovery of solvents from the micelle, degumming, bleaching, deacidification as well as hydrolysis of fats and oils.

MEMBRANE APPLICATION IN DAIRY PROCESSING

Dairy industry is one of the early adaptor of the membrane processes.It has greatly benefited by membranes as a number of applications like removal of water/concentration, liquid-liquid as well as solid-liquid separations are routinely performed. From 1960, the major milestones in membrane development and their application in dairy processing.

APPLICATIONS IN FOOD INDUSTRY

Food industry covers about 20"30% of all worldwide membrane sales out this dairy industry covers nearly 40%. The global market share of membrane separation systems are as follows ultra filtration – 35%, micro filtration- 33%, nano filtration and Reverse osmosis-30%, electrodialysis and pervaporation – 2%. The use of membrane separation techniques in food industry have several advantages like improved product quality, mild thermal treatments, reduced cost of production and less wastage.

Micro filtration

Microfiltration is used as pre-treatment before ultra filtration or reverse osmosis. It removes the suspended particles and colloidal and reduce the rate of fouling in subsequent ultra filtration or reverse osmosis process. Nowadays Microfiltration is commonly used for clarification of beer and wine i.e. for removal of yeasts, microorganism.

Other applications of micro filtration are as follows

- Removal of oil droplets and fat globules
- Removal of bacteria from milk
- Pretreatment of cheese whey.
- Removal of suspended solids and fat present in the brine used in fish processing and cheese making
- Pre filtration of wastewater streams generated in food processing.

Ultra filtration

Ultra filtration is used in dairy industry mainly for two purposes i.e production of protein concentrates from whey and cheese from milk. In ultra filtration of milk, inorganic salts, lactose and water are removed as permeate where as fat globules and proteins are retained over the membrane. This concentrated milk (retantate) is used in the manufacture of many types of cheeses. In ultra filtration of whey, whey protein concentrate is retained over the membrane. This retentate is dried by spray or drum drying to produce whey protein powder.

Ultrafiltration can be used for isolated soybean protein production. In the production of isolated soybean protein, defatted soy flour is extracted with water. This extract is concentrated by ultra filtration. The sugars and solutes are removed partially by repeated steps of dilution and ultra filtration to produce concentrated soy protein extract.

Nano filtration

The important application of nano filtration is water softening where retention of bivalent ions of Ca and Mg produces hardness free water. Nano filtration is also applicable for pre concentration of thin juices in the production of sugar from sugarcane and beet, desalting of lactose, production of protein concentrates, the demineralization and deacidification of whey obtained from ultra filtration.

MEMBRANE TECHNIQUES

- Counter diffusion
- Osmotic distillation
- Electro dialysis
- Ion-exchange

Counter diffusion

- Separate small ions from large molecules.
- Hollow fiber cellulose diffusion tubes.
- Counter diffusion can produce 50 per cent demineralization.
- which may represent 70 per cent removal of monovalents.

Osmotic distillation

- Low pressure and low temperature separation
- Hollow fiber or spiral wound hydrophobic membrane
- Electro dialysis
- Similar to Ion exchange separation.

- Used to separate electrolytes from non-electrolytes and to exchange ions between solutions.
- Direct current is passed through a solution having number of ion exchange plates/membranes.
- The ion selective membranes act as barriers to either cations or anions
- Anions and cations are attracted towards anodes and cathodes, respectively.
- The membranes are arranged alternatively to form ion diluting cells and ion concentrating cells.
- Electro dialysis is used for demineralization of milk products.
- It is also used for desalination of water.
- Also used to de-mineralize whey, to remove potassium and tartaric acid from wines and to de acidify fruit juices.

Ion-exchange

- Food components are separated basing on their electrical properties.
- Metal ions, proteins, amino acids and sugar are transferred from the feed and retained on a solid ion exchange material.
- Two types of ion exchangers (having fixed electrical charges) namely cation exchanger that attracts cations and anion exchanger that attracts anions by a process called electro static adsorption.
- The components are then separated by washing off the ion exchangers.
- They are constructed using porous matrix made of polyacrylamides, polystyrenes, dextrans or silica.
- Generally employed for de-colorization of sugar syrup, protein recovery from whey, softening & demineralization of water and separation of enzymes.

INTEGRATION APPLICATIONS OF DIFFERENT MEMBRANE SEPARATION PROCESSES

From experimental results, researchers proposed a procedure where two steps of UF in combination with NF are employed to obtain a more economical cheese production. In this process, the main target was to utilize whey. The first UF employs membrane of 6-8 kDa MWCO. The retentate of the second UF is introduced back into the cheese process, which can improve both the nutritional and economic values. The permeate solutions of both UF processes can be fed to a NF (400 Da MWCO) to concentrate lactose for sweet industry and the nana filtrate, containing low concentrations of organic material, can be reused in the production or for other purposes, representing a "cleaner" production.

Targeting to consumers having lactose intolerance, reduced-lactose dairy products are produced using lactase, the enzyme which hydrolyzes lactose into glucose and galactose. Vyas et al. developed and tested on a pilot scale of a process using two steps of UF to produce a product called calcium fortified reduced lactose skim milk. Permeate of the first UF is treated by an increase in pH and/or a mild heat treatment to form and precipitate calcium complex which is then retained by the second UF.

The combination of both increase in pH and heat treatment resulted in highest amount of calcium retained by the second UF. The retentate of this UF is then combined with the retentate of the first UF to give a skim milk product which is low in lactose and enriched with calcium). UF offer a very good opportunity to produce other low lactose dairy products such as milk beverages and yogurts.

ISOLATION OF MILK FAT GLOBULE MEMBRANE MATERIALS

Milk fat globule membrane (MFGM), the membrane surrounding the milk fat globules, has received much attention recently due to the fact that it contains many components with health-beneficial properties. This small fraction of milk consists mostly of polar lipids, including phospholipids and sphigolipids, and specific-membrane proteins. Under certain treatments during dairy processing, MFGM is broken into fragments and these are distributed to aqueous streams. Therefore, buttermilk, butter serum, and whey from cheese production contain substantial amount of MFGM fragments. There are several possible ways to isolate these fragments.

Buttermilk and butter serum contain caseins micelles and they should be dissociated into individual species (to be small enough to permeate the filtering membrane) or precipitated to remove before MF or UF in combination with DF can be applied to concentrate MFGM material. MFGM contains high concentration of polar lipids, which are amphiphilic molecules, so it can be used as an emulsifying agent. This material is as well a potential ingredient to develop functional food.

WASTEWATER TREATMENT

The dairy industry is considered as one of the most polluting food industries in term of water consumption and discharged volume of wastewater. Research on applications of membrane separation processes for treatment of wastewater is very active. NF applied to ultrafiltrate of whey has a twofold purpose; recovery of lactose from the retentate and reduction on volume of wastewater required to be treated at the plant since the nanofiltrate can be discharged directly into the sewer., Permeate of a single RO or a NF + RO operation could be reused as heating, cooling, cleaning and boiler feed water., Vourch et al. estimated that to treat 100 m^3/d of wastewater from a dairy plant a 540 m^2 RO unit is required and that the water recovery could reach 95%. Purified water complying with drinking water criteria could be achieved by a two-stage RO + RO process treatment of dairy effluents.

CONCLUSIONS AND PERSPECTIVES

Since the introduction of membrane technology into the food processing industry about four decades back, the number of applications as well as the membrane surface area is increasing very fast. Some of the reasons for this trend include:

- Membrane separation processes can be used as alternatives to conventional processing methods in a way that the formers assist more economic production and better quality products in terms of both technological functionalities and nutrition value.
- Applications of membrane techniques can create products, ingredients with favorable characteristics that conventional techniques are unable to offer.
- Membrane separation processes are considered green technology due to their higher efficiency in energy usage and the winging away from using chemicals and additives, which is both better for environment and human health.
- Membrane technologies create the possibilities to recover valuable components in diluted effluents, by-products, and wastewater. These applications bring more benefits to the producers.
- Membrane separations are a very good technology for wastewater treatment, especially permeated water from certain membrane filtrations can be reused in production activities.

A more comprehensive understanding of membrane technology is continuously acquired and from these, additional techniques for increasing membrane process performance are being developed and optimized. Moreover, more and more membrane producers and scientists are now in search for cheaper and simpler methods of manufacturing membranes. All these factors are impetus in accelerating and extending the scope of membrane applications in the food industry.

CHAPTER

DEHYDRATION 14

INTRODUCTION

Dehydration is one of the oldest methods of food preservation, though methods for drying food have become sophisticated over time. For many people, dehydrating food at home is a convenient way to preserve foods. **Dehydration,** in food processing, means by which many types of food can be preserved for indefinite periods by extracting the moisture, thereby inhibiting the growth of microorganisms. Dehydration is one of the oldest methods of food preservation and was used by prehistoric peoples in sun-drying seeds. The North American Indians preserved meat by sun-drying slices, the Chinese dried eggs, and the Japanese dried fish and rice.

Hot-air dehydration was developed in France in 1795, enabling the commercial production of dehydrated food products, particularly spaghetti and other starch products. Modern dehydration techniques have been largely stimulated by the advantages dehydration gives in compactness; on the average, dehydrated food has about 1/15th bulk of the original or reconstituted product. The need to transport large shipments of food over great distances during World War II provided much of the stimulus to perfect dehydration processes. The advantages of reduced bulk later came to be appreciated by campers and backpackers and also by relief agencies that provide food in times of emergency and disaster.

Dehydration equipment varies in form with different food products and includes tunnel driers, kiln driers, cabinet driers, vacuum driers, and other forms. Compact equipment suitable for home use is also available. A basic aim of design is to shorten the drying time, which helps retain the basic character of the food product. Drying under vacuum is especially beneficial to fruits and vegetables. Freeze-drying benefits heat-sensitive products by dehydrating in the frozen state without intermediate thaw. Freeze-drying of meat yields a product of excellent stability, which on rehydration closely resembles fresh meat.

The preservation of foods by drying is the time honored and most common method used by humankind and the food processing industry. The dehydration of food is one of the most important achievements in human history making our species less dependent upon a daily food supply, even under adverse environmental conditions. Drying in earlier times was done in the sun, but today many types of sophisticated equipment and methods are being used to dehydrate foods. In recent decades, considerable efforts have been made to understand some of the chemical and biochemical changes that occur during dehydration and to develop methods for preventing undesirable quality losses. Foods can be divided into three broad groups based on the value added through processing by drying. In the case of cereals, legumes and root crops, very little value is added per kilogram processed. More value per unit mass is added to foods such as vegetables, fruits and fish; and considerably more to high value crops such as spices, herbs, medicinal plants, nuts, bio-active materials, and enzymes.

BENEFITS

Molds, yeast and bacteria need water to grow. When foods are sufficiently dehydrated, microorganisms cannot grow and foods will not spoil. Dried fruits and fruit leathers may be used as snack foods; dried vegetables may be added to soups, stews or casseroles. Campers and hikers value dried foods for their light weight, keeping qualities and ease of preparation.

NUTRITIONAL VALUE

The nutritive value of food is affected by the dehydration process. Vitamins A and C are destroyed by heat and air. Using a sulfite treatment prevents the loss of some vitamins but causes the destruction of thiamin.

Blanching vegetables before drying, while crucial to a better quality product, results in some loss of vitamin C and B-complex vitamins as well as the loss of some minerals, because these are all water soluble. Yet blanching reduces the loss of thiamin and vitamins A and C during dehydration and storage.

Dried foods have more calories on a weight-for-weight basis because of their nutrient concentration. For example, 100 grams of fresh apricots has 51 calories, whereas 100 grams of dried apricots has 260 calories. Nutritive value, as well as flavor and appearance, is best protected by low temperature and low humidity during storage.

DEHYDRATION METHODS

Foods can be dehydrated by various means: the sun, a conventional oven, an electric dehydrator, a microwave oven (for herbs only), air drying and solar drying. Dehydration, like other preservation methods, requires energy. Unless sun drying is possible where you live, the energy cost of dehydrating foods at home is higher than for canning and, in some cases, more expensive than freezing.

- **Electric dehydrators** produces a better-quality dried product than any other method of drying. Electric dehydrators are self-contained units with a heat source, a ventilation system, and trays to place the food on. Models vary in sophistication and efficiency, so choose options that you desire. Although an electric dehydrator requires a fairly high initial investment, it maintains low temperatures and uses less energy than an oven.
- When using a dehydrator, load food on trays in single layers so that pieces do not overlap. This arrangement allows air to circulate through the trays. Large pieces, such as apricot halves, should be turned halfway through the drying time. Pieces near the sides of the tray should be moved to the center. Stir small pieces every one to two hours, separating bits that stick together. It may be necessary to rotate the trays within the dryer at least once during the drying period to ensure uniform drying. Follow manufacturers' guidance for heat settings for specific foods.
- **Oven drying** requires little initial investment, but continual use of an oven for drying is not recommended because ovens are less energy-efficient than dehydrators, and energy costs tend to be high. Overall oven-dried foods are usually darker, more brittle and less flavorful than foods dried by an electric dehydrator.
- An oven takes 2 to 3 times longer to dry food than a dehydrator. Drying in an oven is slower because ovens do not have built-in fans for the air movement It is difficult to maintain a low drying temperature in an oven, and foods are more susceptible to scorching at the end of the drying period. Check oven settings to see if it has a reading as low as 140° F or a "warm' setting. If the thermostat does not go this low, your food will cook instead of dry. Use an oven thermometer to test the temperature of the oven at its lowest setting.
- Trays upon which the food is placed must be at least 1½ inches narrower than the inside of the oven to allow for air circulation. Allow at least 2½ inches between trays and 3 inches of free space at the top of the oven for good air circulation.
- About 4 to 6 pounds of food can be dried in an oven at one time. Place food on drying trays, or on wire cooling racks covered with cheesecloth or nylon netting, to allow easy removal of dried food. Pieces of food should be in a single layer. Do not place food directly on oven racks. Cookie sheets are acceptable only for fruit leathers, which do not require good air circulation.
- Keep oven door propped open at least 4 inches. Place a fan outside the oven door to aid air circulation. Move it from side to side occasionally. The room

also should be well ventilated. Oven drying, particularly if a fan is used, should be done with caution if small children are around.

- Maintain the temperature at 140° F. Watch the temperature even more carefully toward the end of the drying process. To prevent scorching, lower temperature to 120° F if possible. Examine the food often and turn trays frequently, removing foods as they dry.
- **Microwave drying** is a quick way to dry small quantities of herbs and some leaf vegetables, but it is not successful for most other foods. Food which has been microwave dried often tastes overcooked rather than dried. Follow manufacturers' guidance.
- **Air drying** can be done indoors in a well ventilated attic, room or screened-in porch. Herbs, hot peppers and mushrooms are the most common air-dried items. Herbs and peppers are not pretreated, but simply strung on a string or tied in bundles and suspended until dry. Enclose them in paper bags to protect them from dust or other pollutants.
- **Sun-drying is not recommended in the midwest due to high humidity.** Foods dried in the sun can take 3–4 days to dry; if the humidity is high, as is generally the case in Missouri, the food will mold before it dries. Sun-drying requires constant exposure to direct sunlight during the day and a relative humidity of less than 20%. These conditions are found only in the southwestern states.

DEHYDRATION BASICS

Drying times in conventional ovens or dehydrators vary considerably depending on the amount of food dried, its moisture content, room temperature and humidity level — and in the case of oven drying, the use of fans. Some foods require several hours, and others may take more than a day. General guidelines include:

- Fruits (125 to 135 degrees F)
- Vegetables (115 to 130 degrees F)
- Meat/jerky (155 degrees F)
- Herbs (95 to 105 degrees F)

Interrupting drying time, or prolonging it by using lower temperatures, may result in spoilage.

Different foods that require similar drying times and temperatures can be dried together. Vegetables with strong odors or flavors (garlic, onion and pepper) should be dried separately.

Air temperature and circulation must be controlled during the drying process. If the temperature is too low or the humidity too high (resulting in poor circulation of moist air), the food will dry more slowly than it should and microbial growth can occur. Watch temperatures closely at the beginning and end of the drying period. If the temperature is too high at first, a hard shell may develop on the outside, trapping moisture on the inside. This condition is known as case hardening and leads to a moldy food product. Temperatures that are too high at the end of the drying period may cause food to scorch.

Determining dryness. Judging when food is dry requires experience. It is better to overdry than to underdry. When in doubt, continue drying for an additional 15–30 minutes. Check for doneness. Allow the product to cool before testing.

Vegetables are sufficiently dried when they are leathery or brittle. Leathery vegetables will be pliable and spring back if folded. Edges will be sharp. Corn and peas will shatter when hit with a hammer.

Fruits are adequately dried when moisture cannot be squeezed from them, and if they are tough and pliable when cut. Fruit leathers may be slightly sticky to the touch but should separate easily from the plastic wrap.

Meats should be extremely dry unless they are to be refrigerated or frozen for long-term storage. Meat is sufficiently dried when it is dark-colored and fibrous and forms sharp points when broken.

Herbs are dried when brittle. Their leaves shatter when rubbed together.

After drying, even when a food tests dry, it may not be uniformly dry. Therefore, conditioning and pasteurizing should be done before storing.

Conditioning is the process used to equalize, or evenly distribute, moisture left in the food after drying. It is usually done to fruits, herbs and seeds to improve storage, because it decreases the chance of spoilage, especially by molds. To condition a food, follow these steps:

- Cool foods on trays.
- Pour into a large, nonporous container of food-grade material; fill to about two-thirds full.
- Cover container and shake container daily or stir contents at least once a day for 10–14 days.
- Check for condensation on the lid and any signs of spoilage. If condensation occurs, return food to the dryer to finish the product. Recondition after it is dry.
- Cool thoroughly before packaging.

Freshly dried fruit can be added to the conditioning batch within the first five days. Conditioning time will need to be lengthened to accommodate the additional food.

Pasteurizing is recommended for foods that have been contaminated before or during storage. It can be used as a second treatment for vegetables held in storage if the vegetables do not have any mold on them. Keep in mind that this treatment can cause quality changes. To pasteurize, use one of the two following methods:

- **Freezer**—Seal dried food in a heavy, plastic bag after drying (and conditioning, if necessary). Place in a freezer at zero° F for a minimum of 48 hours.
- **Oven** — Place the food in a single layer on a tray or in a shallow pan. Place in an oven, preheated to 160°F, for 30 minutes. Cool and package for storage.

Package dried foods

Package dried foods in glass jars, food-grade plastic storage containers, or plastic food-storage bags. Make sure the package has an airtight seal. It is a good idea to package dried foods in small amounts, because after the package is opened, the food can absorb moisture from the air and quality deteriorates.

Store containers of dried foods in a cool, dark, dry area, such as a basement or cellar. Exposure to humidity, light or air decreases the shelf life of foods. The lower the temperature, the better: Foods stored at temperatures under 60° F will keep about one year. At 80–90° F, the food begins to deteriorate within several months.

SELECTING AN ELECTRIC DEHYDRATOR

Because an electric dehydrator can be an expensive investment, choose a specific brand or model carefully.

Space requirements

Dehydrators vary in size and can be round, square or rectangular. You will need to find a convenient space to operate it. Place the dehydrator near an electrical outlet so that extension cords are not necessary. You should have room to open the door or lid to easily load and rotate the trays. When deciding where to put the dehydrator, keep these points in mind:

- Food will give off humidity
- Some foods will give off strong odors, for example onions
- The fan will create noise
- The appliance will give off heat

Heat control

The dehydrator should allow you to select heat settings from 85–160° F and should have a thermostat to maintain the selected temperature. The heating element should be enclosed so that it does not present a shock or burn hazard. Some models have timers and automatic shutoffs. Controls should be located on the front of the dehydrator for ease of use. Check the instructions to see how often the trays need to be rotated to ensure even drying.

Air flow

Good air flow is extremely important in removing moisture from food. As much food surface as possible should be exposed to warm, dry, moving air. Some dryers use natural convection only; others use a fan. The addition of a fan will make even air circulation more likely.

A fan with two speeds is desirable. High speed is used at the beginning of the drying period when moisture is readily available at the surface. Low speed is then used to slowly remove the remaining moisture.

The two basic kinds of air flow systems are horizontal and vertical. Horizontal flow moves air across the top and bottom of food and is considered the best system. Vertical air flow is more easily blocked by food trays and requires rotating trays to ensure even drying.

Trays

Tray edges should be a quarter- to a half-inch high to prevent food from spilling or falling off and to allow air to circulate. There should be four to 10 open-mesh trays of sturdy lightweight plastic for drying.

The mesh size of the screening material should be large enough to allow air to circulate yet small enough to prevent food falling through. Finally, the trays should be easy to clean and lightweight.

Door

The door should be easy to open and close. Doors may swing up or down, right or left. Some are removable. You may need both hands to load and unload the food trays so the door should stay in an open position by itself.

It should seal well and have strong hinges and latches that will withstand repeated use as well as heat and humidity.

Cabinet

The cabinet should be built of a material that does not warp or get hot when in use—double-wall construction of metal or plastic. It should be nonflammable and easy to clean.

SAFETY AND WARRANTY

Look for the UL label and a grounded plug. Check the length of the warranty and what it covers. Also, verify that replacement parts are available.

Capacity

Use common sense. If the capacity of the dehydrator does not look large enough to accommodate the size of the loads you will dry frequently, it will not be a good buy.

MODE OF PRESERVATION

Drying reduces the water activity, thus preserving foods by avoiding microbial growth and deteriorative chemical reactions. The effects of heating on the activity of microorganisms and enzymes are also important in the drying of foods. With foods to be preserved by drying, it is important to maximize microorganism and/or enzyme inactivation for preventing spoilage and enhanced safety, and to reduce the components causing the deterioration of dried foods. On the other hand, in the case of drying bacterial cultures, enzymes, or vitamins, minimum inactivation is required. Thus detrimental effects of drying may be desirable or undesirable depending on the purpose of the drying process.

STATE OF WATER IN FOODS

The terms dried and dehydrated are not synonymous. The US Department of Agriculture lists dehydrated foods as those with no more than 2.5% water (dry basis). The concept of bound and free water has been developed from drying principles, and it is important for dried products for its stability during processing and storage. A product containing no water is termed as bone-dry. Water in foods exists in different forms or states. Water in foods having properties different from those of pure water can be defined as bound water. In the literature different forms of bound water are defined, e.g. unfreezeable, immobile, monolayer, and non-solvent water. However the fraction of bound water depends on the definition and measurement techniques. The binding energy of different states of bound water affects the drying process, since it requires more energy to remove bound water than free water.

ENDPOINT OF DRYING

Equilibrium in the drying system is the ultimate endpoint for the process. Water activity is commonly used to estimate the equilibrium point in thermal and osmotic drying processes. In mechanical dewatering, the magnitude of the applied force and rheological properties of the foods affect the equilibrium point. Generally meat, fish, and dairy products are dehydrated to a moisture content of 3% or less; vegetable products usually to 5%; and cereal products frequently to as much as 12%. A maximum moisture level is usually established for each dried product separately, based on desired acceptable quality after drying and during storage. Different attributes of quality can be targeted, thus the end point should be determined from all aspects, such as safety first and then consumer acceptance.

HEATING METHODS IN DRYING

Heating of air by electric heater or flue gas is the conventional heating method used for the drying process. In this case, heat transfer from the gas to the product occurs mainly through convection. The heating method is another important aspect of drying in terms of quality, as well as energy cost. Microwave, infrared, radio frequency, refractance window, and dielectric heating use the electromagnetic wavelength spectrum as a form of energy, which interacts with the materials, thus generating heat and increasing the drying rate. Dielectric drying uses frequencies in the range of 1-100 MHz, whereas microwave drying uses frequencies in the range of 300-300 000 MHz.

Microwave heating is rapid, more uniform in the case of liquids, and more energy efficient than the hot air method. Applying microwave energy under vacuum affords the advantages of vacuum-drying and microwave drying, providing improved energy efficiency and product quality. The energy can be applied in a pulsed or continuous mode. Pulsed microwave drying is more efficient than continuous drying. The development of electrotechnology in drying is becoming a priority in the food industry, to improve drying efficiency as well as food quality.

DRYING METHODS

Drying processes can be broadly classified, based on the water-removing method applied, as:

- thermal drying,
- osmotic dehydration, and
- mechanical dewatering. In thermal drying a gaseous or void medium is used to remove water from the material.

Thermal drying can be divided into three types:

- air drying,
- low air environment drying, and
- modified atmosphere drying.

In osmotic dehydration, a solvent or solution is applied to remove water, whereas in mechanical dewatering physical force is used to remove water. Consideration should be given to many factors before selecting a drying process. These factors include:

- The type of product to be dried,
- Desired properties of the finished product,
- Allowable temperature tolerance,
- The product's susceptibility to heat,
- Pretreatments required,
- Capital and processing cost, and
- Environmental factors. There is no single best technique for drying of all products.

THERMAL DRYING

Air drying

Sun drying

Formerly, sun drying was the only method used for drying food. The main disadvantage is the contamination and product loss by insects and birds. Where the climate is not particularly suitable for air drying or better quality is desired, mechanical air-drying is mainly used. Today, solar and mechanical air-drying is widely used commercially.

Solar drying

In solar drying, radiation energy from the sun is used. Solar drying is a non-polluting process and uses renewable energy. Moreover, it is an abundant energy source that cannot be monopolized. Solar drying has several drawbacks, however, and these limit its use in large-scale production. These are the need for large areas of space and for high labor inputs, the difficulty in controlling the rate of drying, and insect infestation and microbial contamination. More options in designing are now available in the literature.

In-store drying

In-store drying can also be called low-temperature in-bin drying. It may be used where grain remains in store until milled or sold. Weather conditions in tropical climates are less favorable for in-store drying due to high ambient relative humidity values. Twostage drying can produce good quality by preventing discoloration of high moisture grain and reduced cracking of skin-dry kernels.

Convection air-drying

Cabinet and bed type dryers (i.e. kiln, tray, truck tray, rotary flow conveyor, and tunnel) fall into the first generation. This is the simplest drying technique, taking place in an enclosed and heated chamber. The drying medium, hot air, is allowed to pass over the product, which has been placed in open trays. Convection drying is often a continuous process and is most often used for products that are relatively low in value.

Air-drying is usually accomplished by passing air at regulated temperature and humidity over or through the food in a dryer. Factors affect the rates of drying are temperature, humidity, air velocity and distribution pattern, air exchange, product geometry and characteristics, and thickness. The sample is usually placed on mesh trays in one layer or in a bulk bed or hangs from a string for better air circulation. The air circulation can be horizontal or vertical to the layer or bed. The structure and composition, such as fat content, of a product affects the drying rate. In general the hotter the air temperature, the faster is the drying rate; and similarly the higher the velocity the higher the drying rate.

Lower air humidity causes a higher drying rate. The relative humidity (a measure of dryness) falls when air temperature is raised. The drier must expel air to get rid of moisture, thereby allowing new, lower humidity air to enter the system. However, this process causes heat loss from the drier. In many cases, two or multistage dryers are used, with different conditions, e.g. initial drying at 90 °C and then the second or final stage at 60°C. Recirculating exhaust air in dryers is popular because of energy conservation and its effect on grain quality.

Drying is one of the most energy-intensive processes in the food industry. Apart from the rise of energy cost, legislation on pollution and sustainable and environmentally friendly technologies have created greater demand for energy-efficient drying processes in the food industry. Thus, novel thinking could save much money by avoiding costly energy waste.

Heat losses during drying can be grouped as: heat loss with the exhaust air, heat loss with the product, radiation heat loss from the dryer, heat loss due to leakage of air from the dryer, and heat loss due to over drying of products. Changing the dryer design can reduce energy losses while achieving higher product quality, such as by reducing drying time or increasing throughput (better control), avoiding heat losses, and heat recovery from exhaust gas and dried product.

Explosive puff-drying

Explosive puff-drying uses a combination of high temperature and high pressure and a sudden release of the pressure (explosion) to flush superheated water out of a product. This method gives the product good rehydrability. However, the high heat can degrade the food quality, and explosion puffing may compromise the product integrity.

Spray-drying

Spray-drying is used to remove water from a free-flowing liquid mixture, thus transforming it into a powder product. The fluid to be dried is first atomized by pumping it through either a nozzle or a rotary atomizer, thus forming small droplets with large surface areas. The droplets immediately come into contact with a hot drying gas, usually air. The liquid is very rapidly evaporated, thus minimizing contact time and heat damage. Disadvantages include, the size of the equipment required to achieve drying is very large, and very oily materials might require special preparation to remove excessive levels of fat before atomization. Ultra-sonication in the chamber can be used instead of complex atomization to produce small-diameter droplets in spray-drying.

Fluidized bed drying

This technique involves movement of particulate matter in an upward-flowing gas stream, usually hot air. Fluidization mobilizes the solid particulates, thus creating turbulences on the solid surfaces, which increases drying rate. The hot gas is introduced into the bottom of a preloaded cylindrical bed and exits at the top. In some cases, a vibratory mechanism is used to increase the contact of the product with the hot gas. Fluidized bed drying is usually carried out as a batch process and requires relatively small, uniform, and discrete particles that can be readily fluidized. The main advantages of fluidized bed drying are uniform temperature and high drying rates, thus less thermal damage. A rotating chamber is also used with a fluidized bed, thus increasing centrifugal force to further increase the drying rate and mixing. The use of a solid carrier, such as sea sand, and wheat bran could be used to prevent the biomaterial from deterioration due to thermal shock.

Spouted bed drying

In a spouted bed dryer, the heated gas enters the chamber at the center of a conical base as a jet. The particles are rapidly dispersed in the gas, and the drying occurs in an operation similar to flash-drying. This works very well with larger pieces that cannot be dried in a fluidized bed dryer.

Ball drying

In this method the material to be dried is added to the top of the drying chamber through a screw conveyor. The material within the drying chamber comes into direct contact with heated balls made from ceramic or other heat-conductive material. Drying occurs primarily by conduction. Hot drying air is passed through the bottom side of the chamber. When the product arrives at the bottom of the chamber, it is separated from the balls and collected.

Rotary drum drying

Rotary drum dryers are cylindrical shells 1 to 5 m in diameter, 10 to 40 m in length, and rotating at 1 to 8 rpm with a circumferential speed of approximately 0.2 to 0.4 m/s. These conditions depend on the product types used for drying. They are designed to operate at a nearly horizontal position, inclined at only 2 to 60 to maintain the axial advance of solids, which are fed from the upper end of the dryer body.

Drum drying

This technique removes water from a slurry, paste or fluid that has been placed on the surface of a heated drum. The dryer may comprise either a single or a double drum. Drum drying is typically a continuous operation, and care must be taken to ensure that the product that is to be dried adheres well to the drying surface; in some cases, it may be necessary to modify the liquid product by additives to change its surface tension or viscosity.

Low air environment drying

Vacuum-drying

Vacuum-drying of food involves subjecting it to a low pressure and a heating source. The vacuum allows the water to vaporize at a lower temperature than at atmospheric conditions, thus foods can be dried without exposure to high temperature. In addition, the absence of air during drying diminishes oxidation reactions. In general color, texture and flavor of vacuum-dried products are improved compared to air-dried. In some cases the product is comparable to the quality of freeze-dried.

Freeze-drying

In freeze-drying the material that has been frozen is subject to a pressure below the triple point (at 0 °C, a pressure of 610 Pa) and heated to cause ice sublimation to vapor. This method is usually used for high-quality dried products, which contain heatsensitive components such as vitamins, antibiotics, and microbial culture. Virtual absence of air and low temperature prevent deterioration due to oxidation or chemical modification of the product. It also gives very porous products which results in a high rehydration rate. However, freeze-drying is a slow and expensive process. The long processing time requires additional energy to run the compressor and refrigeration units, which makes the process very expensive for commercial use.

Heat pump drying

The heat pump dryer is a further extension of a conventional convection air-dryer with an in-built refrigeration system. The dry heated air is supplied continuously to the product to pick up moisture. The humid air passes through the evaporator of the heat pump where it condenses, thus giving up its latent heat of vaporization to the refrigerant in the evaporator. This heat is used to reheat the cool dry air passing over the hot condenser of the heat pump, thus the latent heat recovered in the process is released at the condenser of the refrigeration circuit and used to reheat the air within the dryer.

The use of the heat pump dryer offers several advantages over conventional hot air dryers for the drying of food products, including higher energy efficiency, better product quality, the ability to operate independently of outside ambient weather conditions, and zero environmental impact. In addition, the condensate can be recovered and disposed of in an appropriate manner, and there is also the potential to recover valuable volatile components from the condensate. One of the main reasons for quality improvements in heat pump dried products is due to its ability to operate at low temperatures. If a heat pump dryer is used at low temperatures (10 to 60 °C) for highly perishable food products, adequate precautions need to be taken. There is also potential to use heat pump modified atmosphere drying for better quality products.

Superheated steam drying

Superheated steam is used as a drying medium. The main advantages of this type of drying are that it can provide an oxygen-free medium for drying and process steam available in the industry can be used without any capital cost. An oxygen-free medium has the potential to give high-quality food products, but it is important to generate more information regarding quality improvement and processing efficiency

Impingement drying

Impingement drying is an old technology that has only recently been applied to food products. An impingement dryer consists of a single gas jet (air or superheated steam) or an array of such jets, impinging normally on a surface. There is a great variety of nozzles that can be used and selection of the nozzle geometry and multi-nozzle configuration have important relevance on the initial and operating costs and product quality. Some characteristics of impingement drying include: rapid drying, popular for convection drying, and the large variety of nozzles available (multizones). Typically, the temperature and jet velocity in impingement drying ranges from 100 to 350 °C and 10 to 100 m/s, respectively.

PRINCIPLE, METHODS AND APPLICATIONS

Introduction

Drying and dehydration of fruits and vegetables is an age old method to preserve these products. Removal of the water (75-90%) present in fresh commodity results in reduction in the water activity and ultimately resistance against most of the deteriorative agents. The removal of water is carried out by the application of heat and this heat is usually supplied in the form of solar energy or artificially generated hot air. Removal of moisture and exposure of heat often results in poor textural attributes, loss in nutritive value (vitamins), discolouration and loss of flavouring components. Although both drying and dehydration are interchangeably used, drying is referred to removal of water to an equilibrium moisture content while dehydration is removal of water to an almost bone dry condition.

A number of processing steps are carefully designed to check all these adverse effects of drying. Some of the new technologies have been introduced in recent years to produce a wholesome and nutritive product. Partial dewatering by osmosis and impregnation soaking process before drying saves energy during drying and improves quality of dried product. Osmotic dehydration is gaining popularity, as the dehydrated product is more stable during storage due to low water activity by solute gain and water loss. The low water activity resulted in fewer rates of chemical reactions avoiding deterioration of the food. Osmotic dehydration in many cases is employed to increase sugar to acid ratio of acidic fruits, thereby to improve the taste, texture and appearance of dried product. The processing steps involve in drying of fruits and vegetables are summarized here.

Why Drying of Foods?

- Water activity is defined as the ratio of vapour pressure of food to that of the vapour pressure of pure water at a constant temperature. Reduction in water activity (a_w) so control/check over chemical and microbiological changes (deterioration).
- Reduction in weight, size and volume of the food material. Hence bulk transportation becomes easier and cheaper.
- Packaging requirements are simple and cheap.
- Facilitate further processing. Example: grain drying for flour.

Water content in foods

Water is present in food as free or/and bound water. **Free water** is defined as water within a food that behaves as pure water. Unbound water is removed during the constant rate period of drying, when the nature of food does not have a great effect on the drying process.

Bound water can be defined as water that exhibits a lower vapour pressure, lower mobility and greatly reduced freezing point. So, bound water molecules have different kinetic and thermodynamic properties than ordinary water molecules. The a_w as affected by the extent of bound water is given in Table 1:

Table 1 Water activity as affected by the extent of bound water

Extent of bound water	Water activity (a_w)
Tightly bound water	< 0.3
Moderately bound water	0.3 to 0.7

Loosely bound water	> 0.7
Free water	~ 1.0

Mechanism of Drying

When hot air is blow over a wet food, heat is transferred to the surface and latent heat of vaporization causes water to evaporate. Water vapours diffuse through a boundary film of air and is carried away by the moving air. This create a region of lower water vapour pressure gradient is established from the moist interior of the food to the dry air. The gradient provides the "driving force" for water removal from the food.

- Liquid movement by capillary force.
- Diffusion of liquids, caused by concentration gradient.
- Diffusion of liquids, which are absorbed in layers at the surfaces of solid components of the food.
- Water vapour diffusion in air spaces within the food caused by vapour pressure gradients.

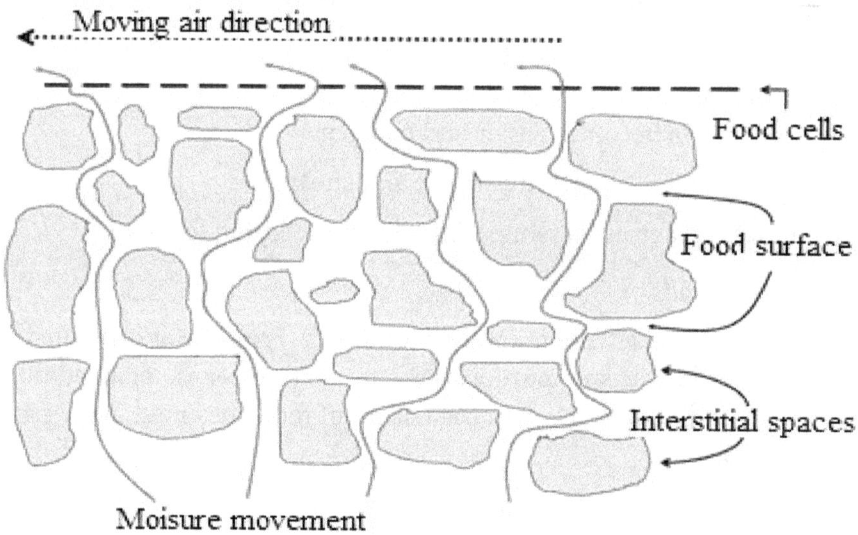

Fig 1 Schematic diagram of movement of moisture in the interstitial spaces of food cells during drying in fruits and vegetables

Phases of drying

- Initial warm up period
- Constant drying rate period
- Falling drying rate period

In hygroscopic food material more than one falling rate period occurs. In the first period plane of evaporation moves inside the food and water diffuses through the dry solids to the drying air. It ends when plane of evaporation reaches to the centre of food and the partial pressure of water falls below the saturated water vapour pressure. Second falling rate period occurs when the partial pressure of water is below the saturated vapour pressure and the drying is by desorption.

Falling rate period is the longest period during drying of food product. **Equilibrium moisture content** (EMC) occurs when dry spots develop at the surface so less area exposed to dry air and evaporation decreases.

There are 4 resistances to heat transfer in drying:

1. Resistance to external heat transfer
2. Resistance to internal heat transfer
3. Resistance to external mass transfer
4. Resistance to internal mass transfer

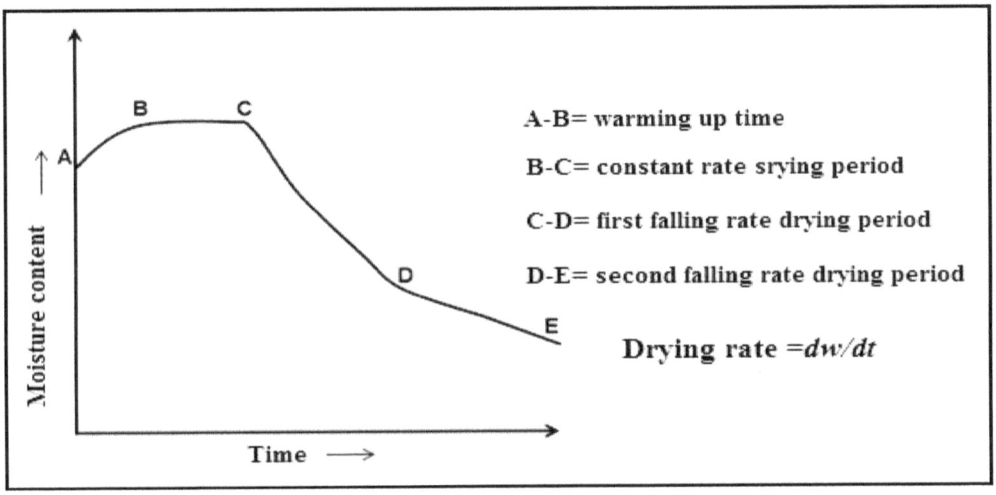

Fig 2 Schematic diagram of change in moisture content with time (drying rate)

Factors Affecting Drying Rate

The factors that affect drying rate are external and internal factors.

The external factors are:

- Dry bulb temperature
- Relative humidity
- Air velocity
- Surface heat transfer coefficient

The Internal factors are:

- Surface to volume ratio

- Surface temperature
- Rate of moisture loss
- Composition i.e. moisture, fat

Effects of Drying on Foods

Shrinkage

During drying as moisture is removed and food material becomes smaller in size. This also affects bulk density (weight per unit volume) of food material. Slow drying results in development of internal stress. These rupture compress and permanently distort the relatively rigid cells, to give the food a shrink / shrivelled appearance. Such food material on rehydration absorbs water more slowly. Gelatinization of starch, denaturation of proteins, and crystallization of cellulose also affect rehydration characteristics. Rapid drying improves textural characteristics such as wettability, sinkability, dispersibility and solubility.

Case hardening

Formation of impervious layer over the surface of a dried food product characterized by inner soft and outer hard layer resulting in inadequate drying. It always occurs in food products rich in solutes and when initial drying temperature is very high. During the initial high temperature solute particles comes out and deposit at the surface resulting of building up of an impermeable layer which prevents further moisture removal. It can be prevented by using lower drying temperature.

Browning

Browning refers to change in the colour of food material to light to dark brown colour. This change in colour may occur by any of the three methods given below.

- Residual enzymatic browning: the residual enzymes especially in vegetables such as polyphenol oxidases cause oxidation that result in the change of colour.
- Maillardýÿs reaction: it is the reaction between the amino group of proteins and reducing sugars of carbohydrates in presence of heat. This type of browning is most common in dried foods.
- Caramelization: it is the conversion of sugars only into dark coloured compounds in presence of heat.

Types of Drying

- **Hot Air drying**: It includes spray drying, tray drying, fluidized bed drying, etc. In spray drying a fine dispersion of pre-concentrated food (40-60% moisture) is ýÿatomizedýÿ to form fine droplets and then sprayed into a co-or counter-current flow of heated air at 150-300ýÿC in a large drying chamber. Tray driers consist of an insulated cabinet fitted with shallow mesh or perforated trays, each of which contains a thin (2-6 cm deep) layer of food. Hot air is blown at 0.5-5.0 m/s through a system of ducts and baffles to promote uniform distribution over and/or through each tray.

- **Microwave drying:** It involves use of microwaves.
- **Freeze drying:** It is also known as ýÿlyophilizationýÿ and is usually used for drying heat sensitive food material by freezing the material and then reducing the surrounding pressure to allow the frozen water to sublimate directly from the solid phase (ice) to gas phase (water vapour).
- Osmotic drying: This is explained in section *Osmotic Dehydration*, below.

Osmotic Dehydration

Osmotic dehydration, also called as dewatering and impregnation soaking (DIS) process, was pioneered by James D. Ponting in 1966. It is a water removal process that involves the soaking of foods - mostly fruits and vegetables in hypertonic salt or sugar or in a combined solution, to reduce the water content while increasing the soluble solid content. Osmotic dehydration is undertaken to reduce the product water activity in minimal processing, which is carried out either at atmospheric pressure or at vacuum conditions. The raw material is placed in concentrated solutions of soluble solids with higher osmotic pressure and lower water activity. Water moves through the selective permeable membrane with much greater ease than in the dissolved substances.

Mechanism of osmotic dehydration

In osmotic dehydration process, a solid product of high moisture content is immersed in a concentrated solution (mainly of sugar or salt), which initiates three types of counter-current mass transfer. The diagrammatic presentation of osmotic dehydration is given in Fig. 3.

- Water outflow from the product to the surroundings solution as a result of osmosis through a semi-permeable membrane.
- Solute transfer, from the solution to the product.
- Leaching out of the water-soluble component along with water from the product to the solution.

The last two mass transfers occur mainly because of diffusion. This counter current mass flow is due to the water and solute activity gradients across the cell's membrane. This process continues till the osmotic potential on two processes reaches equilibrium. In an ideal osmotic solution a semi-permeable membrane would be permeated by the solvent molecules but not by the solute molecules.

In fruits and vegetables, the cell wall membranes are living biological units and selective permeable, which can stretch and expand under the influence of growth and turgor pressure generated inside the cells. These cellular membranes, which are composed mainly of parenchymatous cells, freely allow the solvent molecules to pass through, but they also allow, to a lesser degree, the passage of some of the solute molecules.

Such membranes are called as differentially permeable, rather than semi-permeable. Osmotic dehydration may result in 40-50% decrease in initial volume, if properly performed.

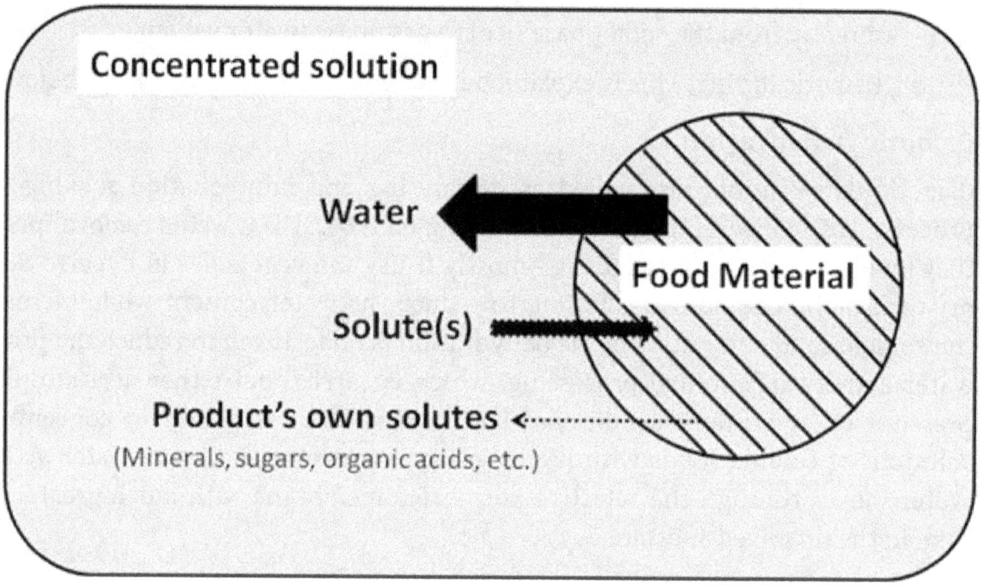

Fig. 3 Diagrammatic presentation of Osmotic Dehydration

Osmotic agents and their requirements

Osmotic agent must have lower water activity (a_w), good solubility, constant concentration during processing, and it should be cheap. However, non-toxicity, inertness to food constituent and good sensory attributes is other added attraction, while selecting any osmotic agent.

These are number of compounds available, satisfying above mentioned criteria, like, sucrose, glucose syrups, invert sugar, corn syrups, honey, and humectants such as sorbitol and glycerol.

The kind of sugar utilized strongly affects the kinetics of water removal, and by increasing the molecular weight of osmotic substance, larger water removal could be achieved with little uptake of solutes. Low molecular weight substances (glucose, fructose, sorbitol etc.) favour the sugar uptake because of the high velocity of penetration of the molecules so that solid enrichment instead of dehydration is the main effect of the process.

Sodium chloride (NaCl) is an excellent osmotic agent for vegetables and other animal derived products, but its use with fruits is restricted because of alteration in taste and bleaching of colour. Addition of NaCl to osmotic solution increased the driving force for drying owing to the a_w lowering capacity of salt.

Synergistic effects between sugar and salt have also been observed. The use of blends comprising two or more solutes seems to be an attractive alternative.

Factors affecting osmotic dehydration

In any type of food product, many mechanisms can be acting at the same time, the relative contribution of which depends upon the following:

- Nature of the product
- Processing temperature
- Operating conditions such as osmosis, diffusion, flux interactions and shrinkage.

Manufacturing Steps Involved In Drying of Fruits And Vegetables

Selection of raw material

The major cost involve in drying operation is the price of raw material. Hence good quality raw material is of prime importance. Immature or over mature fruits and vegetables often result in poor quality product. For example over mature green peas result in less sweet and starchy dried product with poor rehydration characteristics. Similarly high specific gravity fully mature potatoes with low reducing sugar content are desirable because they are large, mealy and less prone to browning or yellowing during dehydration.

Washing

Before further processing raw materials should be thoroughly washed to remove the adhering dirt, dust and other foreign particles. On commercial scale raw material is dumped in large tanks for some time and then sprayed with a jet of water. Alternatively, it may cleaned in a stream of running water. Water for this purpose must be of potable quality. To remove the traces of pesticides, colouring material commonly used nowadays to impart shining; some of the chemicals like dilute acids may also be added in washing water. Addition of sanitizers like chlorine based compounds is permitted to certain level for washing purpose. These sanitizers improve the microbiological quality of the finished product.

Peeling, trimming and sizing

Fruits and vegetables are either dried as whole or undergo size reduction before drying. A number of peeling processes are available on commercial level. Trimming is done to remove unwanted parts. Sizing is done to develop uniform product and it also facilitates subsequent unit operation. Increase in surface area causes faster drying.

Pre-treatments

In order to improve the quality of finished product fruits and vegetables they are invariable subjected to many pre-treatments. Some of the pre-treatments are summarized here.

Checking

Checking is an operation in which fruits specially raisins, prunes are dipped for a brief period in hot solution of alkali. This cause removal of waxy layer and it also improve the drying rate.

Blanching

Blanching is mild heat treatment given to some fruits and majority of vegetables primarily to inactivate the naturally occurring enzymes. These enzymes belong to peroxidase group like polyphenol peroxidases that catalyze the oxidation of phenolic substances, resulting in production of brown coloured compounds. The cell wall degrading enzymes, catalase (off-flavour), and ascorbic acid oxidase also get inactivated during blanching process. Blanching improves the colour of dried products, it aids to rapid reconstitution of dried product, it also increase drying rate, it also expels dissolved oxygen and it also improve bacteriological quality of the finished product.

Sulphiting

Sulphur dioxide and sulphites (usually sodium metasulphite or potassium metabisulphite) are well known anti-browning agents. The sulphur dioxide gas competes with peroxidases for the active sites and prevents the oxidation of phenolic substances. In many products that contain anthocyanin as major pigment, the colour of the product turned blue when exposed to acidic conditions. In such fruits or vegetables sulphur dioxide treatment may prove beneficial.

The blanched or checked fruits or vegetables are exposed to fumes of sulphur dioxide by burning sulphur powder in an enclosed chamber. This is most widely used method of sulphuring. However, they can be dipped in solution of some sulphur salt. Proper penetration of sulphur dioxide is important to prevent internal darkening during drying. In fuming process, the sulphur dioxide absorption is faster as compared dipping. However the absorption is more uniform in later.

Drying

Drying of the fruits and vegetables is usually carried out in cabinet dryer. The material is loaded over perforated aluminium trays and dried using hot air. The temperature, velocity and the humidity of the air are important for drying process. It usually varies from commodity to commodity. A lower drying temperature is generally used for fruits, as there may be chances of case hardening. The temperature varies from 55-80°C. To create perforation initially temperature is maintained towards higher side and them it is reduced. Some vegetables those are light, cylindrical or spherical like peas are dried in fluidized bed dryer. This improves the rehydration characteristic and the nutritive value of the product, as it take lesser time and more uniform drying.

Some Common Processes Used for Drying of Fruits and Vegetables

Drying of some vegetables

Drying of onion

Onion is bulbous crop characterized by pungent flavour, attractive colour and high amount of reducing sugar. It an important seasoning item for traditional Indian cuisines. The demand for onion has increased considerably over the years because of its unique flavour profile and well established medicinal and therapeutic properties. To meet the demand of consumers, the onions are processed in various convenient forms as paste, dried slices powder, etc.

EFFECTS OF DRYING ON PRODUCTS

In an ideal world, drying would have no negative impact on the final product whatsoever. We all know that this is not an ideal world and as a result, we must do whatever we can to minimize the negative effects of drying on the various quality aspects of the product. In cases where we prepare a dehydrated food product that is rehydrated before eating, the quality attributes of the rehydrated food are expected to be as close to that of the original product as possible.

In some cases, the dried product may even be expected to have improved attribute when compared to the initial material. An example of this could be instant mashed potato flakes. The consumer may feel that for the price of the product, not only should they be more convenient, but the resultant mashed potatoes should be lighter or fluffier, or creamier than the mashed potatoes prepared from whole potatoes at home. Not only that, they should be free from any black marks and the like that may be found in home-cooked potatoes.

The following is a brief summary of a number of negative factors that may be introduced into a food product if drying is not properly conducted:

Nutritional Degradation

Vitamins that are present in food products may be unstable and susceptible to heat damage. Overheating due to excessive temperatures or prolonged exposure to heat during drying can cause this problem.

Loss of Structural Integrity:

- On occasion, improper drying can create stresses in the structure of a material that can have serious negative results.

Consider the example of a starch-based cereal grain that is placed in boiling water to "cook" it. When heated sufficiently, the starch can gelatinize to form a rather soft and pliable material. It also expands in water and the water can be taken into the starch network or matrix. Once the starch-based cereal grain is "cooked", it may be necessary to dry it. As the water leaves the inner regions of the cereal grain, the starch matrix may shrink back towards its original size, or it may stay in its expanded state. Once it is dried and cooled, the gelatinized starch changes from a soft, pliable material to a hard glassy or brittle material. If the drying and cooling have been done rapidly, the glassy starch may be under considerable stress. Over time, these stresses may cause the dried product to break apart or shatter.

Reduction in Functionality

"Functionality" is a term used to describe the fact that various components in a food system have a role to perform. For example, ungelatinized starch may be present in a food (e.g., corn or potato starch). If ungelatinized starch enters a dryer, it may become gelatinized if the temperature of the material exceeds the gelatinization temperature of the starch. Typically, potato starch gelatinizes at a temperature of around 60°C.

Once it has gelatinized, the starch behaves much differently than it did before gelatinization. Ungelatinized starch is not soluble in water; but gelatinized starch is able to take up water and swell into a rather wet pasty mass.

When compressed, the starch gel may release much of this moisture. Care must be taken to dry ungelatinized starchy materials at temperatures below their gelatinization temperatures if functionality of the starch is a concern.

Flavour and Aroma Changes

Flavour changes imparted by drying may be desirable or undesirable depending upon the end use of the dried product. During drying, heating may create a toasted flavour in the product along with the associated toasted aroma.

The development of caramel flavours can be attributed to non-enzymatic browning through the Maillard Reaction where reducing sugars in the product react with amino acids. Other flavours and aromas may also develop, depending upon the nature of the material being dried and the drying conditions used.

Colour Changes

- This nonenzymatic browning also produces a characteristic brown colour (hence the name). This is particularly evident in the production of maple syrup from the sap of North American sugar maple trees. In the traditional process, sugars in the tree sap are concentrated from an original level of less than 2% by weight to over 65% by weight. In the original process, water was removed by boiling the sap in open kettles over a fire.

This prolonged heating caused the syrup to turn brown and develop desirable flavours. Some colour changes can be quite undesirable, however. If a white product is being dried, it may become "toasted" and have a light brown colour which consumers find objectionable.

Case Hardening

We have already mentioned case hardening under the heading of "Air Temperature" earlier in this module. It is being repeated here to re-emphasize the negative impact that drying can have on a product if it is not done properly.

Case hardening is a term that refers to the development of a hard shell or crust around the outside of the food particles being dried. It is caused by excessive drying of the surface of the food particle and creates a shell that prevents moisture from being removed from the interior portions of the food particle.

In normal drying, moisture is removed by evaporation from the surface of the food particles after which moisture from the inner portions of the particle travels to the particle surface where it, in turn, is evaporated.

If the surface moisture is removed too rapidly, the surface may become overly dry and form a hard layer that prevents any more moisture from reaching the surface. Since the interior moisture cannot be removed during the drying process, it will be trapped inside the particle where it will remain when the product is packaged. Once in the package, this moisture will slowly and gradually come to the food particle surface where it may promote the growth of mold. Occasionally, the moisture may soften the shell, or create internal stresses which cause the particle to collapse in on itself or to break apart.

Leaching of Soluble Constituents

During the drying process, water from the interior portion of the food particles or pieces travels to the surface where it is evaporated by the hot air inside the dryer. This water may contain dissolved nutrients that it leaves on the food surface when it evaporates. These nutrients are still present on the food surface at the time it is packaged, but if the food has to be rehydrated by soaking it in water or boiling it in water, the nutrients may be solubilized by the water and removed from the food. The nutrients are then not available for the person consuming the cooked food.

CHAPTER

PACKAGING CONCEPTS 16

INTRODUCTION

Food packaging sells your product, extends its shelf life and aids in long-distance transportation. Finding the right type of packaging for your product will help your sales. Currently, if you visit a supermarket or a grocery store they carry almost more than 50,000 products. Consumers can pick one from 15 different options of cereals, chocolates soft drinks etc. For your product to stand out, a strong food packaging design should be implemented with powerful graphics. Food packaging aids and enhances convenience for consumers. Not only that, the right type of packaging will gain additional shelf life to your product if processed and packaged at optimal conditions. The right **food packaging services** should persuade consumers purchase on a shelf as well on a small screen while buying them online.

The food industry has seldom used the word design except as related to packaging and to advertising. The development of the product has usually been called 'product development' and had connotations of laboratory formulation and sensory panel. But today, there may be real benefits in adopting food product design and in associating food product design with other areas of design.

A product has several layers and these are being built up gradually during the product development. There is the company's basic functional product, the total company product (with packaging, aesthetics, brand, price and advertising) and the consumer's product (which relates it to the competitors, the environment, the media, the society, as well as its communication and use) as shown in Fig. 1.

THE VALUE OF PACKAGING TO SOCIETY

The value of food packaging to society has never been greater nor, paradoxically, has packaging attracted so much adverse media publicity and political attention. In response, stakeholders in the food industries need to fully appreciate and actively promote the positive contributions that their packaging makes to the quality of life.

Food packaging is governed by a mass of laws, regulations, codes of practice and guidelines.

Properties

Physical Chemical Microbiological Sensory Nutritional

=

Company's basic functional product

+

Packaging Aesthetics Brand Price Advertising

=

Total company product

+

Competitors Environment Use Media Society Communication

=

Consumer's product

Fig. 1 The total food product

The societal benefits of packaging may include the following:

- prevents or reduces product damage and food spoilage, thereby saving energy and vital nutrients, and protecting the health of the consumer
- requires less municipal solid waste disposal since it promotes processed food residue recycling for use as animal feed or compost. For example, from 454 g (1 lb) of fresh corn-on-the-cob purchased at the supermarket, the customer eats approximately only 170g (six ounces), and the rest ends up in the trash can and, ultimately, in the local landfill (Institute of Packaging Professionals, IOPP, USA). This same amount of edible frozen corn can be packed in a polyethylene bag weighing less than 5 g (less than 0.18 ounce)
- lowers the cost of many foods through economies of scale in mass production and efficiency in bulk distribution. Savings are also derived from reduced product damage
- reduces or eliminates the risk of tampering and adulteration
- presents food in an hygienic and often aesthetically attractive way
- communicates important information about the food and helps consumers make informed purchases

- provides functional convenience in use or preparation, freeing up more time
- promotes goods in a competitive marketplace and increases consumer choice
- facilitates the development of modern retail formats that offer consumers the convenience of the one-stop shop and the availability of food from around the world throughout the year
- extends the shelf life with the benefit of prolonged product use, thereby reducing wastage
- saves energy through the use of ambient packs that do not require refrigeration or frozen distribution and storage.

The food industry is aware of current public concerns related to packaging which include:

- packaging litter and the volume of packaging waste in municipal waste
- cost of disposal and recovery of discarded packaging in municipal waste
- pollution associated with methods of disposal, i.e. landfill and incineration
- ease of opening
- perception of over-packaging due to apparently excessive ullage (free space) resulting from product settlement
- legibility of labels
- integrity of information on labels
- contamination of food due to the packaging itself
- accidents involving packaging. The role of packaging in multiple retail logistics

There are tight constraints on physical distribution and in-store merchandising. The retailer is receptive to packaging that reduces operating costs, increases inventory turnover, transforms to attractive merchandising displays – such as pre-assembled or easyto-assemble aisle displays – and satisfies logistics service levels (reliability, responsiveness and product availability). For example, combined transit and point-of-sale packaging saves store labour through faster shelf loading, provides ease of access to product thereby obviating the need to use potentially dangerous unsafe cutting tools, and presents an opportunity for source reduction.

The total distribution cost affects the total volume of demand through its influence on price. For some fast-moving commodity type products, such as pasteurised milk, the cost of distribution and retail merchandising is usually a sizeable proportion of total product cost representing up to 50 per cent or more of the sales price. The cost of packaging materials and containers also adds slightly to the cost but design of the optimal packaging system can significantly reduce cost in the retail distribution chain. The development of global food supply chains has meant that many points of production have located further away from the points of consumption, often resulting in higher distribution cost.

Controlling distribution cost through improved operational efficiency in the supply chain is a key to competitive advantage for a retailer. The retailer must maximise operational efficiency in the distribution channel. The goal of distribution is to deliver the requisite level of service to customers at the least cost. The identification of the most cost-effective logistical packaging is becoming more crucial.

Cost areas in distribution include storage, inventory, transport, administration and packaging. Storage, inventory, transport and store labour are major cost areas for the retailer while transport, storage and packaging are the main cost areas for the food manufacturer.

Packaging is a means of ensuring the safe delivery of a product to the consumer at the right time in sound condition at optimum cost. The need for safe delivery of products at the right time and optimum cost demands cost-effective protective packaging that facilitates high logistics performance. The objective is to arrive at the optimum protection level that will meet the customer's service requirements at minimal expense. Most organisations build in an expected damage rate, which may be as high as 10% or more for inexpensive and readily replaceable items. Other issues that concern the distributor and store manager are cleanliness and hygiene.

DEFINITION AND REGULATORY ASPECTS

In Europe, active and intelligent packaging must be in agreement with the Legislation about food-contact materials, which includes the evaluation of the overall migration limits (OMLs), specific migration limits (SMLs) and toxicological properties.

The Commission Regulation 450/2009/EC specifically deals with active and intelligent materials and articles intended to come into contact with food. In particular, it is emphasized that intelligent packaging systems must not release their components into food.

The intelligent component is allowed to be placed on the external side of the package and the contact with food may be avoided by a functional barrier that prevents the migration of substances. The functional barrier allows the use of unauthorized substances, under the respect of defined criteria and of the established migration limits.

Furthermore, it is also specified that the risk related to the employment of technologies such as nanoparticles should be analyzed for the specific application until more information is available regarding their use and their possible interactions with food and with human health.

Concerning active packaging, art. 5 indicates that components which may be added to food packaging with active function must be listed in the Community list of authorized substances. On each of the listed substances, the European Food Safety Authority (EFSA) performed a risk and safety assessment including the migration of the active agents, the migration of their reaction products and the evaluation of their toxicological properties.

In general, active packaging can be classified into non-migratory and migratory. The former implies that the action of active agent is carried out without intentional migration of non-volatile or volatile compounds from packaging into food. Nevertheless, the active agent in the packaging may deliberately be intended to be released into food. In this case, the overall migration limit of the packaging may be exceeded, so the Regulation implies two exceptions.

Indeed, if the active function is not a specific feature of the passive material, the amount of released active agent should not be included in the evaluation of overall migration. In addition, the migration of the released active substances can exceed the specific migration limit if its concentration in food is in compliance with the applicable food law.

Furthermore, the active substance intentionally released in food, or that have a technical effect, must comply with the Regulation 1333/2008/EC, concerning direct food additives.

For already approved substances which are incorporated in active materials by means of techniques such as grafting and immobilization, the manufacturer is expected to perform a safety evaluation and to verify the stability of the substances when chemical reactions, degradation or decomposition are expected to occur.

Furthermore, the active substances should not mislead the consumer (e.g., masking spoiled food), and they must be indicated in the label associated with the sentence "do not eat" or a symbol to prevent the nonedible part to be erroneously ingested. An extensive overview about regulation aspects of food active packaging in Europe is reported.

Considering different jurisdictions, a different approach compared to the European one can be observed. The introduction of active and intelligent packaging into the Japanese market is already consolidated, but specific regulations related to these two categories of packaging are still missing.

Indeed, safety of active and intelligent packaging is ensured by the application of the Japanese Food Sanitation Law of 1947, together with the Food Safety Base Law of 2003. Generally, to introduce active and intelligent packaging materials to the Japanese market, risk assessment criteria are adopted including also compliance with foreign regulations, such as the European Union or the United States. For

US regulatory concepts about food-contact materials, toxicological justification is greatly minimized by exposure assessments compared to European legislation. Active and intelligent packagings are considered almost like conventional packaging.

Food-contact materials are subjected to regulatory clearance before placing on the market by the United States Food and Drug Administration if they are considered "food additives" under the Federal Food, Drug, and Cosmetic Act, while as long as the material in the active or intelligent packaging is intended neither to add any substance to the food, nor to have a technical effect in the food (so-called "indirect additives"), there are no specific regulations for these components, which are simply regulated like all other food-contact substances.

The Canadian food packaging regulatory system is based on high safety standards and is similar to the one of US. The crucial difference between the two systems concerns migrating food packaging components, which are not considered to be food additives, legally requiring Health Products and Food Branch (HPFB)preclearance.

ACTIVE PACKAGING

Active food packaging represents an improvement in the function ascribed to the classical packaging, aiming to increase product shelf-life by interacting with food or the environment surrounding food. A thorough definition of "active materials and articles" is given by European Regulation 450/2009/EC.

It defines them as "designed to deliberately incorporate components that would release or absorb substances into or from the packaged food or the environment surrounding the food".

Indeed, conventional food packages are usually intended as passive barriers, which protect food by segregation against environmental contaminations, while active packaging promotes an active action of the packing material providing enhanced food conservation.

This active action may be due to particular features that belong to the packaging material or to active agents, which are added to the passive barrier in order to work as absorbers, emitters or releasers of compounds that are able to play a pivotal role in food preservation.

Thus, the main target of active packaging is to prevent microbial and chemical contamination, as well as to maintain visual and organoleptic properties of food.

Moisture Regulators

Moisture absorbers are active non-migratory packaging, working as controllers of excess moisture. High content of water into the package may be provided by several causes, such as trapping during packaging process, release due to temperature fluctuations, respiration of fresh products, low vapor permeability of the package, etc. Excess water reduces food shelf-life, since promotes microbial proliferation inducing undesired changes in food quality, particularly in dry food. In addition, presence of liquid inside the package of raw fish and meat, due to dripping or formation of foggy film in packaging of fresh fruit and vegetables, results into a low appeal for customers.

The main purpose of moisture absorbers is the reduction of food water activity by using hygroscopic substrates or substances, providing an environment less suitable for mold, yeast and bacteria growth and spoilage.

Some packaging materials used as passive system are inherently moisture scavengers due to their hygroscopicity. According to the Regulation 450/2009/EC, these kind of materials and articles, which work on the basis of the natural constituents only and are not designed to deliberately absorb substances, cannot be defined as active systems (e.g., pads composed of 100% cellulose).

Nevertheless, if they contain components designed for the scope, they fall under the definition of active packaging. Moisture scavengers can be defined as "relative humidity controllers" if they provide a reduction of humidity in the headspace, while "moisture removers" absorb liquids exuded from food and are usually placed on the bottom of fresh products and meat packages.

Examples of moisture absorber commercially available are absorbent pads, sheets, trays and blankets. The structure comprehends two layers of microporous non-woven plastic film in polyethylene, polypropylene, etc. containing the active substance, such as superabsorbent polymer, silica gel-based adsorbents, sodium chloride, starch copolymers, etc.. For package design, it is important that the active agent maintains its properties after the package processing and must not interfere with plastic properties.

In addition, the polymer containing the desiccant has to be water-permeable. Nevertheless, for high water content products, moisture absorber should be carefully designed in order to avoid undesired food dehydration. Sachets containing desiccants as calcium oxide, calcium chloride, molecular sieves, natural clays, and silica gel are usually employed for humidity control in low-level moisture food, for instance in dry food packages. In addition, moisture absorbing material can be dispersed into the polymeric matrix or between two layers of plastic films constituting the package.

Inorganic compounds used as desiccants include materials such as silica gel, bentonite, calcium sulfate, molecular sieves, which are able to retain increasing amounts of water as the humidity increase, or metal oxides, such as calcium, barium, and magnesium oxides, which irreversibly react with water providing the corresponding oxides. Organic-based absorbers are also reported such as fructose, sorbitol and cellulose and modified starch. Researchers reported the application of graphene oxide papers as desiccant for food preservation. They studied the moisture adsorption capacity of graphene oxide papers with different conditions of surface functionalization, showing promising results for applications related to food preservation.

Ethylene Removal Systems

Ethylene is a pure unsaturated hydrocarbon, odorless and colorless. It is a hormone naturally produced by plants and plays an important role in their growth cycle, respiration rate, somatic embryogenesis, seed germination, root growth and development. On the one hand, its action can lead to an appropriate ripening process that prepares fresh produce to the market (e.g., de-greening of lemons), on the other hand the acceleration of ripening and degradation of chlorophyll can cause a deterioration of the quality and reduce the shelf-life during postharvest storage of fresh products.

Climacteric fruit and vegetables produce high amount of ethylene, aldehydes, and other gases during ripening, which further promote the ripening process. The reduction of ethylene in the package atmosphere by active systems can slow undesired effects on these products. The presence of just 1 ppm of ethylene in the package is sufficient to trigger climacteric fruit ripening.

Conversely, the application of ethylene removal systems is not common in packaging of non-climacteric fruit (cherry, strawberry, pineapple, etc.), since they do not require ethylene gas during ripening. Ethylene removal can be carried out by ethylene absorbers, which physically absorb and hold its molecules, and scavengers, acting by chemical reactions.

Ethylene Adsorbents

Ethylene adsorbents include silica, zeolite, montmorillonite, cloisite, activated carbon and Japanese Oya clay. In general, commercial ethylene adsorbents can be provided in sachets or could be incorporated into a plastic film structure, commonly used in fresh produce packaging. Recently, researchers studied alkali-treated halloysite nanotubes as ethylene absorber compared to raw halloysite nanotubes. The alkaline treatment provided an increase of the pore size of natural halloysite allowing a larger and faster adsorption capacity of ethylene than raw halloysite nanotubes.

A novel work provided useful data for the design of new adsorbent based on the interesting properties of Yttrium doped graphene oxide. The adsorption properties of H_2O, CO, and C_2H_4 molecules on the surface of the graphene oxide and Yttrium doped graphene oxide were investigated by the density functional theory calculations. Results showed as the substituted Yttrium atom can affect the electronic properties of graphene oxide molecule, leading to a noticeable increase in the adsorption energies.

Ethylene Scavenger

The most widely used ethylene scavengers is made by potassium permanganate (4–6%) supported on inert matrices, such as alumina or silica gel. This compound is not integrated in surfaces in contact with food because of its toxicity, so it is embedded in minerals or nanoparticles to enhance its scavenging ability and it is usually enclosed in permeable sachets. Moreover, $KMnO_4$-based scavengers are also available in the forms of tube filters, blankets, and films. Mechanism of action involves the conversion of ethylene gas by oxidation into ethylene glycol, carbon dioxide and water.

Due to their high surface activity and reactivity, nanoparticles can be used to oxidize ethylene into water and carbon dioxide by photocatalytic reaction, using silver, titanium dioxide, zinc oxide, copper, and palladium. Chitosan-titanium dioxide nanocomposite film was developed by Siripatrawan and Kaewklin, who aimed to reduce the problems related to the spontaneous agglomeration observed for TiO_2 nanoparticles, and to produce an active packaging with both ethylene scavenger and antimicrobial properties. Chitosan film containing 1% TiO_2 showed suitable water barrier and ethylene degradation properties, as well as antimicrobial activity. Supplementary details about ethylene scavengers can be found elsewhere.

Carbon Dioxide Scavengers

Presence of carbon dioxide into food packages is generally beneficial and this aspect will be discussed in following paragraph. Nevertheless, an excess of carbon dioxide can induce undesirable effects on food and packaging. Excess of carbon dioxide could be due to the metabolism of the product and of microbial contaminant, being CO_2 one of the main results of catabolic reactions that occur in biological systems. For instance, packaged non-pasteurized or non-sterilized fermented foods, such as kimchi, yogurt, cheese, and soy paste, continue the microbial activity during storage and distribution.

The microbial activity produces high levels of CO_2 inside the package, leading to its potential collapse or to undesirable changes in texture and flavor of food, such as discoloration, off-flavor development, and tissue breakdown. Examples of foods, whose quality can be affected by excess of carbon dioxide are potato, lettuce, onion, cucumber, cauliflower, artichoke, apricot, peach, apple, and carrot. Also roasted coffee is able to release high amount of carbon dioxide, which may cause the package to burst. Due to the low permeability of some packaging to CO_2 and to overcome its accumulation, scavengers placed into the food package in form of sachets were developed.

Calcium oxide and hydrating agents such as silica gel enclosed in porous sachets allow the reaction between water and calcium oxide forming calcium hydroxide, which finally reacts with CO_2 producing calcium carbonate. Other absorbers consist of sodium hydroxide, potassium hydroxide in form of sachets or granules or physical absorbers (zeolites, activated carbon) in form of beads and powder.

Antimicrobial Active Packaging

Food shelf-life is strongly affected by the presence of microorganisms. Contamination could occur at every stage of the supply chain, whenever food is exposed to the environment and its outcomes can range from simple alterations of sensory features to serious health hazards for consumers. Among the technologies developed to hinder this issue, antimicrobial packaging can be an effective tool to improve food safety, extend food shelf-life and reduce food waste and economic losses. Active emitters contained in packaging aim to provide a controlled release of compounds, able to ensure a right level of humidity, inhibition of harmful microorganisms and prevent bacteria spoilage.

Carbon Dioxide

Carbon dioxide has a well-known antimicrobial effect on several microorganisms and it is used in modified atmosphere packaging to prolong food shelf-life. Nevertheless, since CO_2 permeability through some packaging materials is higher than oxygen and it can also dissolve in food, carbon dioxide generators are needed to maintain the CO_2 concentration to a desired value and to avoid packaging deformation.

The antimicrobial activity of carbon dioxide is strictly related to its rate of solubility and the amount dissolved in food, which increases as temperature decreases, and in food packaging it is proportional to the gas partial pressure in the headspace.

For fresh products partial pressures of CO_2 higher than 0.1 atm are able to inhibit the respiration rate and microbial growth on food surface. The formation of carbonic acid within the cell, decreasing the intracellular pH and activities, inhibition of decarboxylating enzymes, non-specific inhibition of susceptible non-decarboxylating enzymes, inhibition of membrane functions are all possible mechanisms of CO_2 antimicrobial activity.

Sachets or pads with double action of CO_2 emitters/O_2 scavengers are usually employed. Iron carbonate (II) and metal halide as a catalyst are conventionally used. Ferrous carbonate, combinations of sodium bicarbonate and citric acids, or ascorbic acid and sodium bicarbonate are commercially solutions to obtain CO_2-releasing system.

Ethanol

Ethanol is an antimicrobial agent able to inhibit yeast and bacteria growth. It is particularly effective on molds, and its effect in prolonging shelf-life of bakery products by direct spraying was widely demonstrated. Emitting sachets or films containing food grade ethanol provide the exchange of ethanol with water vapor in the package headspace. To mask the odor of ethanol, sometimes flavors are added to the sachets. The rate of ethanol release depends on carrier water permeability, amount of ethanol into the sachet, food water activity and ethanol permeability through film material. In the latter case, ethanol imbedded films usually require additional layers to provide a sustained release. Recently, researchers proposed an ethanol gel, obtained by gelation reaction between ethanol and sodium stearate, adsorbed on diatomite, in order to enhance the release of ethanol emitter. The main disadvantage of ethanol emitters is its absorption on food, which can be reduced to insignificant values by heating or microwaving the product. Food products consumed without being heated may contain residual ethanol and might cause regulatory problems.

Preservatives

Chlorine dioxide and sulfur dioxide enclosed in sachets and pads attached to the internal part of the package are volatile agent with antimicrobial action. Non-volatile preservatives with antimicrobial activity include weak acids and their salts such as acetate, sorbate, benzoate, and propionate, which probably exert the antimicrobial activity being transported in the undissociated form into the plasma membrane, where the higher pH provide the dissociations in ions that cannot return through the plasma membrane. The antifungal activity of potassium sorbate in films was reported in several studies.

Inorganic Nanoparticles

Moreover, inorganic nanoparticles composed of metal ions of silver, copper, gold, platinum, selenium and metal oxides such as TiO_2, ZnO, MgO, and CuO are used for the production of active packaging, obtained by the incorporation of such materials into adsorbent pads or plastic films. Even if the mechanism of action is still under debate, the antimicrobial activity seems to be due mainly to the release of metal ions and the formation of reactive oxygen species, which provoke deadly damages to the wall and membranes of microbial cells.

The small size of nanoparticles implies a high surface-to-volume ratio, resulting in an enhance of metal reactivity as photocatalysts and improving the interactions between metal nanoparticles and microbial membranes.

Recently a review by researchers explored the advantages of bimetallic and trimetallic nanoparticles compared to monometallic ones and their potential application in food packaging. They are hybrid nanostructured material showing improved thermal, mechanical and gas barrier properties and higher antimicrobial activity compared to monometallic nanoparticles.

However, the use of such inorganic nanoparticles should be carefully selected, basing on the type of food and the properties of packaging film (barrier properties, transparency) that can be affected. Furthermore, some concerns related to potential toxic effects on human health due to migration to food, dermal contact, and inhalation, are presently increasing, and depends on the chemical composition of nanoparticles, size, surface chemistry, solubility, and hydrophobicity.

SYNTHETIC ANTIOXIDANTS AND OXYGEN SCAVENGERS

Food products are generally sensitive to oxygen, which induces undesired changes to the organoleptic properties, such as color modifications, development of off-flavors, as well as deterioration of nutritional properties, and supports microbial growth. Consequently, great importance is given to strategies aiming to minimize oxygen content in the headspace of food packages. Vacuum and modified atmosphere packaging are two methods that proved to be effective in prolonging shelf-life of some foods.

Nevertheless, both of them cannot impede the permeation of oxygen over time from the external environment and are also able to reduce the amount of oxygen in the package up to only 0.5–2 vol.%, while lower values can be achieved by oxygen scavengers. The control of oxygen levels is obtained using synthetic antioxidants and oxygen scavengers.

Oxygen scavengers are used in packaged food products in sachets, bottle crowns, labels, plastic films, and trays. One of the most used solutions available in the market employs metallic scavengers, which carry out oxygen removal by chemical reactions. For instance, in iron-based scavengers (iron powder, activated iron, ferrous oxide, iron salt) the reduced metal oxidation occurs in presence of moisture or Lewis acids.

The rapid rate of oxidation of iron, can be further increased by using nano-iron particles blended with activated carbon, sodium chloride and calcium chloride. However, some disadvantages of iron-based scavenging systems are related to potential contamination of food due to accidental breakage, to interferences with inline metal detectors and to the inhibition of heating by microwave ovens.

Other metals employed are platinum and palladium as efficient catalysts for the conversion of hydrogen in water. For this purpose, the atmosphere needs to be modified in order to hold high pressures of molecular hydrogen and the metal catalyst is used to improve the reaction also in presence of small amount of oxygen.

Nevertheless, as reported by, due to the flammability of hydrogen, there is a maximum of oxygen that can be removed (2.5 vol. %) when hydrogen is introduced in the modified atmosphere. An alternative involves hydrogen-evolving compounds, such as calcium hydride or sodium borohydride, which allow a controlled hydrogen release, and which can be easily inserted in the cap of bottles, while more difficulties are met in their incorporation in films.

In addition, films based on oxidizable transition metals such as copper, zinc, magnesium, manganese, aluminum, or titanium, and nanocrystalline titanium particles are also reported.

The most common synthetic antioxidants comprise phenolic compounds as butylated hydroxytoluene, butylated hydroxyanisole and tert-butylhydroquinone and propyl gallate, which are hydrogen donating free radical scavengers. They are widely used in food active packaging in order to prevent lipid oxidation, thus they are extensively applied to enhance the shelf-life of fat and fat containing products. In addition, although they are generally used as antioxidants, they have shown also antimicrobial activity. However, these compounds are suspected to have potentially toxic and carcinogenic effects, thus the research is more oriented toward the application of natural antioxidants, which will be discussed in this review in the following paragraphs.

AGENTS FOR ACTIVE PACKAGING FROM NATURAL PRODUCTS

The use of naturally occurring agents for the fabrication of active food packaging is the trend to which the research is currently approaching. Natural antioxidant and antimicrobial agents are perceived as safer than synthetic ones by consumers, thus different natural compounds have been proposed for active packaging purposes and the interest of the scientific community in this field is continuously increasing.

Bacteriocins and Enzymes

Bacteriocins, such as Pediocins, Nisins, Enterocins and Sakacins, are peptides produced by bacteria ribosomal synthesis, which gained strong attention by food industry due to their antimicrobial effect, even at low concentrations. The main bacteriocin producers are lactic acid bacteria, so these naturally occurring compounds are present in several fermented food. They are able to inhibit many pathogenic microorganisms, but the mechanism of action is still not completely known. Bacteriocins produced by each strain are active only against their competing bacteria, since the originating bacterial cells possess specific immunity mechanism toward their bacteriocins.

Some hypotheses related to the mechanism of action were evaluated by recent studies and discussed in a recent review by Santos et al.. Such mechanisms include the disturbance of bacterial metabolism, interaction with intracellular targets, inhibition of the synthesis of nucleic acids and proteins, interferences with the formation of cellular components, in which the cationic characters of the antimicrobial peptides play a crucial role in their adsorption on microbial cell surfaces.

Bacteriocins produced by lactic bacteria show properties that make them attractive for active packaging applications. Indeed, they are substances generally recognized as safe and are inactive and nontoxic on eukaryotic cells and on consumers.

In addition, bacteriocins have little influence on consumer's gut microbiota, being inactivated by digestive protease, while show broad antimicrobial activity against many food-borne pathogenic and spoilage bacteria.

Furthermore, they do not provide any alteration of the organoleptic properties of food as they are taste-, odor-, and colorless. Finally, they are active in wide ranges of pH and temperatures. The last aspect plays an important role from an industrial point of view, since standard melting blending technological processes are not suitable for thermosensitive compounds.

Nisin is a commercially acceptable food grade bacteriocin, active against Gram-positive bacteria such as *Listeria monocytogenes*, *Clostridium botulinum*, *Staphylococcus aureus* and *Bacilli*. It is allowed for use in pasteurized cheese and liquid eggs and commercially used in a range of foods including dairy, eggs, vegetables, meat, fish, beverages, and cereal products. Holcapkova et al. found that the antimicrobial activity of nisin against *Micrococcus luteus* is temperature-resistant, indeed they showed that the polylactic acid (PLA)-based films displayed antimicrobial properties even when they were thermally treated above the melting temperature of PLA, but losing the 25% of their activity under treatment at 160 °C and more than 60% at 180 °C for 5 min. Pediocin was also reported to be a remarkable natural biopreservative to overcome the post-processing contamination of meat products against *L. monocytogenes*.

According to literature, nisin and other bacteriocins such as pediocin and lacticin are ineffective against molds, yeasts, and Gram-negative bacteria, with the only exception given by a few strains. Incorporating chelating agents such as Ethylenediaminetetraacetic acid (EDTA) was found to increase the effectiveness of bacteriocins in food packaging application.

However, in a recent work, the bacteriocins extracted from standard *Pediococcus acidilactici* culture and the isolate, *Enterococcus faecium*, showed antibacterial activity against both Gram-positive and Gram-negative bacteria. In addition, Woraprayote et al. reported that Bacteriocin 7293, a novel bacteriocin from *Weissella hellenica* BCC 7293, with polylactic acid/sawdust particle (PLA/SP) film provided both Gram-positive (*Listeria monocytogenes* and *Staphylococcus aureus*) and Gram-negative bacteria (*Pseudomonas aeruginosa*, *Aeromonas hydrophila*, *Escherichia coli* and *Salmonella Typhimurium*) inhibition in pangasius fish fillet. In addition, bacteriocins demonstrated additive or synergistic effects when used in a combination with other bacteriocins or with other preservatives or phenolic compounds.

Enzymes are also currently applied as antimicrobial agents in food packaging. In particular, lysozyme is a naturally occurring enzyme with activity against cellular structure specific to bacteria, since it is able to damage the structural integrity of the cell wall, resulting in the lysis of bacterial cells. This enzyme exhibits antimicrobial activity against Gram-positive bacteria but not on Gram-negative bacteria, but susceptibility of Gram-negative bacteria could be enhanced by the addition of chelating agents such as EDTA.

Another enzyme that can be used as an active agent is Glucose oxidase, whose reaction products, yielding hydrogen peroxide from glucose and oxygen, show antimicrobial power. However, the application of this enzyme in packaging is limited by the cost of enzyme and the requirement of glucose, whose amount in many foods in not sufficient. Both bacteriocins and enzymes were employed as directly incorporated in polymer matrix, as coated onto material's surface or by immobilization in active packaging.

The first method is used with biodegradable polymers such as carbohydrates and proteins, when thermal processing is not used, due to the potential loss of antimicrobial activity. Alternatively, coating by immersing or contacting the film with a peptide solution or by solvent casting can be performed. Immobilization can be achieved by physical or chemical methods and, due to the low tolerance of enzymes to high temperatures, immobilization is the option for the application of enzymes to antimicrobial packaging and can improve the stability of bacteriocins.

Phytochemicals

Among natural food preservatives, essential oils and other extracts from plants, herbs, and spices, deserve particular attention being able to exhibit great antimicrobial and antioxidant potential. Essential oils are volatile substances because of the high content of low molecular weight aromatic compounds. They are biologically produced by plants as secondary metabolites and exert antimicrobial activity due to the high concentration of phenolic compounds (up to 85%), which provoke damages to bacterial enzyme systems, genetic material, and cell membrane phospholipid bilayers.

The main drawback of the use of essential oils and their components as antimicrobial agents is the high concentration required. Most of essential oils extracted from plants and spices are classified as Generally Recognized As Safe but, due to their strong flavor potentially changing organoleptic features of food, their incorporation in films can be preferred to their direct use as food additive. Examples of natural sources of essential oils proposed for active packaging applications are garlic, cinnamon, lemongrass, oregano, rosemary, thyme and bergamot.

Antioxidant and antibacterial packaging materials with essential oils can be obtained by direct mixing into packaging material by blending or solvent casting technologies. Alternatively, adsorption or coating the packaging material can be used for the addition of essential oils. In addition, carriers in which essential oils are adsorbed or embedded can be used for the fabrication of small antibacterial bags continuously releasing the active compound in package headspace, or modified atmosphere packaging can be created by filling the entire packaging with essential oils in gaseous form.

The application of essential oils into hydrophilic materials by casting is mainly obtained by emulsification or homogenization techniques, which allow creating fine emulsion and after drying the lipid droplets remain embedded in the polymer matrix. In addition, biopolymer-stabilized emulsions, also known as Pickering emulsions, are an interesting sustained delivery tool that was used for the same purpose. They involve the stabilization of oil in water O/W or water in oil W/O emulsions by replacing surfactants with solid particles. A solid layer around the essential oils droplets prevents the contact with the aqueous phase, thus allowing higher coalescence stability, higher loading capacity, a more sustained release of the active agent and its protection from the external environment, particularly from oxidation.

Solutions of active packaging with essential oils and examples of applications can be found in recent reviews. However, an important characteristic that should be taken into account is the interaction between essential oils and the packaging material, due to the alterations that the former can induce on the film. For instance, the addition of thyme essential oil to bio-nanocomposites made of whey protein isolate and cellulose nanofiber provided the decrease of water vapor permeability, the reduction of film transparency and the films obtained with thyme essential oil resulted as less strong and elastic than the control samples.

On the other hand, researchers found that the addition of lemongrass essential oil by emulsification into formulations based on glycerol-plasticized cassava starch improved colorimetric attributes, thermal stability, barrier to moisture, and mechanical properties of film.

Active packaging has been developed using several natural extracts from plants and spices. Pure standards of natural antioxidants were used in several works to produce active films by different techniques. For instance, gallic acid and quercetin were added in polyvinyl alcohol (PVA) for the fabrication of active films by solvent casting, quercetin-starch-based complex were incorporated in chitosan-gelatin-based films, quercetin was adsorbed onto maize starch and calcium alginate aerogel through supercritical carbon dioxide adsorption, while α-tocopherol into polyethylene terephthalate/polypropylene (PET/PP) films was added by supercritical CO_2 impregnation, and Poly(L-lactic acid) (PLLA)/antioxidant mixtures containing tocopherol and resveratrol were compression molded to films.

The antioxidant and antimicrobial properties of natural substances such as phenolic compounds are well-known in the literature. Many plants, herbs and spices are considered to be potential antioxidant sources due to the rich amount of phenolic acids, carotenoids, and polyphenols in their extracts, which can be added to packaging materials, resulting in an extension of food shelf-life, as reviewed by researchers.

Free radical scavenging of phenolic compounds is mainly provided by two mechanisms, the former involves an H-atom transfer by the antioxidants, while the second a one-electron transfer, resulting in the formation of a nonreactive phenoxyl radical or a stable radical cation, respectively. The single antioxidant ability to inhibit oxidizing chain reactions depends on the molecular structure, conjugation, and resonance effects.

Very powerful antioxidants are also carotenoids. As reported by researchers, they are most likely involved in the scavenging of singlet molecular oxygen, peroxyl radicals and they are effective deactivators of electronically excited sensitizer molecules, which are involved in the generation of radicals and singlet oxygen by physical quenching.

Antimicrobial activity of phenolic compounds was explained by several mechanisms, such as aggregatory effect on all the bacterial cell, destabilization of cytoplasmic membrane, permeabilization of cell membrane, inhibition of extracellular microbial enzymes, deprivation of substrates required for microbial growth and direct action on microbial metabolism, depending on the subclass considered. Several studies evaluated carotenoids antimicrobial activity against both Gram-positive and Gram-negative bacteria.

Recently, the antimicrobial activity of fucoxanthin, a carotenoid produced by brown algae and diatoms, was evaluated by Karpiñski et al.. However, some studies highlighted synergistic effects of natural compounds, able to increase both their antioxidant and antimicrobial activity when they are simultaneously present in mixtures or in extracts, compared to the same amounts of a single compounds. For this reason, studies on active packaging dealing with the incorporation of natural extracts, rather than a single compound, in polymers for food applications are attracting more and more the attention of the scientific community. In addition, agrifood industry residues, such as spent coffee grounds, apple skins, grape seeds and skins, tomato waste, olive pomace, are still rich of antioxidant compounds, which can be recovered for active packaging purposes, allowing a waste valorization.

PACKAGING CONCEPTS

Make sure your food product packaging checks under the following category. With these values in mind, designing a great food product becomes much easier.

Clarity of message

Consumer buys essential products and would like to try a product if it catches their attention. Shoppers often make their decision in one-third of a second and they can become frustrated if the food packaging design makes it difficult to choose a product. Your product becomes successful when the end consumer can understand the brand, the product and the packaging design.

Integrity

So consumers have bought your product once, and for them to buy it again, they have to trust your brand which is based on the integrity of your intact product. A false expectation should not be created by using dramatically edited images that will not meet the consumer expectations.

Authenticity

Try to bring the common colours and material to suit your product. For instance, if your product is about natural foods, then choose earthen primary colours which will appeal better. Moreover, this authentic style of packaging should be consistent with your brand products. Research has shown colours can be effective to grab customers focus and attention. For instance, red, green, yellow and pink are the most eye-catching colours. This will aid in distinguishing your product from its competition.

Shelf impact

Food retailers also think about food packaging and design which are displayed on the shelves to attract customers and this is called shelf impact. Product placement is done based on the product looks and patterns that they create. Products place next to a certain product can change the perception of consumers leading to both an increase or decrease in their sales.

If your product looks like your competition there is a chance that your product could get blended and get lost in the crowd. However, if the packaging design standout it is more likely to get picked by the consumer.

Versatility

The food packaging design needs to be versatile and the key is to expand the product lines, which may be as simple as adding a new flavour. The best packaging design will be the one that is adaptable to the variety of products in your segment. For instance, show your customers what they can expect to find in the food package. The best template would be to use a consistent or colour.

Usability

Food packaging design not only needs to be attractive but also needs to convenient and usable. For instance, mayonnaise filled in squeeze jars is more usable compared to glass bottles. McDonald's has launched a Mcbike package which allows cyclists to attach the packaging with burger, and chips on to the cycle bar. Dunkin Donut's coffee cup is molded to carry sugar and cream with it. When the product becomes usable, it becomes appealing to consumers and increases the chances of buying again.

CURRENT PACKAGING TRENDS

Technology-enabled smart packaging

This is pretty new and novel for most **food development companies.** Technology can now be used in the packaging to provide consumers with more information for security and confidence. For instance, QR codes and smart labels which when scanned using a smartphone provides information on temperature, pH, special ingredients, allergen if any, compliance with health standard etc. Companies such as Nestle's Maggi soups used patented "granulation based technology' to retain the freshness of its ingredients.

Emotional engagement

Consumers may also expect a story to a product. Emotional content is gaining popularity which then trends in social media with positive or negative sentiments. For instance company such as Paper Boar in association with tetrapak, incorporated the feeling of reminiscence in its packaging.

Another emotion most consumer would love is nostalgic value. Vintage-inspired designs evoke happy emotions in consumers reminding them of simpler times. This could be achieved by thinking about how a certain product would have been packed a decade ago. This concept would be appealing to the younger generation as well as they would have never experience that era.

Transparent and clean labelling

Recent studies on the packaging and consumer behaviours in 2020 have shown 38% of people are keen on buying products with clear information.

As explained in our previous blogs, consumers are well aware and conscious of what they eat. Manufacturers need to adapt to clear labelling to list the contents as it harder for people to read the fine print. This also indicates that the manufacturer is sticking to its promise of delivering healthy packaged product, upscaling the brand identity and integrity.

Personalization

Personalized packaging is gaining momentum as it plays a crucial part in attracting consumers. For instance, Coca-Cola, personalized printing names on the packaging which were initially only meant for gift orders on their online stores. An increase in production speeded up food packaging personalization making it a reality for everyone.

Minimalistic design with colours

Minimalism is another new concept the younger generations can connect with. Manufacturers have shifted from flamboyant designs to adopt simplistic clear packaging. This also aids in highlighting the product's value and keeps the costs low.

Having said that, consumers also love the colour, shape and design of the most attractive aspects of packaging. Using this information, manufacturers are using multiple areas of design, gradient colours, from logo to packaging to show a level of depth and intricacy.

Sustainability

Consumers are willing to pay extra to support green and sustainable packaging. The ecofriendly options have driven big multi-level corporations to ditch single-use plastics and provide biodegradable and recyclable packaging options. To add, single-use food cutlery is also switched for plant-based materials for improved sustainability.

NEW CONCEPTS IN FOOD PACKAGING

In recent years there has been a surge of criticism against plastic packaging and the damage it does to the planet.

As fruit and vegetable processors well know, though, packaging plays an important role in protecting product and increasing shelf life, while doubling as a marketing and branding tool. The following companies have designed new packaging solutions that offer the best of both worlds — functionality and environmental protection.

Fox Packaging's Eco Stand-Up bag

Texas-based company Fox Packaging showcased two new products at the United Fresh Produce Association expo in Chicago earlier this year: the Eco Stand-up bag and the HD Stand-Up bag.

The HD Stand-Up bag is made of high-density polyethylene, which creates a more stiff and robust packaging solution for fresh produce.

"The Fox HD Stand-Up bag significantly reduces waste, energy use and CO_2 emissions by %20," Fox said. "Upholding the similar stiffness of a polypropylene, the HD Stand-Up bag holds an RIC (Resin Identification Code) of 2#, compared to polypropylene's 5#."

The Fox Eco Stand-up bag (RIC #4) offers the same convenience as the stand-up format, but features an eco-starch resin, which reduces plastic consumption. The biopolymer is both sustainable and durable.

Not only do the new HD Stand-Up bag and Eco Stand-Up bag offer various on-shelf displays, they are also our most sustainable and eco-friendly packages yet.

The HD Stand-Up bag becomes the Fox Stand-Up Combo bag when you add a mesh window to it. This is a truly unique aspect to our pouch line of offerings, besides, of course, the recyclability. This ability to customize not only caters to our customer's brands, but specifically serves a commodity's needs.

Fig. 2. Fox Packaging's Eco Pouch contains potato resin to make it reusable and recyclable

SEAWEED-BASED PACKAGING

According to Indonesia-based packaging company Evoware, Indonesia is the world's second largest contributor of plastic waste in the ocean and it is estimated that 25% of its fish are contaminated with plastic waste. Most of that waste comes from plastic food wrapping.

Poverty is also a very real issue in Indonesia where five of the six poorest provinces produce seaweed, but farmers are not well paid due to lengthy supply chains and overproduction. Working in collaboration with local seaweed farmers, the company has been able to address both concerns through the creation of high-quality seaweed-based packaging.

The packaging has a two-year shelf life and dissolves in warm water, which makes it a zero-waste product. It is 100% biodegradable and can be used as a fertilizer for plants. It is Halal certified, safe to eat and produced in compliance with HACCP standards. According to the company, it can be customized to add flavor, color and even a brand logo.

Eco-Lok

Small changes can amount to big change. At least, that's the thought behind Kwik Lok's new bag closure design. Using proprietary technology from BioLogiQ, Eco-Lok was designed with environmental stewardship in mind.

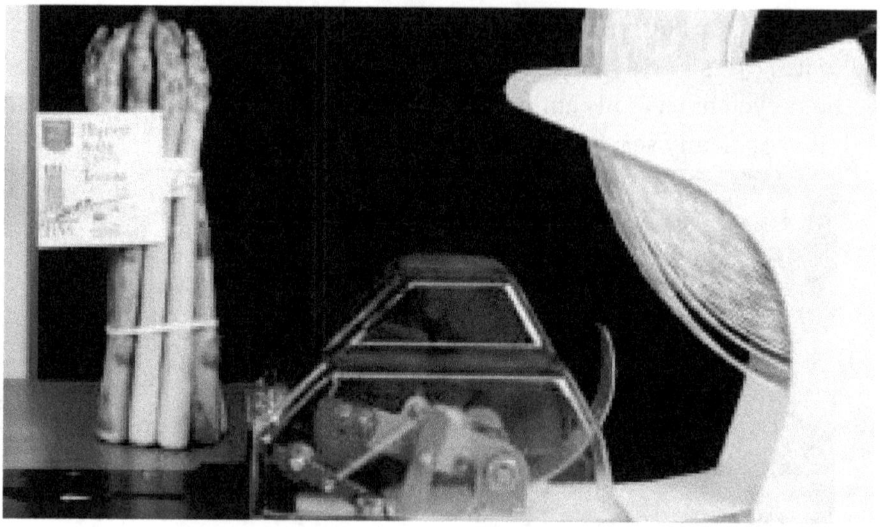

Fig. 3. Eco-Lok

Schur Star Microwaveable Bag

The unique functionality of the Schur Star Bags in combination with the multiple filling positions on the Schur Star filling line, makes the concept suitable for ready-meal production.

The demand for convenience food is steadily growing and there is a trend for smaller single-servings meals. Therefore, Schur has developed flexible solutions for this market.

Fig. 4. SchurStar's microwavable bag

Aimed at the convenience market, Schur Star now offers an easy-to-open microwaveable bag. The company says the bags can be filled quickly and easily, and the attractive packaging increases visibility in supermarkets. Steam valves have been applied to the bag, so the contents can be heated directly in the packaging.

Another product worth mentioning is Branding By Banding. It's a concept by Bandall and specialist printer MaxAarts. The printed band is designed to wrap around product, and includes all relevant information, such as bar code, expiration date, nutritional value and ingredients. The aim of the concept is to reduce packaging, overall waste, energy consumption and handling.

INTELLIGENT PACKAGING

In the literature there are many definitions of the concept of intelligent packaging. European Commission defines intelligent packaging as "materials and articles that monitor the condition of packaged food or the environment surrounding the food". Intelligent packaging has also been defined as science and technology that uses the communication function to facilitate decision making to extend shelf-life, improve safety, guarantee quality, provide information, and warn of any problems by monitoring changes in the internal and external environment of the packages.

More in details, intelligent packaging is a packaging system that uses the internal (e.g., metabolites) or external (e.g., temperature) package environment as "information" to monitor the status of product quality, in order to improve not only product safety, but also to track the product for automatic identification and traceability.

The aim of intelligent packaging lies in its communication ability; the package can be considered the best companion of the product, since they constantly move together along the entire supply chain and therefore the package can, in principle, constantly communicate the conditions of the product. The development of new intelligent packaging capable of continuously providing information on the conditions of both the food product and the packaging integrity, allows for a safer and more efficient supply chain, avoiding unnecessary transports and logistics and reducing food waste.

Products are regularly subjected to microbiological and chemical tests, which are carried out during production, but ensuring adequate control after delivery to the supermarket it is often not possible. Intelligent packaging can fill this gap. Intelligent packaging can also contribute to improving 'Hazard Analysis and Critical Control Point' (HACCP) and 'Quality Analysis and Critical Control Point' (QACCP) methodologies, used to control, detect, prevent, reduce, and eliminate any possible criticism that could compromise the food product and its final quality.

The main technologies for intelligent packaging system are three: indicators, sensors, and data carriers. Indicators and sensors have the main function of providing information relating to product quality, while the class of data carriers is more involved in the management of the supply chain logistics. These systems can be placed on the primary packaging, inside or outside, on the secondary or tertiary packaging.

Indicators

The main function of indicators is to transmit to the consumer information related to the presence or absence of a specific substance, to indicate whether a reaction is taking place between two or more components or to monitor the concentration of a certain substance. This information is translated into signals usually in the form of immediate visual changes (e.g., different color intensities or diffusion of a dye along a straight path), providing qualitative or semiquantitative information. In most of the cases, the basic requirement of an indicator is that these changes in color or intensity are irreversible.

The indicators represent a very consistent class of intelligent packaging; the most used classification, also for this review, is based on the type of variable controlled, so they can be grouped into three macro categories: time-temperature indicators, freshness indicators, and gas indicators. For a more general classification, the indicators can be classified as external or internal, based on their positioning on the packaging.

Time-Temperature Indicators

Temperature is a fundamental factor in monitoring and determining the shelf-life of a food product, especially for perishable products. One of the main challenges of the perishable food industry is to ensure the quality of food by protecting it from unwanted temperature fluctuations, a challenge that is not simple at all. Temperature monitoring and control along all the supply chain are essential to maintaining the quality and safety of perishable foods. Incorrect or uncontrolled temperature management can cause a large amount of food waste during production, as well as during the distribution chain. Studies report that poorly managed temperature in perishable food logistics can cause up to 35% loss of products. To meet these needs, time-temperature indicators (TTI) provide valuable support.

Time-temperature indicators find application majorly for temperature-sensitive foods such as chilled and frozen products. Furthermore, TTIs have also been applied to control the pasteurization and sterilization process. The classification of TTI varies slightly from one author to another, the clearest grouping them into two categories: partial or full history indicators. A partial history indicator does not respond unless a certain predetermined threshold temperature is exceeded, and therefore identifies abusive temperature conditions, warning consumers of the potential survival of microorganisms and protein denaturation during, for example, freezing or defrosting.

A full history indicator responds continuously to all temperatures along the food supply chain, so as to provide a measure relating to the entire life of the product. The TTIs currently available on the market, have operating mechanisms based on different principle that mainly include physical change or chemical reaction such as polymerization, melting or acid-based reaction, generally expressed through a visible response in the form of mechanical deformation, color development or color movement. These types are the most common, but there is another class to mention, which is the enzymatic one, based on a biological response. In this case, TTI is sensitive to the change in biological activity of microorganisms, spores or enzymes. The rate of change is directly proportional to the change in temperature similarly to the reactions responsible for the product deterioration.

However, enzymatic TTIs still have limitations in terms of high cost and enzymatic instability. Some authors have developed an immobilized laccase-based TTI by electrospinning zein fiber to increase the stability of the enzyme. Immobilization onto solid carriers is the most used strategy to improve stability and fibrous membranes made by electrospinning have a great potential to be used for enzyme immobilization.

Usually, TTIs are made of small, self-adhesive labels attached onto containers or individual consumer packages, so as to be clearly visible and interpretable. These systems must represent time-temperature dependent changes in an easily measurable manner that can be quickly correlated with the extent of deterioration and the remaining shelf-life of the food.

Freshness Indicators

Freshness indicators are intelligent devices that allow the monitoring of the quality of food products during storage and transport. Freshness decay can be due both to exposure to harmful conditions and to a shelf-life longer than that established. Freshness indicators, unlike temperature indicators, provide direct information on the quality of the product, analyzing the chemical reactions of food deterioration due to target microorganisms. Changes in the concentration of metabolites such as glucose, organic acids, ethanol, carbon dioxide, biogenic amines, volatile nitrogen compounds or sulfur derivatives are indicators of microbial growth and their presence can therefore be exploited in the freshness indicators.

In most cases, freshness indicators are based on the use of dyes sensitive to pH variations caused by the deterioration of the product, which lead to the visible change in the color of the indicator. These indicators find applications in various products including fresh food, fruit, and seafood, but their marketing is still rare, especially in Europe, primarily for legislative issues related to substances that can be in contact with food.

In most cases, various polymeric matrices (e.g., polypropylene, nylon, or cellulose films) are impregnated with pH-sensitive dyes that are typically: bromocresol purple, bromocresol green, bromophenol blue, methyl red, and cresol red. In recent years, the use of natural dyes has also been studied. For instance, curcumin, grape peel, and beetroot extracts have been explored to detect the aging of cod flesh, or a system based on corn starch, chitosan and red cabbage extract was tested during fish spoilage.

Freshness indicators for seafood are based on the total volatile basic nitrogen content (TVB-N), as volatile amines, which are formed as the food spoils. Instead, hydrogen sulfide indicators can be used to determine the quality of meat products; in fact, hydrogen sulfide is correlated with the color of myoglobin, which is considered a quality attribute for meat products.

Most of these sensing mechanisms are based on color changing dyes, but penetration of the dyes into the package can affect the organoleptic properties of the packed food. To avoid this issue, some researchers have developed a freshness indicator consisting of a triple layer: a layer in contact with the food in non-woven high-density polyethylene, a layer containing a dye sensitive to pH variations, bromocresol green, immobilized with the binding polymer, and an external layer in low-density polyethylene. This indicator is able to detect chicken spoilage as a function of the change in concentration of volatile basic nitrogen, CO_2, and bacterial count, without migration of the dye inside the package, thus offering a safe packaging solution, simple and reliable.

Normally, freshness indicators are printed on the packaging film or are incorporated, in the form of labels, inside the packaging film, reacting with agents produced during the storage. A common problem with the category of indicators is related to false negatives/positives. In the first case, there would be samples that seem safe but that are not, constituting a danger to the health of the consumer, while in the case of false positives the problem would be connected to the useless food waste induced by samples that seem dangerous or damaged, but which are actually healthy. In this sense, it would be necessary to develop standard protocols to accelerate the development of indicators, in particular those of freshness, also on an industrial scale.

Gas Indicators

The activity of the food, the nature of the package, and the environmental conditions to which the package is subjected lead to the variation of the gas composition in the headspace of the package. Gas indicators are used to monitor these changes and are also often used to evaluate the effectiveness of active packaging components (e.g., O_2 and CO_2 scavengers) or to detect the occurrence of leaks across the packaging, and therefore to detect package integrity, which is an essential requirement, especially when the food is stored in a modified atmosphere packaging (MAP). The most well-known gas indicators are used to control oxygen and carbon dioxide concentrations.

Oxygen is responsible for the microbial and biochemical deterioration of food and is therefore removed from the packaging and replaced with gas such as nitrogen or using oxygen scavenger. However, during transport and storage, oxygen may be able to penetrate the package due to poor sealing, defects, or damage. For this reason, the use of visible oxygen indicators becomes essential to quickly and easily verify the presence of oxygen inside the package, without using professional equipment or laboratory analysis.

The most commonly used oxygen indicator is a colorimetric indicator based on a redox dye, generally methylene blue and a reducing agent, such as glucose, in an alkaline solution such as sodium or potassium hydroxide. If no oxygen is present, the glucose in the alkaline solution reduces the methylene blue which is, in this case, colorless. If oxygen is present in the monitored environment, the dye is oxidized and intensely colored. The reducing agent is sensitive to oxygen, therefore these indicators must be created and maintained in anaerobic conditions to avoid the rapid exhaustion of the reducing agent, due to the reaction with oxygen, and to the consequent malfunction of the indicator.

Furthermore, there may be safety problems due to the contamination of the food product by the synthetic and, often harmful, chemical components used in the indicator. A solution was proposed by researchers, who developed an indicator by physically separating the components via a barrier. Activation of the indicator occurs by hand pressing and breaking the barrier, so components can come into contact.

The rate of the color change is related to the oxygen concentration. In addition, the same authors have redesigned the indicator using natural organic compounds such as cysteine and laccase, thus finding a possible solution to the second issue mentioned. The other predominant class of gas indicators is related to CO_2. The microbial growth responsible for the deterioration of food usually begins immediately after packaging, and therefore there is a production of CO_2 due to the metabolic activity of microorganisms. The rate of carbon dioxide production depends on the type of food, time, and conditions of storage, and on packaging materials.

Carbon dioxide is also used, in combination with oxygen and nitrogen, in protective atmosphere technology to inhibit microbial metabolism over time. Again, most of the carbon dioxide indicators are colorimetric labels, whose color changes as a function of the pH of the product. Researchers have, for example, studied the performance of a colorimetric indicator composed of a mixture of lysine, polylysine and anthocyanins extracted from red cabbage, tested on cold-preserved foods such as poultry meat.

The reversible reaction of amino group of lysine in presence of CO_2, leads to the formation of the corresponding carbamic acid derivative, and to a consequent variation in the pH of the solution, which is shown by a significant variation in the color of the anthocyanin dye. In recent years, anthocyanins have attracted the attention of many researchers who have investigated their behavior as natural dyes, in combination with natural-based polymer matrices, for the production of safe, non-toxic, biodegradable, and suitable colorimetric indicators to be use in the food sector.

Gas indicators, mainly in the form of labels or printed directly on the packaging films, are positioned inside the package to monitor changes in the internal gas composition, thus providing a strategy for monitoring the quality and safety of food products.

PACKAGING SPECIFICATIONS AND STANDARDS

The packaging assessment must include a definition of the optimum quality standards and these standards should not be compromised by cost. Ideally, packaging supplier selection is a techno-commercial decision agreed during discussions between the purchasing function and packaging technologists. Buyers are becoming more discerning and now expect suppliers to have quality assurance schemes that are accredited by a third party. Widely used quality management systems are those based on ISO 9000.

Total quality management (TQM) is a technique that examines the overall quality image as perceived by suppliers and customers. There has been a change in focus from inspecting production outputs to monitoring process variations. TQM can be regarded as a source of competitive advantage, especially where quality is perceived to be a problem by the customer.

Quality assurance on production and packaging lines has been facilitated by the use of integrated computerised micro-electronic control systems that can detect a range of defects and automatically eject reject packs. For example, there are automatic check-weighers, metal detectors, fill-level sensors, pack-leak detectors, label detectors, pack-dimension sensors, light-transmission sensors and odour detectors.

A retailer's own brand products, delivered by the manufacturer to the retailer's centralised distribution system, may have packaging specifications that are tailored to meet the rigours of that retailer's specific system. The packaging specifications for a national brand, however, may require packaging that needs to provide a higher degree of protection due to the wide variation in storage conditions and distribution hazards experienced during a pack's delivery through several retailers' distribution channels.

Thus, the rigours of the distribution system and a lack of control over it by food manufacturers often lead to specifications that generate extra cost and use up resources. Packaging specifications are geared to ensuring that a very high percentage of products arrive in pristine and safe condition, despite the rigours of the journey. However, this approach to packaging is in conflict with the significant pressure to minimise packaging.

The main testing methods for materials are available from a wide range of sources including ISO, American Society for Testing and Materials (ASTM), British Standards Institute (BSI), DIN etc.

CONCLUSION

This chapter has introduced the subject of packaging design and innovation which is driven by the need of industry to create and sustain competitive advantage, changes in consumer behaviour, industry's cost sensitivity, shorter product life cycles, new legislation, growing environmental pressures, availability of new materials and technological processes.

A framework has been used to analyse and describe key considerations in the design and development of packaging. It has emphasised the importance of an integrated approach to packaging by those involved in product design and development, materials and machinery development, production processes, logistics and distribution, quality assurance, marketing and purchasing. An integrated approach to packaging design and innovation requires consideration of key issues and values throughout the packaging chain.

GLOSSARY

Acetic acid. A compound found in vinegar, amongst other foods, that gives a sour and pungent smell

Acid. A substance with pH below 7 that can break down to release hydrogen ions

Acrylamide. A by-product made when foods high in starch are fried, roasted or baked at a high temperature

Additive. Ingredients added to foodstuffs to alter them in order to preserve or enhance specific qualities eg taste, colour

Aflatoxin. Poisonous, carcinogenic chemicals produced by certain moulds; more widespread in hot, humid environments with poor storage facilities

Aftertaste. The after-effect of flavour leaving a coating in the mouth after chewing food

Alkaline. Basic; a pH of greater than 7.

Amino acids. Simple organic compounds containing an amino group and a carboxyl group: the building blocks of proteins

Amylopectin. Part of the starch molecule made up of glucose molecules in a branched structure

Anthocyanin. A blue, violet, or red flavonoid pigment found in plants.

Antioxidant. A substance that inhibits oxidation, particularly important in the preservation of stored food products

Aroma. The smell given off by food (eg baked bread) and drink which is detected by receptors in the nose

Ascorbic acid. The scientific name for vitamin C; essential for growth and defence against infection

Astringency. A sensory phenomenon characterized by a dry, puckery feeling in the mouth. Often present in tea and red wine.

Baking Powder. A chemical leavener consisting of a mixture of baking soda, acid(s) and an inert filler such as cornstarch.

Baking Soda. A white chemical leavening powder consisting of sodium bicarbonate.

Blanch. To dip a food briefly into boiling water.

Calorie (kcal). The amount of energy required to raise 1 gram of water °1C (measured between °14.5 and °15.5C at normal atmospheric pressure).

Casein. The primary protein (80 percent) found in milk; it can be precipitated (solidified out of solution) with acid or certain enzymes.

Chemical. A substance obtained through chemistry or that produces a chemical effect. The term "chemical" is used here in its scientific sense. Chemicals and chemical processes can be natural or artificial.

Chlorophyll. A green pigment, present in all green plants and in cyanobacteria, responsible for the absorption of light to provide energy for photosynthesis. Its molecule contains a magnesium atom held in a porphyrin ring.

Chlorophyllins. Bright green water-soluble salts that are derivatives of chlorophyll. They differ from chlorophyll in that their phytyl and methyl groups have been removed due to exposure to alkaline conditions.

Coagulate. To clot or become semisolid. In milk, denatured proteins often separate from the liquid by coagulation.

Colloidal dispersion. A solvent containing particles that are too large to go into solution (dissolve), but not large enough to precipitate out.

Complete protein. A protein, usually from animal sources, that contains all the essential amino acids in sufficient amounts for the body's maintenance and growth.

Conduction. The direct transfer of heat from one substance to another with which it is in contact. For example, a hot skillet transferring heat to a steak that has been placed on its surface.

Convection. The transfer of heat by moving hot air or liquid (usually water or fat) through or around food. For example, a convection oven transfers heat by moving hot air around food using a fan. Deep fried and boiled foods are also heated by convection via the heated liquid around them.

Crumb. The cell structure appearing when a baked product is sliced. Evaluation is based on cell size (medium to large cells are described as open, while small cells are described as closed), cell shape and cell thickness (thin walls are described as a fine crumb, while thick walls are described as a coarse crumb).

Curd. The coagulated or thickened part of milk that is formed in the initial stages of cheese-making. This is the non-whey portion of milk and is comprised mainly of the milk's protein (casein) and fat.

Denaturation. The irreversible process in which the structure of a protein is disrupted, resulting in a partial or complete loss of function. This can be caused by a number of things, the most common of which are heating and exposure to acid. When exposed to heat, egg whites will change from clear and goopy to white and hard as their proteins denature. In ceviche, acidic lime juice will denature proteins in seafood and give it the opaque appearance of having been cooked, although it never was exposed to heat. Different proteins will have varying sensitivities to disruptors.

Dietary Fiber. Food material, particularly plant material, that is not hydrolyzed by enzymes secreted by the human digestive tract but that may be digested by microflora in the gut. Plant components that fall within this definition include non-starch polysaccharides (NSP) such as celluloses, some hemi-celluloses, gums and pectins, as well as lignin, resistant dextrins and resistant starches. Sources of dietary fiber include vegetables, wheat and most other grains. Foods rich in soluble fiber include fruits, oats, barley and beans.

Emulsifier. A compound that possesses both water-loving (hydrophilic) and water-fearing (hydrophobic) properties and so is able to disperse in either water or oil.

Emulsion. A liquid dispersed in another liquid with which it is usually immiscible (incapable of being mixed, such as oil and water).

Enriched. Foods that have had certain nutrients, which were lost through processing, added back to levels established by federal standards. For example, enriched wheat flour has had vitamins and minerals added to it that were present in the original grain, but lost during processing.

Enzymatic browning. A reaction in which an enzyme acts on a phenolic compound in the presence of oxygen to produce brown -colored products. The browning of raw apple and potato slices is caused by enzymatic browning.

Enzyme. A protein the catalyzes (facilitates) a chemical reaction without itself being altered in the process. That means that enzymes can continue to facilitate that same reaction without restraint until either they are denatured or the chemicals used in the reaction run out. For example, sliced raw apples brown due to enzymatic browning, but this reaction can be stopped by the addition of lemon juice, an acid which denatures the enzymes and renders them non-functional.

Essential amino acids. The essential amino acids for humans are: phenylalanine, valine, threonine, methionine, arginine (required for the young, but not for adults), tryptophan, histidine, isoleucine, leucine and lysine. Acronym often used for remembering the essential amino acids: PVT. MAT HILL. Yes, Private Mat spells his name with only one T.

Essential nutrients. Nutrients that the body cannot synthesize at all or in necessary amounts to meet the body's needs. These nutrients must be acquired through the diet.

Fat bloom (chocolate). A layer of fat particles that have migrated to the surface of a piece of chocolate and crystallized there, creating a light brown or whitish appearance. Usually occurs due to the fluctuations in temperature during storage (melting and re-solidifying) or because the chocolate was not properly tempered during manufacture. For the less common cause of whitish looking chocolate, see sugar bloom (chocolate).

Fermentation. The conversion of carbohydrates to carbon dioxide and alcohol by yeast or bacteria.

Flavonoid. A plant pigment having a structure based on or similar to that of flavone. Found in citrus, tea, wine and dark chocolate. A subject of interest in the scientific community for their in vitro antioxidant properties and potential health benefits. So far, research has not shown conclusive evidence that a diet high in flavonoid compounds is beneficial.

Foam. A colloidal dispersion of a gas in a liquid.

Fortified. Foods that have had nutrients added that were not present in the original food. For example, orange juice is often fortified with calcium, and milk is often fortified with vitamin D.

Freeze-dry. To remove water from food when it is in a frozen state, usually under a vacuum.

Freezer burn. White or grayish patches on frozen food caused by water evaporation. Vacuum packing or otherwise preventing the exposure of frozen food to air will help prevent freezer burn.

Fructose. A simple sugar (or monosaccharide) that cannot be broken down into any smaller carbohydrate. Most commonly found in fruits and honey. It is the sweetest of the natural sugars at 170-150 percent of the sweetness of sucrose (table sugar). It is also known as fruit sugar or levulose.

Functional food. A food or beverage that imparts a physiological benefit that enhances overall health, helps prevent or treat a disease or condition, or improves physical/mental performance. For example, drinking orange juice (rich in vitamin C) to prevent, or help speed recovery from, a cold.

Galactose. A simple sugar (or monosaccharide) that cannot be broken down into any smaller carbohydrate. It possesses only 30 percent of the sweetness of sucrose (table sugar).

Gelatinization. The increase in volume, viscosity, and translucency of starch granules when they are heated in a liquid. For example, adding flour or starch to cold chicken broth would produce a cloudy thin liquid, but with heat, the starch granules would gelatinize, resulting in a thicker and more translucent liquid. This is one of the principles behind gravy making.

Genetically modified organisms (GMOs). Plants, animals, or microorganisms that have had their genes altered through genetic engineering using the application of recombinant deoxyribonucleic acid (rDNA) technology.

Glucose. A simple sugar (or monosaccharide) that cannot be broken down into any smaller carbohydrate. It is commonly found throughout nature, and is necessary for many biological functions. It has only 80-70 percent of the sweetness of sucrose (table sugar), but is still sweet. It is also known as dextrose.

Gluten. A group of proteins (most commonly found in wheat) which bond to form an elastic network that gives necessary structure and function to doughs, batters and baked goods.

GRAS list. A list of compounds that are exempt from the "food additive" definition because they are "generally recognized as safe" based on "a reasonable certainty of no harm from a product under the intended conditions of use." Many food items on this list are those that have been in use for so long with no adverse effects that the government does not see a need to prove their safety through testing. Salt is an example of a GRAS food. However, some newer additions to the GRAS list have been more controversial.

Hemicellulose. Any of a class of substances that occur as constituents of the cell walls of plants and are polysaccharides of simpler structure than cellulose.

Homogenization. A mechanical process that breaks up the fat globules in milk into much smaller globules that do not clump together and are permanently dispersed in a fine emulsion. This process allows milk to be a uniform liquid, without the cream (fat) portion rising to the top.

Hydrogenation. A commercial process in which hydrogen atoms are added to the double bonds in unsaturated fatty acids to make them more saturated. Hydrogenation can change liquid fats into solid fats. Commercial shortenings are made from vegetable oil that has been partially hydrogenated.

Fun fact. Partial hydrogenation can lead to the artificial formation of trans fats, but complete hydrogenation (fully saturating every bond) cannot yield trans fats. Commercial shortenings (historically very high in trans fats) have decreased their levels of trans fat by using a combination of almost completely hydrogenated fat and completely unhydrogenated oil. Mixing the two creates a shortening with similar properties to the partially hydrogenated, but minimizes the formation of trans fats.

Hydrolysis. The chemical breakdown of a compound due to a reaction with water.

Hydrophilic. A term describing "water-loving" or water-soluble substances.

Hydrophobic. A term describing "water-fearing" or non-water-soluble substances.

Incomplete protein. A protein, usually from plant sources, that does not provide all the essential amino acids.

Induction. The transfer of heat energy to a neighboring material without contact. For example, an induction cooktop works by creating an electromagnetic field in its immediate vicinity. Any ferrous (magnetic) cookware within that field will begin to heat up and cook the food. The source of the energy (induction coil) and the cookware never touch.

Inert. Non-reactive.

Invert sugar. An equal mixture of glucose and fructose, created by hydrolyzing sucrose (table sugar). For example, many candy recipes call for adding cream of tartar (an acid) to a sugar solution. The purpose of adding the acid is to hydrolyze the sucrose molecule so it will break down into its two more basic sugar components—one glucose and one fructose.

Julienne. To cut food lengthwise into thin strips.

Knead. To work a dough by pushing, stretching and folding it. Important in the formation of gluten in baked goods.

Kosher. Food that is prepared and eaten according to Jewish dietary laws.

Lactose. A disaccharide comprised of glucose and galactose. It is commonly found in milk and dairy products and is the least sweet of the common natural sugars with only 15 percent of the sweetness of sucrose (table sugar). It is also known as milk sugar.

Maillard reaction. The reaction between a reducing sugar (most commonly glucose, fructose, lactose and maltose) and a protein in the presence of heat, resulting in the formation of brown complexes. For example, the browning of meat and toast are the result of the Maillard reaction, which introduces important flavors and aromas as well as an attractive appearance to cooked foods.

Maltose. A disaccharide comprised of two glucose molecules. It is found in barley and is often produced when amylase (an digestive enzyme found in saliva) breaks down starch. It is approximately 40 percent as sweet as sucrose (table sugar). It is also known as malt sugar and plays an important role in the brewing of beer and adds flavor to malted milkshakes and malted milk ball candy.

Marbling. In meat, this is fat deposited in the muscle that can be seen as white streaks or spots.

Melanin. A class of back or brown pigments naturally occurring in many plant and animal tissues.

Meniscus. Vital to accurate and precise measurement of liquids, this is the bottom of the concave arc at the water's surface. Liquids should be measured at eye level in a liquid measuring cup, and the meniscus should line up with the desired measurement.

Modified starch. A starch that has been chemically or physically modified to create unique functional characteristics. Can be made from a variety of grains such as wheat, rice, tapioca, or, most commonly, corn. Industrially, modified starches are available for use in natural, organic or gluten free foods in addition to conventional foods, so you may see it on the labels of all sorts of pre-packaged goods.

Monounsaturated fat. A fatty acid that has one double bond along its chain; the remaining bonds are single.

Muffin Method. A method of combining ingredients where the liquids are added to the dry ingredients and then stirred only enough to hydrate the dry ingredients. The batter should remain lumpy. Continued stirring beyond what is needed to wet the dry ingredients will result in the overproduction of gluten, which will yield a less tender product with large tunnels (tunneling) or air cells. The muffin method is used in the preparation of muffins, pancakes, coffee cakes, waffles, and cake-like cookies. It is also sometimes called the quick-bread method, blending method or stirring method.

Nitrate. A salt or ester of nitric acid, containing the anion NO_3^- or the group $-NO_3$. Historically, nitrates were often added to sausages and cured meats in the form of potassium nitrate (KNO_3), which was discovered in the Middle Ages and named saltpeter, for its salt-like appearance. It serves to brighten meat color and improve its flavor, safety and shelf life. Scientists have since discovered that these benefits are actually imbued by nitrites produced by small amounts of salt-tolerant bacteria that break down the nitrates that were originally added to the meat. Although added nitrites are more common in today's rapidly produced cured meats, long-cured and aged cured meats (like prosciutto) still uses nitrates, usually in the form of saltpeter or those naturally occurring in sea salts. In the US, residual nitrate and nitrite content is limited to 200 parts per million (0.02%) due to the small risk of the formation of nitrosamine in food, a carcinogen.

Nitrite. A salt or ester of nitrous acid, containing the anion NO_2^- or the group -NO2. Today, very small amounts of nitrites are often added to sausages and cured meats. It serves to brighten meat color and improve its flavor, safety, shelf life and prevent the growth of *Clostridium botulinum*, a bacteria that could otherwise produce a lethal botulism-causing toxin when present in an oxygen-free environment (like the inside of a sausage). In the US, residual nitrate and nitrite content is limited to 200 parts per million (0.02%) due to the small risk of the formation of nitrosamine in food, a carcinogen.

Nutraceutical. A bioactive compound (nutritive or non-nutritive) that has health benefits.

Nutrients. Food components that nourish the body to provide growth, maintenance and/or repair. Foods components that do not meet this definition are often referred to as non-nutritive, as in artificial sweeteners and Calorie-free natural sweeteners like stevia. Non-nutritive components are still considered food, but do not provide any nutritional value.

Olfactory. Relating to the sense of smell.

Osmosis. The movement of a solvent through a semi-permeable membrane to the side with the higher solute concentration, equalizing the solute concentration on both sides of the membrane.

Overrun (ice cream). The volume over and above the volume of the original frozen dessert mix, caused by the incorporation of air during freezing.

Oxidation. The addition of an oxygen molecule to a given compound. Alternately defined as the loss of electrons. Oxidation has many implications in chemistry, but with regards to food, it is often responsible for off-flavors and rancidity in fats. Antioxidants are chemical compounds, such as vitamin E, (aka tocopherols) that prevent harmful oxidation reactions, usually by attracting the oxygen (or losing electrons) themselves.

Parboil. To partially boil, but not fully cook, a food.

Pasteurization. A food preservation process that heats liquids to °160F (°71C) for 15 seconds, or °143F (°62C) for 30 minutes, in order to kill bacteria, yeasts and molds. Commercially available milk, juices and liquid egg products are pasteurized. Raw milk is not pasteurized, is not commercially available, and is generally considered to be a health risk by the scientific community.

pH scale. Measures the degree of acidity or alkalinity of a substance, with 1 the most acidic, 7 neutral, and 14 the most alkaline (basic).

Phenolic compounds. Also called phenols or phenolics, are a class of chemicals which are comprised of an aromatic (circular) ring attached to one or more hydroxyl (-OH) groups.

Pheophytin. A chlorophyll molecule lacking a central Mg_2+ ion. It can be produced from chlorophyll by treatment with a weak acid, producing a drab brown pigment.

Photosynthesis. The process by which green plants and some other organisms use sunlight to synthesize foods from carbon dioxide and water. Photosynthesis in plants generally involves the green pigment chlorophyll and generates oxygen as a byproduct.

Pith. The spongy white inner rind of citrus fruits. Also known as the *albedo*, it is most known for its uncommonly powerful bitterness. Its flavor is generally considered undesirable, with the notable exceptions of candied citrus peel and certain cocktails where the addition of bitterness improves the overall flavor profile.

Plasticity. The ability of a substance to be shaped or molded. Solid fat is an example of a food with plastic characteristics. (Think butter sculpture!)

Polymers. A molecule, compound or mixture of compounds whose structure is composed of multiple repeating units.

Polyunsaturated fat. A fatty acid that has two or more double bonds along its chain; the remaining bonds are single.

Prebiotics. Non-digestible food ingredients (generally fibers such as fructo-oligosaccharides (FOS) and inulin) that support the growth of probiotics.

Probiotics. Live microbial food ingredients (i.e., bacteria) that have a beneficial effect on human health.

Protein complementation. Two incomplete-protein foods, each of which supplies the essential amino acids missing in the other, combined to yield a complete protein profile. Some foods historically eaten together have been found to complement each other, such as rice and beans. This technique of combining incomplete-protein foods is very important in meat- or animal-protein-free diets.

Protein. A large molecule composed of one or more chains of amino acids. An essential component of the human diet. In addition to providing amino acids needed for the assembly of proteins in our bodies, a gram of protein also yields approximately 4 Calories.

Quickbread. Bread leavened with air, steam, carbon dioxide from baking soda or baking powder or other non-yeast source. Muffins, pancakes, coffee cakes, waffles, banana bread, and cake-like cookies are all examples of quickbreads.

Rancid. The breakdown of the polyunsaturated fatty acids in fats that results in disagreeable odors and flavors.

Reducing sugar. Sugars such as glucose, fructose, lactose, maltose and others that have a reactive aldehyde or ketone group. These sugars play a crucial role in the Maillard browning reaction. Sucrose is not a reducing sugar.

Rennin. An enzyme originally obtained from the lining of a calf's stomach and sold commercially as rennet. A non-animal origin alternative is now also available. This enzyme is key in cheese-making.

Retrogradation. The seepage of water out of an aging gel due to the contraction of the gel (bonds tighten between the amylase molecules). Also known as syneresis or weeping.

Roux. A thickener made by cooking equal parts or flour and fat (usually butter). Short cooking times lead to a blonde or light-colored roux with powerful thickening properties that is historically used in sauce-making in French cuisine. It is also useful in thickening gravy, stews and soups. In Creole cuisine, roux is traditionally cooked for much longer until it develops dark, complex flavor and color. However, most of a roux's thickening power is lost with longer cooking.

Saturated fat. A fatty acid that has only single bonds along its chain; there are no double bonds along its structure.

Saturated solution. A solution holding the maximum amount of dissolved solute at room temperature.

Sear. To brown the surface of meat by brief exposure to high heat.

Simple syrup. A basic mixture of boiled sugar and water. Usually a 1:1 ratio.

Smoke point. The temperature at which a fat or oil begins to smoke.

Solubility. The ability of one substance to blend uniformly with another. For example, salt is soluble in water. Salt has greater solubility in hot water than in cold water.

Solute. Solid, liquid or gas compounds dissolved in another substance. For example, in saltwater salt is the solute and water is the solvent. In chocolate milk, chocolate syrup is the solute and milk is the solvent. In seltzer water, carbon dioxide is the solute and water is the solvent.

Solution. A completely homogeneous mixture of a solute (usually a solid) dissolved in a solvent (usually a liquid). For example, in saltwater salt is the solute and water is the solvent. In chocolate milk, chocolate syrup is the solute and milk is the solvent. In seltzer water, carbon dioxide is the solute and water is the solvent.

Solvent. A substance, usually a liquid, in which another substance is dissolved. For example, in saltwater salt is the solute and water is the solvent. In chocolate milk, chocolate syrup is the solute and milk is the solvent. In seltzer water, carbon dioxide is the solute and water is the solvent.

Standards of Identity. Requirements for the type and amount of ingredients a food should contain in order to be labeled as that food. For example, for something to be labeled as "orange juice" it must contain a certain percentage of juice from oranges; if it falls below the requirement in any way, it must be labeled something other than "orange juice", usually orange drink or something similar. Not every type of food has a Standard of Identity.

Structure/function claims. Statements identifying relationships between nutrients or dietary ingredients and body function. These health claims are held to the standard of "significant scientific agreement." Cheerios' claim to be a heart healthy food is an example of a structure/function claim. Note that the FDA does not typically evaluate claims made by supplement manufacturers, so use caution!

Sublimation. The process in which a solid changes directly to a vapor without passing through the liquid phase. Dry ice is an example of this phenomenon.

Sucrose. A disaccharide comprised of glucose and fructose. It is commonly known as table sugar or saccharose and is plays an important role in many foods. Sucrose is one of the only sugars that will not contribute to the Maillard browning reaction, as it is non-reducing.

Sugar bloom (chocolate). A layer of sugar particles that have migrated to the surface of a piece of chocolate and dried there, creating a dusty white appearance. Usually occurs due to the presence of moisture or humidity. For the more common cause of whitish looking chocolate, see fat bloom (chocolate).

Supersaturated solution. An unstable solution created when more than the maximum solute is dissolved in solution. This is usually made possible by raising the temperature of the solution, thereby increasing solubility, then allowing the mixture to cool slowly. Without agitation, the extra solute can remain in solution for a time; during that time, the solution is considered supersaturated. This technique is often used in candy-making.

Suspension. A mixture in which particles too large to go into solution remain suspended in the solvent.

Tannins. Polymers of various flavonoid compounds, some of which yield reddish and brown pigments. Present in tea and red wine.

Tempering (chocolate). To heat and cool chocolate to specific temperatures, making it more resistant to melting and resulting in a smooth, glossy, hard finish.

Trans fat. An unsaturated fat molecule with a trans-isomer fatty acid. "Trans" describes the molecular formation in which a double bond on an unsaturated fatty acid has carbons attached on opposite sides of the molecule, as opposed to "cis" formation, where carbons are attached on the same side. Trans formation yields a linear structure, much like a saturated fatty acid, unlike a cis fatty acid, which is bent. Trans fats can be mono- or polyunsaturated, but never saturated. They do occur rarely in nature, but are more commonly introduced in the diet via foods that contain partially hydrogenated fats. There are ways to reduce trans fats in partially hydrogenated products, such as vegetable shortening, by increasing the level of hydrogenation (thereby reducing the remaining unsaturated fats) and then adding back in pure unsaturated fats in the form of vegetable oil. Trans fats have been implicated in various health problems, including an increased risk for heart disease.

Ultrahigh-temperature (UHT) milk. Milk that has been pasteurized using very high temperatures, is aseptically sealed, and is capable of being safely stored unrefrigerated for up to three months.

Unsaturated fat. A fatty acid that has one or more double bonds along its chain.

Viscosity. The resistance of a fluid to flowing freely, caused by the friction of its molecules against a surface. In simplified terms, viscosity is used to describe the thickness a food, usually a sauce or other fluid. Something with a high viscosity (like sour cream) will move slowly and is considered "thick", something with a low viscosity (like water) will move rapidly and is considered "thin". Viscosity measurements must be taken at a constant temperature. After all, everyone knows molasses moves slower in winter, right?

Water activity (aw). Measures the amount of available (free) water in foods. Water activity ranges from 0 to 1.00, which is pure water. Water activity is one of the single most important concepts in food safety today. Limiting the amount of available water prevents the growth of microbes, which need water to survive and thrive. In addition to keeping food safe, a low water activity is one way to allow for extended shelf life without refrigeration for many foods.

Whey. The liquid portion of milk, consisting primarily of water, lactose and whey proteins. It is the watery component removed from the curd in cheese manufacture. Leftover whey from cheesemaking is often dried and used as filler in many pre-packaged dry goods and seasonings, especially those with a dairy flavor element. Check the label on dry mac and cheese or Doritos next time you pick some up. You might be surprised at how many places you find whey.

Xanthan gum. A water-soluble natural gum produced by bacterial fermentation and used as a thickener, gelling agent, binder, or stabilizer in food.

Yeast (active). A fungus (a plant that lacks chlorophyll) that is able to ferment sugars and that is used for producing food products such as bread and alcohol.

Zest. Shavings from the colorful, outermost layer of a citrus fruit. Also known as the *flavedo*, it contains a high concentration of flavorful essential oils and can imbue powerful citrus flavors without adding unwanted acidity or moisture to recipes.

BIBLIOGRAPHY

Abbas, A.; Waseem, M.; Yang, M. An ensemble approach for assessment of energy efficiency of agriculture system in Pakistan. Energy Effic. 2020, 13, 683–696.

Achir, N., Vitrac, O., and Trystram, G., 2008, "Simulation and Ability to Control the Surface Thermal History and Reactions During Deep Fat Frying," Chem. Eng. Process.: Process Intensification, 47(11), pp. 1953–1967.

Alqdah, K. Prospects of energy savings in the national meat processing factory. Int. J. Sustain. Energy 2013, 32, 670–681.

Alvarez, S.; Carballo-Penela, A.; Mateo-Mantecón, I.; Rubio, A. Strengths-Weaknesses-Opportunities-Threats analysis of carbon footprint indicator and derived recommendations. J. Clean. Prod. 2016, 121, 238–247.

Arrieta, E.M.; Cuchietti, A.; Cabrol, D.; González, A.D. Greenhouse gas emissions and energy efficiencies for soybeans and maize cultivated in different agronomic zones: A case study of Argentina. Sci. Total Environ. 2018, 625, 199–208.

Asteriadou, K., Hasting, A. P. M., Bird, M. R., and Melrose, J., 2006, "Computational Fluid Dynamics for the Prediction of Temperature Profiles and Hygienic Design in the Food Industry," Food Bioproducts Process., 84(C2), pp. 157–163.

Asteriadou, K., Hasting, T., Bird, M., and Melrose, J., 2007, "Predicting Cleaning of Equipment Using Computational Fluid Dynamics," J. Food Process Eng., 30(1), pp. 88–105.

Ayappa, K. G., Davis, H. T., Davis, E. A., and Gordon, J., 1992, "2-Dimensional Finite-Element Analysis of Microwave-Heating," AIChE J., 38(10), pp. 1577–1592.

Aydin, B.; Aktürk, D. Energy use efficiency and economic analysis of peach and cherry production regarding good agricultural practices in Turkey: A case study in Çanakkale province. Energy 2018, 158, 967–974.

Baer-Nawrocka, A.; Mrówczyñska-Kamiñska, A. Material and import intensity in the agriculture of the European Union input-output analysis. Probl. Agric. Econ. 2019, 358, 3–21.

Baer-Nawrocka, A.; Sadowski, A. Food security and food self-sufficiency around the world: A typology of countries. PLoS ONE 2019, 14.

Baier, R. E., 1980, "Substrate Influences on Adhesion of Microorganisms and Their Resultant New Surface Properties," Adsorption of Microorganisms to Surfaces, G. Bitton and K. S. Marshall, eds., John Wiley & Sons, Inc., New York, p. 59.

Bajan, B.; Mrówczyñska-Kamiñska, A. Carbon footprint and environmental performance of agribusiness production in selected countries around the world. J. Clean. Prod. 2020, 276, 123389.

Bajan, B.; Mrówczyñska-Kamiñska, A. Measuring the agribusiness GDP in European Union countries. Econ. Sci. Agribus. Rural Econ. 2019, 3, 15–21.

Ball, C. O., and Olson, F. C. W., 1957, Sterilization in Food Technology, McGraw-Hill, New York.

Banaeian, N.; Omid, M.; Ahmadi, H. Greenhouse strawberry production in Iran, efficient or inefficient in energy. Energy Effic. 2012, 5, 201–209.

Baran, M.F.; Gokdogan, O. Determination of energy balance of sugar beet production in Turkey: A case study of Kýrklareli Province. Energy Effic. 2016, 9, 487–494.

Beckett, S. T., 2002, The Science of Chocolate, RSC, London.

Benning, R., Petermeier, H., Delgado, A., Hinrichs, J., Kulozik, U., and Becker, T., 2003, "Process Design for Improved Fouling Behaviour in Dairy Heat Exchangers Using a Hybrid Modelling Approach," Food Bioproducts Process., 81(C3), pp. 266–274.

Bhadbhade, N.; Patel, M.K. Analysis of energy efficiency improvement and carbon dioxide abatement potentials for Swiss Food and Beverage sector. Resour. Conserv. Recycl. 2020, 161, 104967.

Bird, M. R., and Fryer, P. J., 1991, "An Experimental Study of the Cleaning of Surfaces Fouled by Whey Proteins," Food Bioproducts Process., 69, pp. 13–21.

Blel, W., Le Gentil-Lelievre, C., Benezech, T., Legrand, J., and Legentilhomme, P., 2009, "Application of Turbulent Pulsating Flows to the Bacterial Removal During a Cleaning in Place Procedure. Part 1: Experimental Analysis of Wall Shear Stress in a Cylindrical Pipe," J. Food Eng., 90(4), pp. 422–432.

Blel, W., Legentilhomme, P., Benezech, T., Legrand, J., and Le Gentil-Lelievre, C., 2009, "Application of Turbulent Pulsating Flows to the Bacterial Removal During a Cleaning in Place Procedure. Part 2: Effects on Cleaning Efficiency," J. Food Eng., 90(4), pp. 433–440.

Bolandnazar, E.; Keyhani, A.; Omid, M. Determination of efficient and inefficient greenhouse cucumber producers using Data Envelopment Analysis approach, a case study: Jiroft city in Iran. J. Clean. Prod. 2014, 79, 108–115.

Boulange-Petermann, L., 1996, "Processes of Bioadhesion on Stainless Steel Surfaces and Cleanability: A Review with Special Reference to the Food Industry," Biofouling, 10(4), pp. 275–300.

Bows, J. R., Patrick, M. L., Janes, R., Metaxas, A. C. R., and Dibben, D. C., 1999, "Microwave Phase Control Heating," Int. J. Food Sci. Technol., 34(4), pp. 295–304.

Briam, R.; Walker, M.E.; Masanet, E. A comparison of product-based energy intensity metrics for cheese and whey processing. J. Food Eng. 2015, 151, 25–33.

Brondino, G. Productivity growth and structural change in China (1995–2009): A subsystems analysis. Struct. Chang. Econ. Dyn. 2019, 49, 183–191.

Burfoot, D., and Middleton, K., 2009, "Effects of Operating Conditions of High Pressure Washing on the Removal of Biofilms from Stainless Steel Surfaces," J. Food Eng., 90(3), pp. 350–357.

Canning, P. Energy Use in the US Food System; Diane Publishing: Darby, DE, USA, 2011.

Cao, S.; Xie, G.; Zhen, L. Total embodied energy requirements and its decomposition in China's agricultural sector. Ecol. Econ. 2010, 69, 1396–1404.

Carlsson-Kanyama, A.; Ekstrom, M.; Shanahan, H. Food and life cycle energy inputs: Consequences of diet and ways to increase efficiency. Ecol. Econ. 2003, 44, 293–307.

Changani, S. D., Belmar-Beiny, M. T., and Fryer, P. J., 1997, "Engineering and Chemical Factors Associated with Fouling and Cleaning in Milk Processing," Exp. Therm. Fluid Sci., 14(4), pp. 392–406.

Changani, S. D., Belmar-Beiny, M. T., and Fryer, P. J., 1997, "Engineering and Chemical Factors Associated with Fouling and Cleaning in Milk Processing," Exp. Therm. Fluid Sci., 14, pp. 392–406.

Chen, C. R., and Ramaswamy, H. S., 2004, "Multiple Ramp-Variable Retort Temperature Control for Optimal Thermal Processing," Food Bioproducts Process., 82(1), pp. 78–88.

Chen, C., Abdelrahim, K., and Beckerich, I., 2010, "Sensitivity Analysis of Continuous Ohmic Heating Process for Multiphase Foods," J. Food Eng., 98(2), pp. 257–265.

Chen, H., Tang, J., and Liu, F., 2008, "Simulation Model for Moving Food Packages in Microwave Heating Processes Using Conformal Fdtd Method," J. Food Eng., 88(3), pp. 294–305.

Christian, G. K., and Fryer, P. J., 2006, "The Effect of Pulsing Cleaning Chemicals on the Cleaning of Whey Protein Deposits," Food Bioproducts Process., 84(C4), pp. 320–328.

Christian, G. K., Changani, S. D., and Fryer, P. J., 2002, "The Effect of Adding Minerals on Fouling From Whey Protein Concentrate: Development of a Model Fouling Fluid for a Plate Heat Exchanger," Food Bioproducts Process., 80(4), pp. 231–239.

Clairand, J.M.; Briceño-León, M.; Escrivá-Escrivá, G.; Pantaleo, A.M. Review of energy efficiency technologies in the food industry: Trends, barriers and opportunities. IEEE Access 2020, 8, 48015–48029.

Cole, P. A., Asteriadou, K., Robbins, P. T., Owen, E. G., Montague, G. A., and Fryer, P. J., "Comparison of Cleaning of Toothpaste from Surfaces and Pilot Scale Pipework," Food Bioproducts Process., 88(4), pp. 392–400.

Connor, D.J. Organic agriculture cannot feed the world. Field Crop. Res. 2008, 106, 187–190.

Corsatea, T.D.; Lindner, S.; Arto, I.; Román, M.V.; Rueda-Cantuche, J.M.; Velázquez, A.; Amores, A.F.; Neuwahl, F. World Input-Output Database Environmental Accounts; Publications Office of the European Union: Brussels, Belgium, 2019.

Cox, P. W., Bakalis, S., Ismail, H., Forster, R., Parker, D. J., and Fryer, P. J., 2003, "Visualisation of Three-Dimensional Flows in Rotating Cans Using Positron Emission Particle Tracking (Pept)," J. Food Eng., 60(3), pp. 229–240.

Curet, S., Rouaud, O., and Boillereaux, L., 2009, "Effect of Sample Size on Microwave Power Absorption within Dielectric Materials: 2D Numerical Results Versus Closed-Form Expressions," AIChE J., 55(6), pp. 1569–1583.

Datta, A. K., Teixeira, A. A., and Manson, J. E., 1986, "Computer-Based Retort Control Logic for on-Line Correction of Process Deviations," J. Food. Sci., 51(2), pp. 480–483.

Davies, L. J., Kemp, M. R., and Fryer, P. J., 1999, "The Geometry of Shadows: Effects of Inhomogeneities in Electrical Field Processing," J. Food Eng., 40(4), pp. 245–258.

De Alwis, A. A. P., and Fryer, P. J., 1990, "A Finite-Element Analysis of Heat-Generation and Transfer During Ohmic Heating of Food," Chem. Eng. Sci., 45(6), pp. 1547–1559.

De Alwis, A. A. P., and Fryer, P. J., 1990, "The Use of Direct Resistance Heating in the Food Industry," J. Food Eng., 11(1), pp. 3–27.

De Alwis, A. A. P., and Fryer, P. J., 1992, "Operability of the Ohmic Heating Process - Electrical-Conductivity Effects," J. Food Eng., 15(1), pp. 21–48.

De Alwis, A. A. P., Halden, K., and Fryer, P. J., 1989, "Shape and Conductivity Effects in the Ohmic Heating of Foods," Chem. Eng. Res. Des., 67(2), pp. 159–168.

De Jong, P., Te Giffel, M. C., and Kiezebrink, E. A., 2002, "Prediction of the Adherence, Growth and Release of Microorganisms in Production Chains," Int. J. Food Microbiol., 74(1–2), pp. 13–25.

De Jong, P., Te Giffel, M. C., Straatsma, H., and Vissers, M. M. M., 2002, "Reduction of Fouling and Contamination by Predictive Kinetic Models," Int. Dairy J., 12(2–3), pp. 285–292.

De Roeck, A., Mols, J., Duvetter, T., Van Loey, A., and Hendrickx, M., 2010, "Carrot Texture Degradation Kinetics and Pectin Changes During Thermal Versus High-Pressure/High-Temperature Processing: A Comparative Study," Food Chem., 120(4), pp. 1104–1112.

Dietzenbacher, E.; Los, B.; Stehrer, R.; Timmer, M.P.; de Vries, G.J. The construction of world input-output tables in the WIOD project. Econ. Syst. Res. 2013, 25, 71–98.

Epstein, N., 1981, "Fouling in Heat Exchangers," Fouling of Heat Transfer Equipment, E. F. C. Somerscales and J. G. Knudsen, eds., Hemisphere, Washington, DC, pp. 235–253.

Epstein, N., 1981, "Technical Aspects", Fouling of Heat Transfer Equipment, E. F. C. Somerscales and J. G. Knudsen, eds., Hemisphere, Washington, DC.

Epstein, N., 1981, "Thinking About Heat Transfer Fouling: A 5 × 5 Matrix," Heat Transfer Eng., 4(1), pp. 43–56.

Fennema, O. R., 1996, Food Chemistry, CRC Press, Boca Raton, FL.

Fernandez-Torres, M. J., Fitzgerald, A. M., Paterson, W. R., and Wilson, D. I., 2001, "A Theoretical Study of Freezing Fouling: Limiting Behaviour Based on a Heat and Mass Transfer Analysis," Chem. Process Eng., 40(4), pp. 335–344.

Fessas, D., Signorelli, M., and Schiraldi, A., 2005, "Polymorphous Transitions in Cocoa Butter—A Quantitative Dsc Study," J. Therm Anal. Calorim., 82(3), pp. 691–702.

Finnveden, G.; Hauschild, M.Z.; Ekvall, T.; Guinée, J.; Heijungs, R.; Hellweg, S.; Koehler, A.; Pennington, D.; Suh, S. Recent developments in life cycle assessment. J. Environ. Manag. 2009, 91, 1–21.

Frasch-Melnik, S., Spyropoulos, F., and Norton, I. T., 2010, "W-1/O/W-2 Double Emulsions Stabilised by Fat Crystals—Formulation, Stability, and Salt Release," J. Colloid Interface Sci., 350(1), pp. 178–185.

Fryer, P. J., and Asteriadou, K., 2009, "A Prototype Cleaning Map: A Classification of Industrial Cleaning Processes," Trends Food Sci. Technol., 20(6–7), pp. 255–262.

Fryer, P. J., and Robbins, P. T., 2005, "Heat Transfer in Food Processing: Ensuring Product Quality and Safety," Appl. Therm. Eng., 25(16), pp. 2499–2510.

Fryer, P. J., de Alwis, A. A. P., Koury, E., Stapley, A. G. F., and Zhang, L., 1993, "Ohmic Processing of Solid Liquid-Mixtures - Heat-Generation and Convection Effects," J. Food Eng., 18(2), pp. 101–125.

Fryer, P. J., Porres-Parral, G., and Bakalis, S., 2011, "Multiphysics Modelling of Ohmic Heating," Innovative Food Processing Technologies: Advances in Multiphysics Simulation, K. Knoerzer, P. Juliano, P. Roupas, and C. Versteeg, eds., IFT Press and John Wiley and Sons, Chichester, pp. 155–170.

Gallaher, M.; Delhotal, K.; Petrusa, J. Estimating the potential CO2 mitigation from agricultural energy efficiency in the United States. Energy Effic. 2009, 2, 207–220.

Garcìa, M. -S. G. G., Balsa-Canto, E., Alonso, A. A., and Banga, J. R., 2006, "Computing Optimal Operating Policies for the Food Industry," J. Food Eng., 74(1), pp. 13–23.

Geedipalli, S. S. R., Rakesh, V., and Datta, A. K., 2007, "Modeling the Heating Uniformity Contributed by a Rotating Turntable in Microwave Ovens," J. Food Eng., 82(3), pp. 359–368.

Genty, A.; Arto, I.; Neuwahl, F. Final Database of Environmental Satellite Accounts: Technical Report on Their Compilation. WIOD Deliverable. Available online: http://www.wiod.org/publications/source_docs/Environmental_Sources.pdf. (accessed on 5 August 2020).

Ghnimi, S., Zaid, I., Maingonnat, J. F., and Delaplace, G., 2009, "Axial Temperature Profile of Ohmically Heated Fluid Jet: Analytical Model and Experimental Validation," Chem. Eng. Sci., 64(13), pp. 3188–3196.

Gillham, C. R., Fryer, P. J., Hasting, A. P. M., and Wilson, D. I., 1999, "Cleaning-in-Place of Whey Protein Fouling Deposits: Mechanisms Controlling Cleaning," Food Bioproducts Process., 77(C2), pp. 127–136.

Gillham, C. R., Fryer, P. J., Hasting, A. P. M., and Wilson, D. I., 2000, "Enhanced Cleaning of Whey Protein Soils Using Pulsed Flows," J. Food Eng., 46(3), pp. 199–209.

Goyal, S.K.; Jogdand, S.V.; Agrawal, A.K. Energy use pattern in rice milling industries—A critical appraisal. J. Food Sci. Technol. 2014, 51, 2907–2916.

Grassi, B., Strazza, D., and Poesio, P., 2008, "Experimental Validation of Theoretical Models in Two-Phase High-Viscosity Ratio Liquid-Liquid Flows in Horizontal and Slightly Inclined Pipes," Int. J. Multiphase Flow, 34(10), pp. 950–965.

Grauwet, T., Van Der Plancken, I., Vervoort, L., Hendrickx, M. E., and Van Loey, A., 2010, "Protein-Based Indicator System for Detection of Temperature Differences in High Pressure High Temperature Processing," Food Res. Int., 43(3), pp. 862–871.

Guiavarc'h, Y. P., Deli, V., Van Loey, A. M., and Hendrickx, M. E., 2002, "Development of an Enzymic Time Temperature Integrator for Sterilization Processes Based on Bacillus Licheniformis Alpha-Amylase at Reduced Water Content," J. Food. Sci., 67(1), pp. 285–291.

Guiavarc'h, Y. P., Dintwa, E., Van Loey, A. M., Zuber, F. T., and Hendrickx, M. E., 2002, "Validation and Use of an Enzymic Time-Temperature Integrator to Monitor Thermal Impacts Inside a Solid/Liquid Model Food," Biotechnol. Prog., 18(5), pp. 1087–1094.

Guiavarc'h, Y., Van Loey, A., Zuber, F., and Hendrickx, M., 2004, "Development Characterization and Use of a High-Performance Enzymatic Time-Temperature Integrator for the Control of Sterilization Process' Impacts," Biotechnol. Bioeng., 88(1), pp. 15–25.

Guiavarc'h, Y., Van Loey, A., Zuber, F., and Hendrickx, M., 2004, "Bacillus Licheniformis [Alpha]-Amylase Immobilized on Glass Beads and Equilibrated at Low Moisture Content: Potentials as a Time-Temperature Integrator for Sterilisation Processes," Innovative Food Sci. Emerging Technol., 5(3), pp. 317–325.

Halden, K., De Alwis, A. A. P., and Fryer, P. J., 1990, "Changes in the Electrical Conductivity of Foods During Ohmic Heating," Int. J. Food Sci. Technol., 25(1), pp. 9–25.

Halder, A., Black, D. G., Davidson, P. M., and Datta, A., 2010, "Development of Associations and Kinetic Models for Microbiological Data to be Used in Comprehensive Food Safety Prediction Software," J. Food. Sci., 75(6), pp. R107–R120.

Hamilton, A.; Balogh, S.B.; Maxwell, A.; Hall, C.A. Efficiency of edible agriculture in Canada and the US over the past three and four decades. Energies 2013, 6, 1764–1793.

Hayakawa, K.-I., 1977, "Mathematical Methods for Estimating Proper Thermal Processes and Their Computer Implementation," Adv. Food Res., 23, pp. 75–141.

Hedau, N.K.; Tuti, M.D.; Stanley, J.; Mina, B.L.; Agrawal, P.K.; Bisht, J.K.; Bhatt, J.C. Energy-use efficiency and economic analysis of vegetable crop-ping sequences under greenhouse condition. Energy Effic. 2014, 7, 507-515.

Hendrickson, C.T.; Lave, L.B.; Matthews, H.S. Environmental Life Cycle Assessment of Goods and Services: An Input-Output Approach; RFF Press: Washington, DC, USA, 2006.

Hendrickx, M. E., and Knorr, D., 2001, Ultra-High Pressure Treatments of Foods, Kluwer, New York.

Henningsson, M., Regner, M., Ostergren, K., Tragardh, C., and Dejmek, P., 2007, "Cfd Simulation and Ert Visualization of the Displacement of Yoghurt by Water on Industrial Scale," J. Food Eng., 80(1), pp. 166–175.

Holdsworth, S. D., and Simpson, R., 2008, Thermal Packaging of Processed Foods, Springer, New York.

Hooper, R. J., Paterson, W. R., and Wilson, D. I., 2006, "Comparison of Whey Protein Model Foulants for Studying Cleaning of Milk Fouling Deposits," Food Bioproducts Process., 84(4), pp. 329–337.

Huijbregts, M.; Hellweg, S.; Frischnecht, R.; Hungerbuhler, K.; Hendriks, A. Cumulative Energy demand as predictor for the environmental burden of commodity production. Environ. Sci. Technol. 2010, 44, 2189–2196.

Icier, F., Yildiz, H., and Baysal, T., 2006, "Peroxidase Inactivation and Colour Changes During Ohmic Blanching of Pea Puree," J. Food Eng., 74(3), pp. 424–429.

Ishiyama, E. M., Paterson, W. R., and Wilson, D. I., 2009, "The Effect of Fouling on Heat Transfer, Pressure Drop, and Throughput in Refinery Preheat Trains: Optimization of Cleaning Schedules," Heat Transfer Eng., 30(10–11), pp. 805–814.

Jensen, B. B. B., Friis, A., Benezech, T., Legentilhomme, P., and Lelievre, C.L., 2005, "Local Wall Shear Stress Variations Predicted by Computational Fluid Dynamics for Hygienic Design," Food Bioproducts Process., 83(C1), pp. 53–60.

Jousse, F., Jongen, T., Agterof, W., Russell, S., and Braat, P., 2002, "Simplified Kinetic Scheme of Flavor Formation by the Maillard Reaction," J. Food. Sci., 67(7), pp. 2534–2542.

Juliano, P., Knoerzer, K., Fryer, P. J., and Versteeg, C., 2009, "C-Botulinum Inactivation Kinetics Implemented in a Computational Model of a High-Pressure Sterilization Process," Biotechnol. Prog., 25(1), pp. 163–175.

Jun, S., and Sastry, S., 2007, "Reusable Pouch Development for Long Term Space Missions: A 3D Ohmic Model for Verification of Sterilization Efficacy," J. Food Eng., 80(4), pp. 1199–1205.

Kananeh, A. B., Scharnbeck, E., Kück, U. D., and Räbiger, N., 2010, "Reduction of Milk Fouling Inside Gasketed Plate Heat Exchanger Using Nano-Coatings," Food Bioproducts Process., 88(4), pp. 349–356.

Kazemi, H.; Shokrgozar, M.; Kamkar, B.; Soltani, A. Analysis of cotton production by energy indicators in two different climatic regions. J. Clean. Prod. 2018, 190, 729–736.

Kern, D. Q., and Seaton, R. E., 1959, "A Theoretical Analysis of Thermal Surface Fouling," Br. Chem. Eng., 4, pp. 258–262.

Khan, S.; Hanjra, M. Footprints of water and energy inputs in food production—Global perspectives. Food Policy 2009, 34, 130–140.

Khan, S.; Khan, M.A.; Hanjra, M.A.; Mu, J. Pathways to reduce the environmental footprints of water and energy inputs in food production. Food Policy 2009, 34, 141–149.

Khonpikul, S.; Jakrawatana, N.; Sangkaew, P.; Gheewala, S.H. Resource use and improvement strategy analysis of the livestock and feed production supply chain in Thailand. Int. J. Life Cycle Assess. 2017, 22, 1692–1704.

Kinta, Y., and Hartel, R. W., 2010, "Bloom Formation on Poorly-Tempered Chocolate and Effects of Seed Addition," J. Am. Oil Chem. Soc., 87(1), pp. 19–27.

Kiziltas, S., Erdogdu, F., and Koray Palazoglu, T., 2010, "Simulation of Heat Transfer for Solid-Liquid Food Mixtures in Cans and Model Vali-dation Under Pasteurization Conditions," J. Food Eng., 97(4), pp. 449–456.

Knoerzer, K., Buckow, R., and Versteeg, C., 2010, "Adiabatic Compression Heating Coefficients for High-Pressure Processing—A Study of Some Insulating Polymer Materials," J. Food Eng., 98(1), pp. 110–119.

Knoerzer, K., Buckow, R., Chapman, B., Juliano, P., and Versteeg, C., 2010, "Carrier Optimisation in a Pilot-Scale High Pressure Sterilisation Plant—An Iterative CFD Approach Employing an Integrated Temperature Distributor (Itd)," J. Food Eng., 97(2), pp. 199–207.

Knoerzer, K., Buckow, R., Sanguansri, P., and Versteeg, C., 2010, "Adiabatic Compression Heating Coefficients for High-Pressure Processing of Water, Propylene-Glycol and Mixtures—A Combined Experimental and Numerical Approach," J. Food Eng., 96(2), pp. 229–238.

Liu, W., Zhang, Z., and Fryer, P. J., 2006, "Identification and Modelling of Different Removal Modes in the Cleaning of a Model Food Deposit," Chem. Eng. Sci., 61(22), pp. 7528–7534.

Loisel, C., Keller, G., Lecq, G., Bourgaux, C., and Ollivon, M., 1998, "Phase Transitions and Polymorphism of Cocoa Butter," J. Am. Oil Chem. Soc., 75(4), pp. 425–439.

Lonchampt, P., and Hartel, R. W., 2004, "Fat Bloom in Chocolate and Compound Coatings," Eur. J. Lipid Sci. Tech., 106(4), pp. 241–274.

Lynch, D.; MacRae, R.; Martin, R. The carbon and global warming potential impacts of organic farming: Does it have a significant role in an energy constrained world? Sustainability 2011, 3, 322–362.

Maesmans, G., Hendrickx, M., De Cordt, S., Van Loey, A., Noronha, J., and Tobback, P., 1994, "Evaluation of Process Value Distribution With Time Temperature Integrators," Food Res. Int., 27(5), pp. 413–423.

Majeau-Bettez, G.; Strømman, A.H.; Hertwich, E.G. Evaluation of process- and input-output-based life cycle inventory data with regard to truncation and aggregation issues. Environ. Sci. Technol. 2011, 45, 10170–10177.

Margosch, D., Ganzle, M. G., Ehrmann, M. A., and Vogel, R. F., 2004, "Pressure Inactivation of Bacillus Endospores," Appl. Environ. Microbiol., 70(12), pp. 7321–7328.

Marra, F., Zell, M., Lyng, J. G., Morgan, D. J., and Cronin, D. A., 2009, "Analysis of Heat Transfer During Ohmic Processing of a Solid Food," J. Food Eng., 91(1), pp. 56–63.

Martínez, C.I.P. Analysis of energy efficiency development in the German and Colombian food industries. Int. J. Energy Sect. Manag. 2010, 4, 113–136.

Matser, A.A., Krebbers, B., Van Den Berg, R.W., and Bartels, P. V., 2004, "Advantages of High Pressure Sterilisation on Quality of Food Products," Trends Food Sci. Technol., 15(2), pp. 79–85.

Mehauden, K., Bakalis, S., Cox, P. W., Fryer, P. J., and Simmons, M. J. H., 2008, "Use of Time Temperature Integrators for Determining Process Uniformity in Agitated Vessels," Innovative Food Sci. Emerging Technol., 9(3), pp. 385–395.

Nasseri, A. Energy use and economic analysis for wheat production by conservation tillage along with sprinkler irrigation. Sci. Total Environ. 2019, 648, 450–459.

Navarro, D.M.; Iocola, I.; Persiani, A.; Blanco, J.M.M.; Lakkenborg, H.K.; Hefner, M.; Tamm, K.; Bender, I.; Védie, H.; Willekens, K.; et al. Energy flows in European organic vegetable systems: Effects of the introduction and management of agroecological service crops. Energy 2019, 188, 116096.

Nemecek, T.; Hugenin, O.; Dubois, D.; Gaillard, G.; Schaller, B.; Chervet, A. Life cycle assessment of Swiss farming systems: II. Extensive and intensive production. Agric. Syst. 2010, 104, 233–245.

Ni, H., Datta, A. K., and Torrance, K. E., 1999, "Moisture Transport in Intensive Microwave Heating of Biomaterials: A Multiphase Porous Media Model," Int. J. Heat Mass Transfer, 42(8), pp. 1501–1512.

Norton, I. T., Fryer, P. J., and Moore, S., 2006, "Product/Process Integration in Food Manufacture: Engineering Sustained Health," AIChE J., 52(5), pp. 1632–1640.

Palaniappan, S., and Sastry, S. K., 1991, "Electrical Conductivities of Selected Solid Foods During Ohmic Heating," J. Food Process Eng., 14(3), pp. 221–236.

Peleg, M., 2006, Advanced Quantitative Microbiology for Foods and Biosystems. Models for Predicting Growth and Inactivation, CRC Press, Boca Raton, FL.

Pelletier, N.; Arsenault, N.; Tyedmers, P. Scenario-modeling potential eco-efficiency gains from a transition to organic agriculture: Life cycle perspectives on Canadian canola, corn, soy and wheat production. Environ. Manag. 2008, 42, 989–1001.

Pelletier, N.; Audsley, E.; Brodt, S.; Garnett, T.; Henriksson, P.; Kendall, A.; Kramer, K.J.; Murphy, D.; Nemecek, T.; Troell, M. Energy intensity of agriculture and food systems. Annu. Rev. Environ. Resour. 2011, 36, 223–246.

Petermeier, H., Benning, R., Delgado, A., Kulozik, U., Hinrichs, J., and Becker, T., 2002, "Hybrid Model of the Fouling Process in Tubular Heat Exchangers for the Dairy Industry," J. Food Eng., 55(1), pp. 9–17.

Pimentel, D.; Williamson, S.; Alexander, C.; Gonzalez-Pagan, O.; Kontak, C.; Mulkey, S. Reducing energy inputs in the US food system. Hum. Ecol. 2008, 36, 459–471.

Rahman, S.; Rahman, M.S. Energy productivity and efficiency of maize accounting for the choice of growing season and environmental factors: An empirical analysis from Bangladesh. Energy 2013, 49, 329–336.

Rakesh, V., Datta, A. K., Amin, M. H. G., and Hall, L. D., 2009, "Heating Uniformity and Rates in a Domestic Microwave Combination Oven," J. Food Process Eng., 32(3), pp. 398–424.

Ramirez, C.A.; Patel, M.; Blok, K. How much energy to process one pound of meat? A comparison of energy use and specific energy consumption in the meat industry of four European countries. Energy 2006, 31, 2047–2063.

Robbins, P. T., and Fryer, P. J., 2003, "The Spouted-Bed Roasting of Barley: Development of a Predictive Model for Moisture and Temperature," J. Food Eng., 59(2–3), pp. 199–208.

Rosenhahn, A., Ederth, T., and Pettitt, M. E., 2008, "Advanced Nanostructures for the Control of Biofouling: The FP6 EU Integrated Project Ambio," BioInterphases, 3(1), pp. IR1–IR5.

175. Rosmaninho, R., Rizzo, G., Muller-Steinhagen, H., and Melo, L. F., 2008, "Deposition From a Milk Mineral Solution on Novel Heat Transfer Surfaces under Turbulent Flow Conditions," J. Food Eng., 85(1), pp. 29–41.

Ruan, R., Chen, P., Chang, K., Kim, H. J., and Taub, I. A., 1999, "Rapid Food Particle Temperature Mapping During Ohmic Heating Using Flash Mri," J. Food Sci., 64(6), pp. 1024–1026.

Sahu, K. C., Valluri, P., Spelt, P. D. M., and Matar, O. K., 2007, "Linear Instability of Pressure-Driven Channel Flow of a Newtonian and a Herschel-Bulkley Fluid," Phys. Fluids, 19(12), 122107.

Saikhwan, P., Geddert, T., Augustin, W., Scholl, S., Paterson, W. R., and Wilson, D. I., 2006, "Effect of Surface Treatment on Cleaning of a Model Food Soil," Surf. Coat. Technol., 201(3–4), pp. 943–951.

Sarang, S., Sastry, S. K., and Knipe, L., 2008, "Electrical Conductivity of Fruits and Meats During Ohmic Heating," J. Food Eng., 87(3), pp. 351–356.

Sastry, S. K., 1992, "A Model for Heating of Liquid-Particle Mixtures in a Continuous Flow Ohmic Heater," J. Food Process Eng., 15(4), pp. 263–278.

Sastry, S. K., and Palaniappan, S., 1992, "Mathematical Modeling and Experimental Studies on Ohmic Heating of Liquid-Particle Mixtures in a Static Heater," J. Food Process Eng., 15(4), pp. 241–261.

Sims, R.E.H. Global energy resources, supply and demand, energy security and on-farm energy efficiency. In Sustainable Energy Solutions in Agriculture; Bundschuh, J., Chen, G., Eds.; CRC Press: London, UK, 2014; Volume 1, pp. 19–52. [Google Scholar]

Singh, P.; Singh, G.; Sodhi, G.P.S. Energy auditing and optimization approach for improving energy efficiency of rice cultivation in south-western Punjab, India. Energy 2019, 174, 269–279.

Smith, L.G.; Williams, A.G.; Pearce, B.D. The energy efficiency of organic agriculture: A review. Renew. Agric. Food Syst. 2015, 30, 280–301.

Soni, P.; Soe, M.N. Energy balance and energy economic analyses of rice production systems in Ayeyarwaddy region of Myanmar. Energy Effic. 2016, 9, 223–237.

Sun, K. H., Pyle, D. L., Fitt, A. D., Please, C. P., Baines, M. J., and Hall-Taylor, N., 2004, "Numerical Study of 2D Heat Transfer in a Scraped Surface Heat Exchanger," Comput. Fluids, 33(5–6), pp. 869–880.

Svanberg, L., Ahrnè, L., Lorèn, N., Windhab, E., 2011, "Effect of Pre-Crystallization Process and Solid Particle Addition on Microstructure in Chocolate Model Systems," Food Res. Int., 44(5), pp. 1339–1350.

Taborek, J., Palen, J. W., Aoki, T., Ritter, R. B., and Knudsen, J. G., 1972, "Fouling—The Major Unresolved Problem in Heat Transfer," Chem. Eng. Prog., 68(2), pp. 59–67.

Tamime, A. V., 2008, Cleaning-in-Place: Dairy, Food and Beverage Operations Wiley-Blackwell, London.

Taoukis, P. S., and Labuza, T. P., 1989, "Applicability of Time-Temperature Indicators as Shelf Life Monitors of Food Products," J. Food Sci., 54(4), pp. 783–788.

Tewkesbury, H., Stapley, A. G. F., and Fryer, P. J., 2000, "Modelling Temperature Distributions in Cooling Chocolate Moulds," Chem. Eng. Sci., 55(16), pp. 3123–3132.

Tharakan, A., Norton, I. T., Fryer, P. J., and Bakalis, S., 2010, "Mass Transfer and Nutrient Absorption in a Simulated Model of Small Intestine," J. Food. Sci., 75(6), pp. E339–E346.

Timmer, M.P.; Dietzenbacher, E.; Los, B.; Stehrer, R.; de Vries, G.J. An illustrated user guide to the world input–output database: The case of global automotive production. Rev. Int. Econ. 2015, 23, 575–605.

Timmer, M.P.; Los, B.; Stehrer, R.; de Vries, G.J. An Anatomy of the Global Trade Slowdown Based on the WIOD. 2016. Available online: https://www.rug.nl/ggdc/html_publications/memorandum/gd162.pdf (accessed on 5 August 2020).

Tucker, G. S., Brown, H. M., Fryer, P. J., Cox, P. W., Poole Ii, F. L., Lee, H. S., and Adams, M. W. W., 2007, "A Sterilisation Time-Temperature Integrator Based on Amylase From the Hyperthermophilic Organism Pyrococcus Furiosus," Innovative Food Sci. Emerging Technol., 8(1), pp. 63–72.

United Nations (UN). World Population Prospects 2019. Available online: https://population.un.org/wpp/DataQuery (accessed on 5 August 2020).

Van Boekel, M. A., 2008, "Kinetic Modeling of Food Quality: A Critical Review," Compr. Rev. Food Sci. Food Saf., 7(1), pp. 144–158.

Van Der Plancken, I., Grauwet, T., Oey, I., Van Loey, A., and Hendrickx, M., 2008, "Impact Evaluation of High Pressure Treatment on Foods: Considerations on the Development of Pressure-Temperature-Time Integrators (PTTIs)," Trends Food Sci. Technol., 19(6), pp. 337–348.

Van Der Sman, R. G. M., and Van Der Goot, A. J., 2009, "The Science of Food Structuring," Soft Matter, 5(3), pp. 501–510.

Wang, L. Energy efficiency technologies for sustainable food processing. Energy Effic. 2014, 7, 791–810.

Wang, W.-C., and Sastry, S. K., 1993, "Salt Diffusion Into Vegetable Tissue as a Pretreatment for Ohmic Heating: Electrical Conductivity Profiles and Vacuum Infusion Studies," J. Food Eng., 20(4), pp. 299–309.

Ye, X., Ruan, R., Chen, P., Doona, C., and Taub, I. A., 2003, "Mri Temperature Mapping and Determination of Liquid-Particulate Heat Transfer Coefficient in an Ohmically Heated Food System," J. Food Sci., 68(4), pp. 1341–1346.

Yoo, J. Y., Chen, X. D., Mercade-Prieto, R., and Wilson, D. I., 2007, "Dissolving Heat-Induced Protein Gel Cubes in Alkaline Solutions Under Natural and Forced Convection Conditions," J. Food Eng., 79(4), pp. 1315–1321.

Yuan, S.; Peng, S.; Wang, D.; Man, J. Evaluation of the energy budget and energy use efficiency in wheat production under various crop management practices in China. Energy 2018, 160, 184–191.

Zhang, J., Datta, A. K., and Mukherjee, S., 2005, "Transport Processes and Large Deformation During Baking of Bread," AIChE J., 51(9), pp. 2569–2580.

Zhang, L., and Fryer, P. J., 1994, "Food Sterilization by Electrical Heating: Sensitivity to Process Parameters," AIChE J., 40(5), pp. 888–898.

Zhao, Q., Liu, Y., Wang, C., Wang, S., and Muller-Steinhagen, H., 2005, "Effect of Surface Free Energy on the Adhesion of Biofouling and Crystalline Fouling," Chem. Eng. Sci., 60(17), pp. 4858–4865.

Zorrilla, S. E., and Rubiolo, A. C., 2005, "Mathematical Modeling for Immersion Chilling and Freezing of Foods.: Part I: Model Development," J. Food Eng., 66(3), pp. 329–338.

INDEX

www.ingramcontent.com/pod-product-compliance
Lightning Source LLC
LaVergne TN
LVHW062113200926
848484LV00007B/1656